Mersenne Numbers
and Fermat Numbers

Selected Chapters of Number Theory: Special Numbers

Series Editor: Elena Deza (*Moscow State Pedagogical University, Russia*)

Published

Vol. 1   *Mersenne Numbers and Fermat Numbers*
      by Elena Deza

Selected Chapters of Number Theory: Special Numbers

Mersenne Numbers
and Fermat Numbers

Elena Deza

Moscow State Pedagogical University, Russia

 **World Scientific**

NEW JERSEY · LONDON · SINGAPORE · BEIJING · SHANGHAI · HONG KONG · TAIPEI · CHENNAI · TOKYO

Published by

World Scientific Publishing Co. Pte. Ltd.

5 Toh Tuck Link, Singapore 596224

USA office: 27 Warren Street, Suite 401-402, Hackensack, NJ 07601

UK office: 57 Shelton Street, Covent Garden, London WC2H 9HE

Library of Congress Control Number: 2021033420

British Library Cataloguing-in-Publication Data
A catalogue record for this book is available from the British Library.

Selected Chapters of Number Theory: Special Numbers — Vol. 1
MERSENNE NUMBERS AND FERMAT NUMBERS

Copyright © 2022 by World Scientific Publishing Co. Pte. Ltd.

ISBN 978-981-123-031-8 (hardcover)
ISBN 978-981-123-032-5 (ebook for institutions)
ISBN 978-981-123-033-2 (ebook for individuals)

For any available supplementary material, please visit
https://www.worldscientific.com/worldscibooks/10.1142/12100#t=suppl

Typeset by Stallion Press
Email: enquiries@stallionpress.com

Printed in Singapore

Mathematics is the queen of the Sciences and Number Theory is the queen of Mathematics.
— Gauss

God invented the integers; all else is the work of man.
— Kronecker

I know numbers are beautiful. If they aren't beautiful, nothing is.
— Erdős

Contents

Notations

- X — a set; $X = \{x_1, ..., x_n\}$ — a *finite set*; $|X| = n$ — the number n of elements in the finite set $X = \{x_1, ..., x_n\}$.

- $\mathbb{N} = \{1, 2, 3, ...\}$ — the set of *positive integers* (or *natural numbers*).

- $\mathbb{Z} = \{..., -3, -2, -1, 0, 1, 2, 3, ...\}$ — the set of *integers*.

- $\mathbb{Q}$ — the set of *rational numbers*.

- $\mathbb{R}$ — the set of *real numbers*.

- $\mathbb{C} = \{a + b \cdot i : a, b \in \mathbb{R}, i^2 = -1\}$ — the set of *complex numbers*.

- $b|a$ — a non-zero integer b divides an integer a: $a = bc$, where $c \in \mathbb{Z}$.

- $gcd(a_1, ..., a_n)$ — *greatest common divisor* of integers $a_1, ..., a_n$, at least one of which is non-equal to 0, i.e., the greatest integer, dividing $a_1, ..., a_n$. If $gcd(a_1, ..., a_n) = 1$, the numbers $a_1, ..., a_n$ are called *relatively prime*, or *coprime*; if $gcd(a_i, a_j) = 1$ for any distinct $i, j \in \{1, ..., n\}$, the numbers $a_1, ..., a_n$ are called *pairwise relatively prime*.

- $lcm(a_1, ..., a_n)$ — *least common multiple* of non-zero integers $a_1, ..., a_n$, i.e., the least positive integer divided by $a_1, ..., a_n$.

- $rest(a, b)$ — the *remainder* of an integer a after its *integer division* by a positive integer b: $a = b \cdot q + rest(a, b)$, where $q, rest(a, b) \in \mathbb{Z}$, and $0 \leq rest(a, b) < b$.

- $P = \{2, 3, 5, 7, 11, 13, 17, 19, ...\}$ — the set of *prime numbers*, i.e., the positive integers, having exactly two positive integer divisors; $p, q, p_1, p_2, ..., p_k, ..., q_1, q_2, ..., q_s, ...$ — prime numbers.

- $n = p_1^{\alpha_1} \cdot ... \cdot p_k^{\alpha_k}$, $p_1 < \cdots < p_k \in P$, $\alpha_1, ..., \alpha_k \in \mathbb{N}$ — the *prime factorization* of a positive integer $n > 1$, i.e., its representation as a product of positive integer powers of different prime numbers $p_1, ..., p_k$.

- $S = \{4, 6, 8, 9, 10, 12, 14, 15, 16, 18, ...\}$ — the set of *composite numbers*, i.e., the positive integers, having more than two positive integer divisors.

- $n = (c_s c_{s-1} ... c_1 c_0)_g = c_s g^s + c_{s-1} g^{s-1} + \cdots + c_1 g + c_0$, $g \in \mathbb{N} \backslash \{1\}$, $0 \le c_i \le g - 1$, $c_s \ne 0$ — representation of a positive integer n in *base* g.

- $\lfloor x \rfloor$, $x \in \mathbb{R}$ — *floor function*: the largest integer less than or equal to x.

- $\{x\}$, $x \in \mathbb{R}$ — *fractional value* of x: $\{x\} = x - \lfloor x \rfloor$.

- $\lceil x \rceil$, $x \in \mathbb{R}$ — *ceiling function*: the smallest integer greater than or equal to x.

- $\phi(n)$, $n \in \mathbb{N}$ — *Euler's totient function*: the number of positive integers that are relatively prime to n: $\phi(n) = |\{x \in \mathbb{N} : x \le n, gcd(x, n) = 1\}|$.

- $\lambda(n)$, $n \in \mathbb{N}$ — *Carmichael function*: for prime $p \ge 3$ and positive integer α, $\lambda(p^\alpha) = \phi(p^\alpha)$; for positive integer $\alpha \ge 3$, $\lambda(2^\alpha) = 2^{\alpha-2}$, while $\lambda(2) = 1$, and $\lambda(4) = 2$; moreover, $\lambda(p_1^{\alpha_1} ... p_k^{\alpha_k}) = lcm(\lambda(p_1^{\alpha_1}), ..., \lambda(p_k^{\alpha_k}))$.

- $\mu(n)$, $n \in \mathbb{N}$ — *Möbius function*: $\mu(1) = 1$, $\mu(n) = (-1)^k$, if $n = p_1 \cdot ... \cdot p_k$ is a product of k distinct primes, and $\mu(n) = 0$, if n has repeated prime factors.

- $\pi(x) = \sum_{p \le x} 1$, $x \in \mathbb{R}, x \ge 0$ — *prime counting function*: the number of primes less than or equal to a given non-negative real number x.

- $\tau(n) = \sum_{d|n} 1$, $n \in \mathbb{N}$ — *tau function* (or *number of divisors' function*): the number of positive integer divisors of a positive integer n.

- $\sigma(n) = \sum_{d|n} d$, $n \in \mathbb{N}$ — *sigma function* (or *sum of divisors' function*): the sum of positive integer divisors of a positive integer n.

- $\sigma_k(n) = \sum_{d|n} d^k$, $n \in \mathbb{N}, k \in \mathbb{C}$ — *divisor function*: the sum of the k-th powers of positive integer divisors of a positive integer n. In particular, $\sigma_0(n) = \tau(n)$, and $\sigma_1(n) = \sigma(n)$.

- $\nu(n)$ — the number of distinct prime divisors of a positive integer n.

- $a \equiv b(mod\ n)$ — an integer a is *congruent* to an integer b modulo n, $n \in \mathbb{N}$, i.e., $n|(a-b)$.

- $[a]_n = \mathbf{a}_n = \{x \in \mathbb{Z} : x \equiv a(mod\ n)\} = \{..., a - 2n, a - n, a, a + n, a + 2n, a + 3n, ...\}$ — *residue class* (*of* a) *modulo* n: the set of all integers, congruent to a modulo n. Any representative r_a of $[a]_n$ is called *residue* of a modulo n. The smallest non-negative representative of $[a]_n$ is called *smallest non-negative residue* of a modulo n; it is the *remainder* $rest(a, n)$ of a after its *integer division* by n. The smallest in absolute value representative of $[a]_n$ is called *minimal residue* of a modulo n.

- $(Z/nZ, +, \cdot)$, where $Z/nZ = \{[0]_n, [1]_n, [2]_n, ..., [n-1]_n\}$ — the *factor ring* (or *quotient ring*) modulo n; for a prime number p, the ring $(Z/pZ, +, \cdot)$ is a (finite) field.

- $(Z/nZ^*, \cdot)$, where $Z/nZ^* = \{[1]_n, [2]_n, ..., [n-1]_n\}$ — the *multiplicative group of the factor ring* $(Z/nZ, +, \cdot)$.

- $\left(\frac{a}{p}\right)$ — *Legendre symbol*: $\left(\frac{a}{p}\right) = 1$, if an integer a, relatively prime to an odd prime number p, is a *quadratic residue modulo* p (i.e., the congruence $x^2 \equiv a(mod\ p)$ has a solution $x_0 \in \mathbb{Z}$), $\left(\frac{a}{p}\right) = -1$, if an integer a, relatively prime to an odd prime number p, is a *quadratic nonresidue modulo* p (i.e., the congruence $x^2 \equiv a(mod\ p)$ does not have a solution), and $\left(\frac{a}{p}\right) = 0$, if $p|a$.

- $\left(\frac{a}{n}\right) = \left(\frac{a}{p_1}\right)^{\alpha_1} \cdot \dots \cdot \left(\frac{a}{p_k}\right)^{\alpha_k}$ for odd positive integer $n = p_1^{\alpha_1} \cdot \dots \cdot p_k^{\alpha_k}$ — *Jacobi symbol.* When n is a prime, the Jacobi symbol $\left(\frac{a}{n}\right)$ reduces to the Legendre symbol.

- $ord_n\, a$ — *multiplicative order* (or *modulo order*) of an integer a, relatively prime to n, modulo n: the smallest positive integer exponent γ for which $a^\gamma \equiv 1 (mod\ n)$.

- $ind_g\, a$ — *index* (or *discrete logarithm*) of an integer a with respect to the base g modulo n, i.e., the only integer $\beta \in [0, \phi(n) - 1]$ for which $a \equiv g^\beta (mod\ n)$. Here $n \in \{2, 4, p^\alpha, 2p^\alpha\}$, where $p \in P\backslash\{2\}$, $\alpha \in \mathbb{N}$, g — *primitive root modulo* n (i.e., $ord_n\, g = \phi(n)$), and a — an integer, relatively prime to n.

- $[a_0, a_1, ..., a_n, ...] = a_0 + \cfrac{1}{a_1 + \cfrac{1}{\cdots \cfrac{1}{a_n + \cdots}}}$ — *continued fraction:* here $a_0 \in \mathbb{Z}$, $a_1, ..., a_n \in \mathbb{N}$, and the last term a_n, if it exists, is greater than 1.

- $\delta_k = [a_0, a_1, ..., a_k] = \frac{P_k}{Q_k}$, $k = 0, 1, ..., n, ...$ — the *convergants* of the continued fraction $[a_0, a_1, ..., a_n, ...]$.

- a_1, $a_2 = a_1 + d$, $a_3 = a_2 + d = a_1 + 2d$, ..., $a_n = a_{n-1} + d = a_1 + (n-1)d$, ..., where $a_1, d \in \mathbb{R}$ — *arithmetic progression* with *difference* d; $a_1 + \cdots + a_n = \frac{n(a_1 + a_n)}{2}$.

- b_1, $b_2 = b_1 \cdot q$, $b_3 = b_2 \cdot q = b_1 \cdot q^2$, ..., $b_n = b_{n-1} \cdot q = b_1 \cdot q^{n-1}$, ..., where $b_1, q \in \mathbb{R}\backslash\{0\}$ — *geometric progression* with *common ratio* q; $b_1 + \cdots + b_n = \frac{b_1(q^n - 1)}{q - 1}$ for $q \neq 1$.

- $f(x) = c_0 + c_1 x + c_2 x^2 + \cdots + c_n x^n + ...$, $|x| < r$ — *generating function* of the sequence $c_0, c_1, c_2, ..., c_n, ...$.

- $b_0 c_{n+k} + b_1 c_{n+k-1} + \cdots + b_n c_k = 0$, $b_0, ..., b_n \in \mathbb{R}$, $b_0 \neq 0$ — *linear recurrent equation* of n-th order for a sequence $c_0, c_1, c_2, ..., c_{n-1}$, $c_n = -\frac{b_1}{b_0} c_{n-1} - ... - \frac{b_n}{b_0} c_0$, $c_{n+1} = -\frac{b_1}{b_0} c_n - ... - \frac{b_n}{b_0} c_1$, ..., $c_{n+k} = -\frac{b_1}{b_0} c_{n+k-1} - ... - \frac{b_n}{b_0} c_k$, ... with the *initial values* $c_0, c_1, ..., c_{n-1}$.

- $A = ((a_{ij}))$, $1 \leq i, j \leq n$ — square $n \times n$ matrix with real entries a_{ij}. The matrix A is called *symmetric*, if $a_{ij} = a_{ji}$ for all $i, j \in \{1, 2, ..., n\}$. The matrix A is called *identity matrix* and denoted by I_n, if $a_{ii} = 1$ and $a_{ij} = 0$ for $i \neq j$. The matrix $A^T = ((a_{ji}))$ is called *transpose* of A. The matrix A^{-1}, such that $A \cdot A^{-1} = I_n$, is called *inverse* of A.

- $det\, A$ or $|A|$ — *determinant* of a given $n \times n$ matrix $A = ((a_{ij}))$. For $n = 2$ it holds $det\, A = a_{11}a_{22} - a_{21}a_{12}$; for any $n \geq 3$ it holds $det\, A = \sum_{j=1}^{n}(-1)^{1+j}a_{1j}\cdot det\, A_{1j}$, where A_{1j} is the $(n-1)\times(n-1)$ matrix, obtained by deletion the first row and the j-th column of A.

- $n!$ — *factorial* of a non-negative integer n: $n! = 1 \cdot 2 \cdot ... \cdot n$, $n \in \mathbb{N}$; $0! = 1$.

- $n\sharp$ — *primorial* of a positive integer n: the product of all primes, less than or equal to n.

- $x^{\underline{n}}$, $x \in \mathbb{R}$, $n \in \mathbb{N}$ — *falling factorial* of x: $x^{\underline{n}} = x \cdot (x-1) \cdot ... \cdot (x - n + 1)$.

- $x^{\overline{n}}$, $x \in \mathbb{R}$, $n \in \mathbb{N}$ — *raising factorial* of x: $x^{\overline{n}} = x \cdot (x+1) \cdot ... \cdot (x + n - 1)$.

- $\binom{n}{m}$ — *binomial coefficients*; C_n^m — the number of *combinations*, i.e., the ways to choose an unordered subset of m elements from a fixed set of n elements; $\binom{n}{m} = C_n^m = \frac{n!}{m!(n-m)!}$, $0 \leq m \leq n$. They form the *Pascal's triangle*

$$\binom{0}{0} = 1$$
$$\binom{1}{0} = 1 \quad \binom{1}{1} = 1$$
$$\binom{2}{0} = 1 \quad \binom{2}{1} = 2 \quad \binom{2}{2} = 1$$
$$\binom{3}{0} = 1 \quad \binom{3}{1} = 3 \quad \binom{3}{2} = 3 \quad \binom{3}{3} = 1$$
$$\binom{4}{0} = 1 \quad \binom{4}{1} = 4 \quad \binom{4}{2} = 6 \quad \binom{4}{3} = 4 \quad \binom{4}{4} = 1$$
$$\binom{5}{0} = 1 \quad \binom{5}{1} = 5 \quad \binom{5}{2} = 10 \quad \binom{5}{3} = 10 \quad \binom{5}{4} = 5 \quad \binom{5}{5} = 1$$
$$\binom{6}{0} = 1 \quad \binom{6}{1} = 6 \quad \binom{6}{2} = 15 \quad \binom{6}{3} = 20 \quad \binom{6}{4} = 15 \quad \binom{6}{5} = 6 \quad \binom{6}{6} = 1$$

$$\cdots \quad \cdots \quad \cdots$$

— number triangle, the sides of which are formed by 1, and any inner entry is obtained by adding the two entries diagonally above.

- $f(x) = O(g(x))$ — *big O notation*: for a complex valued function f and a real valued function g, $g(x) > 0$ for all large enough values of x, there exists a positive real number C and a real number x_0 such that $|f(x)| \leq C \cdot g(x)$ for all $x \geq x_0$.

- $f(x) = o(g(x))$ — *small o notation*: for a complex valued function f and a real valued function g, $g(x) > 0$ for all large enough values of x, $\lim_{x \to \infty} \frac{f(x)}{g(x)} = 0$.

- $f(x) \sim g(x)$: for a complex valued function f and a real valued function g, $g(x) > 0$ for all large enough values of x, $\lim_{x \to \infty} \frac{f(x)}{g(x)} = 1$.

- $e = 2.718281828...$ — the *Euler's number*: $e = \lim_{n \to \infty} (1 + \frac{1}{n})^n$.

- $\gamma = 0.5772156649...$ — the *Euler–Mascheroni constant*: $\gamma = \lim_{n \to \infty} \left(\sum_{k=1}^{n} \frac{1}{k} - \log n \right)$.

- $\log x = \ln x = \log_e x$ — *natural logarithm* of a positive real number x: its logarithm to the base e.

- $Li(n)$ — *logarithmic integral*: $Li(n) = \int_2^n \frac{dx}{\log x}$.

Preface

Special numbers, amongst integers, are important part of Number Theory, general Mathematics and several applied areas (such as Cryptography).

While the "names" of some of them (Fermat numbers, Mersenne numbers, Fibonacci numbers, Catalan numbers, Bernoulli numbers, Stirling numbers etc.) are known to every mathematician, actual information on them is often scattered in the special literature. There are only a few books, which systematically present their history and properties, while giving main proofs.

Moreover, if such books exist (see, for example, [Bond93], [CoGu96], [Hogg69], [Line86], [LiNi96], [KLS01], [Madd05], [SlPl95], [Uspe76], [Voro61]), many new results appeared during several decades after their publication; so, this material should be updated and reworked.

We should also mention several classical monographs on Number theory ([Apos86], [Dick05], [HaWr79], [MSC96], [Ribe89], [Ribe96], [Sier64], etc.). Those are basic books of fundamental importance. But they treat Number Theory as a whole and special numbers got only a small place there.

Our first book on special numbers, the monograph "Figurate numbers" [DeDe12], decides all such problems for polygonal numbers and their different generalizations, giving full retrospective of the history of most known class of special integers.

In the presented Series "Selected chapters of Number Theory: Special numbers" we are going to consider, by similar structure, the history and properties of several other famous classes of special numbers. The offered books of the Series are: "Mersenne and Fermat numbers"; "Perfect and amicable numbers"; "Stirling numbers", "Catalan numbers"; "Bernoulli numbers"; "Euler numbers".

So, the main purpose of the Series is to give a complete presentation of the theory of several classes of classical special numbers, considering much of their properties, facts and theorems with full proofs.

In particular, we expect:

- to find and to organize much of scattered material;

- present updated material with all details, in clear and unified way;

- consider all ranges of well-known and hidden connections of a given set number with different mathematical problems;

- draw up a system of multilevel tasks.

Mersenne and Fermat numbers, as well as a majority of classes of special numbers, have long and rich history, connected with the names of many famous mathematicians. This history is mainly an important part of the history of prime numbers, since the interest in Mersenne and Fermat numbers is related, above all, to the questions of their primality.

A *Mersenne number* is a positive integer of the form $M_n = 2^n - 1$, $n \in \mathbb{N}$.

Mersenne numbers were considered by Euclid (IV-th century BC) as an attempt to find all perfect numbers. The *Euclid–Euler theorem* states, that an even positive integer is a perfect number if and only if it can be represented as a product $2^{k-1}(2^k - 1)$, where $2^k - 1$ is a prime. In fact, it is a Mersenne prime, so, there exists an one-to-one relationship between even perfect numbers and Mersenne primes; each Mersenne prime generates one even perfect number, and vice versa. Finally the numbers of the form $2^n - 1$ were named after Marin Mersenne (1588–1648), a French monk and mathematician, who discussed them in his work "Cogita physico mathematics" (1644) and stated some conjectures about the number's occurrence.

A *Fermat number* F_n is a positive integer of the form $F_n = 2^{2^n} + 1$, $n \in \mathbb{Z}$, $n \geq 0$.

Fermat numbers were studied by the XVII-th century French lawyer and mathematician Pierre de Fermat (1607–1665), who conjectured (1650) that these numbers would always give a prime for $n = 0, 1, 2, \dots$. In fact, the first five Fermat numbers are primes: $F_0 = 3$, $F_1 = 5$, $F_2 = 17$, $F_3 = 257$, $F_4 = 65537$.

But in 1732 Leonhard Euler (1707–1783), a Swiss-Russian mathematician, physicist, astronomer, geographer, logician and engineer, found that F_5 is a composite number. So, Fermat's conjecture that F_n is always prime is clearly false. Moreover, beyond F_4, no further Fermat primes have been found.

The discovery that established further interest in these numbers came in 1796 when Johann Carl Friedrich Gauss (1777–1855), a German mathematician and physicist, proved the constructibility of the regular 17-gon. The *Gauss–Wantzel theorem* states that a regular n-gon can be constructed with compass and straightedge if and only if n is the product of a power of 2 and any number of distinct Fermat primes (including none).

Nowadays, the theory of Mersenne and Fermat numbers contains many interesting mathematical facts and theorems, as well as a lot of important applications.

Mathematical part of this theory is closely connected with classical Arithmetics and Number Theory. It contains many information about divisibility properties of Mersenne and Fermat numbers, their connections with other classes of special numbers, etc.

Moreover, Mersenne and Fermat numbers are used in the construction of new large primes, and have numerous applications in contemporary Cryptography. For these applications, one should study well-known deterministic and probabilistic primality tests, standard algorithms of integer factorization, the questions and open problems of Computational Complexity Theory.

The book deals in detail with all these aspects.

In Chapter 1 we consider some important questions of elementary Number Theory, which will be used later. In elementary Number Theory, integers are studied without use of techniques from other mathematical fields. Questions of divisibility, use of the Euclidean algorithm to compute greatest common divisors, factorization of integers into prime numbers, investigation of behaviour of prime numbers and congruences, properties and applications of continued fractions belong to this part of elementary Mathematics. Several important discoveries of this field are the Fermat's little theorem, the Euler's theorem, the Chinese remainder theorem and the law of quadratic reciprocity. The properties of multiplicative functions such as the Möbius function, Euler's totient function, divisor function integer sequences and factorials all also fall into this area. In this Chapter of the book we collect: the basic facts of the Theory of divisibility of integers; main questions of modular Arithmetics; examples of important arithmetic functions; algorithms of solutions of congruences; a base of the Theory of quadratic residues (including properties of Legendre symbol and Jacobi symbol); definitions and properties of multiplicative order, primitive roots and indexes (discrete logarithms) modulo n; some important elementary properties of continued fractions. We represent all needed definitions, give many examples, collect the main properties of considered mathematical objects without proofs, but with large list of possible references.

As the main part of the theory of Mersenne and Fermat numbers focuses on the questions of their primality, in Chapter 2 we give the basic Theory of prime numbers, including the history of the question, elementary properties of primes, the well-known classical algorithms of primality testing, examples of formulas of primes, large collection of the famous and hidden connections of prime numbers with other classes of special numbers. In the end of this Chapter the full list of open problems of the Theory of prime numbers is represented.

Chapter 3 contains a rich review of the mathematical Theory of Mersenne numbers, with all needed proofs and many examples. Starting with the history of the question, we consider elementary properties of Mersenne numbers, including the questions of their divisibility and features of their recurrent behaviour. As the main part of the Theory of Mersenne numbers is connected with problems of their possible primality, we consider the well-known properties of prime divisors of Mersenne numbers, as well as prove the algorithm of Lucas-Lehmer primality test for Mersenne numbers. Moreover, we give the large collection of possible connections of Mersenne numbers with other classes of special numbers, including perfect numbers and elements of Pascal's triangle. At last, we represent a rich list of open problems in Number Theory, connected with Mersenne numbers.

In Chapter 4 we consider similarly main questions of the Theory of Fermat numbers. The structure of the Chapter 4 is the same, as the structure of the Chapter 3. Starting with the history of the question, we consider elementary properties of Fermat numbers, give the list of possible prime divisors of Fermat numbers, prove the Pépin's primality test for Fermat numbers. After we give a large collection of the possible connections of Fermat numbers with other classes of special numbers, and present a rich list of open problems in Number Theory, connected with Fermat numbers.

Chapter 5 contains some important questions, concerning to the classical and contemporary applications of Mersenne and Fermat numbers. First of all, we prove the Euclid–Euler theorem and give a detailed sketch of the classical proof of the Gauss–Wantzel theorem. After we consider many practical questions and interesting facts, connected with the research of large prime numbers, first of all, Mersenne primes. At last, we collect the most important number-theoretical aspects of applications of Mersenne and Fermat numbers in Cryptography.

In a small Chapter 6, we collected some remarkable individual numbers, related to the Theory of Mersenne and Fermat numbers, as well as to the general Theory of prime numbers.

In Chapter 7, a huge Mini Dictionary lists all classes of special numbers mentioned in the text.

The Chapter 7 "Exercises" contain a big exclusive collection of problems, connected with the Theory of Mersenne and Fermat numbers, with full solutions or, at least, with short hints of proofs.

The target audience of the book consists of university professors and students (especially graduate ones) interested in Number Theory, General Algebra, Cryptography and related fields, as well as school teachers and general audience of amateurs of Mathematics.

Specifically, Chapters 2, 3 and 4 are accessible to undergraduate students and general readers of Mathematics, while Chapters 5 is slightly more difficult; it is more involved and requires some (still basic) mathematical and computational culture.

The book is so organized that it can be used as a source material for individual scientific work of university students and even, partially, of school students. In fact, the material of the book was already used for the preparation of bachelor's and master's theses of many students of Moscow State Pedagogical University.

Chapter 1

Preliminaries

1.1 Divisibility of integers

By positive integer (natural) numbers we mean the numbers $1, 2, 3, \ldots$, by integers we mean the positive integers, the number zero, and the negative integers: $\ldots, -3, -2, -1, 0, 1, 2, 3, \ldots$

Division algorithm

1.1.1. The *division algorithm* (or *integer division algorithm*) is a theorem in Mathematics which precisely expresses the outcome of the usual process of division of integers.

Theorem (division algorithm). *For any $a \in \mathbb{Z}$ and $b \in \mathbb{N}$, there exist unique integers q and r, such that $a = bq + r$, and $0 \le r < b$.*

The number q is called the *quotient*, r is called the *remainder* and is denoted by $rest(a, b)$, b is called the *divisor*, a is called the *dividend*.

For example, $-10 = 3 \cdot (-4) + 2$; $22 = 3 \cdot 7 + 1$; $100 = 20 \cdot 5 + 0$.

Divisibility

1.1.2. We say that *an integer a is divisible by an integer b, $b \ne 0$,* and write $b|a$, *if there is an integer c such that*

$$a = b \cdot c.$$

In this case b is called a *divisor* of a, and a is called a *multiply* of b.

For example, $2|10$, as $10 = 2 \cdot 5$; $(-6)|42$, as $42 = (-6) \cdot (-7)$; but $14 \nmid 25$, as there exists no integer c, such that $14 = 25 \cdot c$.

Divisors can be negative as well as positive, although often we restrict our attention to positive divisors.

1 and -1 divide (are divisors of) every integer, every integer (except 0) is a divisor of itself, and every integer is a divisor of 0, except by convention 0 itself.

Numbers divisible by 2 are called *even*; numbers not divisible by 2 are called *odd*.

A divisor of n that is not 1, -1, n or $-n$ is known as a *non-trivial divisor*.

For example, 6 is an even number, as it is divisible by 2; it has four trivial divisors ± 1, ± 6, and four non-trivial divisors ± 2, ± 3; 3 is an odd number, as it is not divisible by 2; it has only trivial divisors ± 1, ± 3; 2 is an even number; it has only trivial divisors ± 1, ± 2; 1 is an odd number, as it is not divisible by 2; it has only trivial divisors ± 1.

A positive divisor of n which is different from n is called a *proper divisor* (or *aliquot part*) of n.

For example, 6 has three proper divisors: 1, 2, and 3.

Properties of divisibility

1. If $a|b$ and $a|c$, then $a|(b+c)$; in fact, $a|(mb+nc)$ for all integers m and n.

2. If $a|b$ and $b|c$, then $a|c$.

3. If $a|b$ and $b|a$, then $a = b$, or $a = -b$.

The proof of these properties and some additional information see, for example, in [Buch09], [DeKo13].

Prime and composite numbers

1.1.3. Positive integers with non-trivial divisors are known as *composite numbers*, while *prime numbers* have no non-trivial divisors.

Formally, a positive integer p is called *prime*, if it has exactly two positive integer divisors (which are 1 and the prime number itself).

For example, 2 is a prime number, as it has exactly two positive divisors, 1 and 2; 3 is a prime number, as it has exactly two positive divisors, 1 and 3.

The first 30 prime numbers are 2, 3, 5, 7, 11, 13, 17, 19, 23, 29, 31, 37, 41, 43, 47, 53, 59, 61, 67, 71, 73, 79, 83, 89, 97, 101, 103, 107, 109, and 113 (see sequence A000040 in OEIS).

The property of being a prime is called *primality*.

A positive integer n is called *composite*, if it has more than two positive integer divisors.

For example, 4 is a composite number, as it has three positive divisors, 1, 2, 4; 6 is a composite number, as it has four positive divisors, 1, 2, 3, 6; 15 is a composite number, as it has four positive divisors, 1, 3, 5, 15.

The first 30 composite numbers are 4, 6, 8, 9, 10, 12, 14, 15, 16, 18, 20, 21, 22, 24, 25, 26, 27, 28, 30 32, 33, 34, 35, 36, 38, 39, 40, 42, 44, and 45 (see the sequence A002808 in the OEIS).

The number 1 has exactly one positive divisor, 1 itself. It is the only positive integer, which is neither prime, nor composite.

So, any positive integer is or prime number, or composite number, or is equal to 1.

Any composite number n can be represented as $n = a \cdot b$, where $1 < a \leq b < n$.

For example, $4 = 2 \cdot 2$; $6 = 2 \cdot 3$; $20 = 2 \cdot 10 = 4 \cdot 5$.

For any prime number, if $p = a \cdot b$, then $a, b \in \{\pm 1, \pm p\}$.

Any natural number n greater than one has a prime divisor; in fact, it is the smallest positive divisor of n, greater than 1.

For example, $2|4$; $3|15$; $5|5$.

The *fundamental theorem of Arithmetic* (or *unique factorization theorem*) states that *every positive integer larger than 1 can be written as a product of one or more primes in an unique way, i.e., unique except for the order.*

For example, $4 = 2^2$; $5 = 5$; $6 = 2 \cdot 3$; $600 = 2^3 \cdot 3 \cdot 5^2$.

For additional information see, for example, [Buch09], [DeKo13], [Ribe89], [Sier64].

Greastest common divisor and least common multiple

1.1.4. Given integers $a_1, \ldots, a_n$, not all equal to zero, the *greatest common divisor* $gcd(a_1, \ldots, a_n)$ of $a_1, \ldots, a_n$ is defined as the largest positive integer that divides all numbers $a_1, \ldots, a_n$.

For example, the set of divisors of the number 4 is $\{\pm 1, \pm 2, \pm 4\}$; the set of divisors of the number 6 is $\{\pm 1, \pm 2, \pm 3, \pm 6\}$; so, the set of common divisors of the numbers 4 and 6 is $\{\pm 1, \pm 2\}$, and $gcd(4, 6) = 2$.

Given non-zero integers $a_1, \ldots, a_n$, the *least common multiple* $lcm(a_1, \ldots, a_n)$ of $a_1, \ldots, a_n$ is defined as the smallest positive integer that is a multiple of all numbers $a_1, \ldots, a_n$.

For example, the set of multiplies of the number 4 is $\{\pm 4, \pm 8, \pm 12, \pm 16, \pm 20, \pm 24, \ldots\}$; the set of multiplies of the number 6 is $\{\pm 6, \pm 12, \pm 18, \pm 24, \pm 30, \pm 36, \ldots\}$; so, the set of common multiplies of the numbers 4 and 6 is $\{\pm 12, \pm 24, \pm 36, \ldots\}$, and $lcm(4, 6) = 12$.

Properties of *gcd* and *lcm*

1. $gcd(p_1^{\alpha_1} \cdot \ldots \cdot p_k^{\alpha_k}, p_1^{\beta_1} \cdot \ldots \cdot p_k^{\beta_k}) = p_1^{\gamma_1} \cdot \ldots \cdot p_k^{\gamma_k}$, where $\alpha_i, \beta_i \geq 0$, and $\gamma_i = \min\{\alpha_i, \beta_i\}$, $i = 1, 2, \ldots, k$.

2. $lcm(p_1^{\alpha_1} \cdot \ldots \cdot p_k^{\alpha_k}, p_1^{\beta_1} \cdot \ldots p_k^{\beta_k}) = p_1^{\delta_1} \cdot \ldots \cdot p_k^{\delta_k}$, where $\alpha_i, \beta_i \geq 0$, and $\delta_i = \max\{\alpha_i, \beta_i\}$, $i = 1, 2, \ldots, k$.

3. Every common divisor of a and b is a divisor of $gcd(a, b)$.

4. If $gcd(a, b) = d$, then there exist integers x and y, such that $ax + by = d$ (*Bézout's identity*).

5. If $a|bc$, and $gcd(a, b) = d$, then $\frac{a}{d}|c$.

6. $gcd(m \cdot a, m \cdot b) = m \cdot gcd(a, b)$ for any $m \in \mathbb{N}$.

7. $gcd(a + m \cdot b, b) = gcd(a, b)$ for any $m \in \mathbb{Z}$.

8. If $m|a$ and $m|b$, where $m \in \mathbb{N}$, then $gcd(\frac{a}{m}, \frac{b}{m}) = \frac{gcd(a,b)}{m}$.

9. $gcd(a, b) \cdot lcm(a, b) = a \cdot b$ for any positive integers a, b.

It is useful to define $\gcd(0, 0) = 0$ and $\text{lcm}(0, 0) = 0$ because then the set of natural numbers becomes a complete distributive lattice with *gcd* as meet and *lcm* as join operation.

The greatest common divisor can more generally be defined for elements of an arbitrary commutative ring.

For a proof of these properties and for some additional information see, for example, [Buch09], [DeKo13], [Kost82].

Euclidean Algorithm

1.1.5. The *Euclidean algorithm* (also called *Euclid's algorithm*) is an algorithm to determine the greatest common divisor of two integers by repeatedly dividing the two numbers and the remainders in turns.

Theorem (Euclidean Algorithm). *For any integer a and any natural b which does not divide a, the greatest common divisor of a and b is equal to the last non-zero remainder r_s in the following algorithm:*

$$a = bq_1 + r_1,$$

$$b = r_1q_2 + r_2,$$

$$\ldots,$$

$$r_{s-2} = r_{s-1}q_s + r_s,$$

$$r_{s-1} = r_sq_s + 0,$$

where $r_1 > r_2 > \cdots > r_s > 0$.

For example, as

$$1071 = 1029 \cdot 1 + 42,$$

$$1029 = 42 \cdot 24 + 21,$$

$$42 = 21 \cdot 2 + 0,$$

we obtain, that $gcd(1071, 1029) = 21$.

The Euclidean algorithm is one of the oldest known algorithms, since it appeared in Euclid's *Elements* around 300 BC. Euclid originally formulated the problem geometrically, as the problem of finding a common "measure" for two line lengths, and his algorithm proceeded by repeated subtraction of the shorter from the longer segment.

However, the algorithm was probably not discovered by Euclid and it may have been known up to 200 years earlier. It was almost certainly known by Eudoxus of Cnidus (about 375 BC); Aristotle (about 330 BC) hinted at it in his *Topics*.

This algorithm can be used in any context where division with remainder is possible. This includes rings of polynomials over a field as well as the ring of Gaussian integers, and in general all Euclidean domains.

For a proof of the theorem and for some additional information see, for example, [Buch09], [DeKo13], [DeKo18], [Kost82], [Stra16].

Coprime numbers

1.1.6. Two integers a and b are called *coprime* (or *relatively prime*) if their greatest common divisor is equal to 1, i.e., they have no common factors other than 1 and -1.

For example, 6 and 35 are coprime, as $gcd(6, 35) = 1$, but 6 and 27 are not because $gcd(6, 27) = 3$.

The number 1 is coprime to every integer; 0 is coprime only to 1 and -1.

Properties of coprime numbers

1. a and b are coprime if and only if there exist integers x and y, such that $ax + by = 1$.

2. If $b|ac$, and $gcd(b, c) = 1$, then $b|a$.

3. If $b|a, c|a$, and $gcd(b, c) = 1$, then $bc|a$.

4. $gcd(a, b) = 1$ if and only if $gcd(a^n, b^m) = 1$ for any non-negative integers m, n.

For the first property, giving an criterion of coprime numbers, see *Bézout's identity*. The second property can be viewed as a generalization of *Euclid's lemma*, which states that if p is a prime, and $p|bc$, then either $p|b$, or $p|c$.

In fact, we can easily check all properties of coprime numbers using the fundamental theorem of Arithmetics, as two coprime numbers have no common prime divisors.

For a proof of these properties and for some additional information see, for example, [Arno38], [BiBa70], [Buch09], [CoGu96], [DeKo13], [DeKo18], [Dick05], [Gelf98], [IrRo90], [Knut68], [Kost82], [Lege79], [Moze09], [Ore48], [Stra16], [Wiki20].

Exercises

1. Prove, that any integer number can be represented as:

   (a) $6k$, or $6k \pm 1$, or $6k \pm 2$, or $6k + 3$, $k \in \mathbb{Z}$;
   (b) $7k$, or $7k \pm 1$, or $7k \pm 2$, or $7k \pm 3$, $k \in \mathbb{Z}$;
   (c) $10k$, or $10k \pm 1$, or $10k \pm 2$, or $10k \pm 3$, or $10k \pm 4$, or $10k + 5$, $k \in \mathbb{Z}$;
   (d) $12k$, or $12k \pm 1$, or $12k \pm 2$, or $12k \pm 3$, or $12k \pm 4$, or $12k \pm 5$, or $12k + 6$, $k \in \mathbb{Z}$.

2. Prove, that a perfect square cannot be represented as:

   (a) $3k - 1$; (c) $5k + 2$; (e) $6k + 2$;

   (b) $4k - 1$; (d) $5k - 2$; (f) $6k - 1$.

3. Prove, that $a^2 + b^2 \neq c^2$ for any odd integers a and b.

4. Prove, that $a^2 + b^2 \neq c^3$ for any odd integers a and b.

5. Prove, that for any integer a it holds:

   (a) $a^{10} - 9a + 8$ is divisible by 2;

   (b) $a^5 + 3a^3 - 12$ is divisible by 4;

   (c) $a^3 - 7a + 18$ is divisible by 6;

   (d) $a^7 - a - 56$ is divisible by 7;

   (e) $a^5 - 17a^3 + 24$ is divisible by 8;

   (f) $a^7 + 17a - 18$ is divisible by 9.

6. Prove that $n \cdot (n^2 + 1) \cdot (n^2 + 4)$ is divisible by 5 for any integer n.

7. Prove that $5^{2n+1} \cdot 2^{n+2} + 3^{n+2} \cdot 2^{2n+1}$ is divisible by 19 for any $n \geq 0$.

8. Find greatest common divisor of 11111111 and $\underbrace{111......111}_{100 \text{ unities}}$.

9. For any positive integer n find greatest common divisor of:

   (a) $n^3 + 11n$ and 6;   (c) $n^5 + 4n$ and 10;

   (b) $(n^2 - 1) \cdot n^2 \cdot (n^2 + 1)$ and   (d) $n^5 - 5n^4 + 4n$ and 10.
       60;

10. Prove, that for any positive integer n it holds:

   (a) $n^6 - n$ is divisible by 60;   (c) $10^n + 5$ is divisible by 15;

   (b) $n^7 - n^3$ is divisible by 120; (d) $7^{6n} - 1$ is divisible by 18.

1.2 Modular arithmetics

Congruence relation

1.2.1. *Modular Arithmetic* (sometimes called *modulo Arithmetic*) is a system of Arithmetic for integers, where numbers "wrap around" after they reach a certain value — the *modulus*. Modular Arithmetic was introduced by C.F. Gauss in his book *Disquisitiones Arithmeticae*, published in 1801.

Two integers a and b are called *congruent modulo n*, $n \in \mathbb{N}$, if a and b have the same remainder when divided by n, or, equivalently, if $n|(a - b)$. In this case, it is expressed as

$$a \equiv b (mod\ n).$$

For example, $6 \equiv 22 (mod\ 4)$, as $4|(22 - 6)$; $2 \equiv -7 (mod\ 3)$, as $3|(2 - (-7))$; but $5 \not\equiv 12 (mod\ 6)$, as $6 \nmid (5 - 12)$.

Properties of congruence relation

1. $a \equiv a (mod\ n)$ (*reflexivity*);

2. if $a \equiv b (mod\ n)$, then $b \equiv a (mod\ n)$ (*symmetry*);

3. if $a \equiv b(mod\ n)$, and $b \equiv c(mod\ n)$, then $a \equiv c(mod\ n)$ (*transitivity*);

4. If $a \equiv b(mod\ n)$, then $f(a) \equiv f(b)(mod\ n)$, where $f(x) = a_m x^m + a_{m-1} x^{m-1} + \cdots + a_1 x + a_0$ is a polynomial with integer coefficients $a_m, \ldots, a_1, a_0$.

5. $a \equiv b(mod\ n)$ if and only if $k \cdot a \equiv k \cdot b(mod\ k \cdot n)$ for any positive integer k.

6. $a \equiv b(mod\ n)$ if and only if $k \cdot a \equiv k \cdot b(mod\ n)$ for any integer k, such that $gcd(k, n) = 1$.

7. $a \equiv b(mod\ n_i)$, $i = 1, \ldots, k$, if and only if $a \equiv b(mod\ M)$, where $M = lcm(n_1, \ldots, n_k)$.

Easy to see (properties 1, 2, 3), that the congruence relation $\equiv$ modulo n is an *equivalence relation* on the set $\mathbb{Z}$ of integers.

The proof of these properties and some additional information see, for example, in [Buch09], [DeKo13], [Gaus01].

Congruence classes

1.2.2. The set

$$[a]_n = \{x \in \mathbb{Z} : x \equiv a(mod\ n)\}$$
$$= \{\ldots, a - 3n, a - 2n, a - n, a, a + n, a + 2n, a + 3n, \ldots\}$$

of all integers congruent to a given number a modulo n is called the *congruence class* (or *residue class*) *of a modulo n*.

The smallest non-negative representative of $[a]_n$ is called *smallest non-negative residue* of a modulo n; it is the remainder $rest(a, n)$ of a after its integer division by n. The smallest in absolute value representative of $[a]_n$ is called *minimal residue* of a modulo n.

For example,

$$[2]_5 = \{x \in \mathbb{Z} : x \equiv 2(mod\ 5)\}$$
$$= \{\ldots, 2 - 3 \cdot 5, 2 - 2 \cdot 5, 2 - 5, 2, 2 + 5, 2 + 2 \cdot 5, 2 + 3 \cdot 5, \ldots\}$$
$$= \{\ldots, -13, -8, -3, 2, 7, 12, 17, \ldots\}.$$

The smallest non-negative residue of the class $[2]_5$, as well as its minimal residue, is 2. On the other hand,

$$[19]_5 = \{x \in \mathbb{Z} : x \equiv 19(mod\ 5)\} = \{\ldots, -6, -1, 4, 9, 14, 19, 24, \ldots\}.$$

The smallest non-negative residue of the class $[19]_5$ is 4, while its minimal residue is -1.

Properties of residue classes modulo n

1. $[a]_n = \{a + mt : t \in \mathbb{Z}\}$.

2. $[a]_n = [b]_n$ if and only if $a \equiv b(mod\ n)$.

3. There are exactly n residue classes modulo n.

4. If $x \in [a]_n$, then $gcd(x, n) = gcd(a, n)$.

5. A residue class $[a]_n$ modulo n gives k residue classes $[a]_{kn}$, $[a + n]_{kn}$, $[a + 2n]_{kn}, \ldots, [a + (k - 1)n]_{kn}$ modulo kn, $k \in \mathbb{N}$.

One can define an addition and multiplication on the set

$$Z/nZ = \{[0]_n, [1]_n, [2]_n, \ldots, [n - 1]_n\}$$

of all equivalence classes modulo n by the following rules:

$$[a]_n + [b]_n = [a + b]_n, \quad \text{and} \quad [a]_n \cdot [b]_n = [a \cdot b]_n.$$

On this way, Z/nZ becomes a commutative ring $(Z/nZ, +, \cdot)$ with n elements. For a prime number p, the ring $(Z/pZ, +, \cdot)$ forms a field.

The proof of all these facts and some additional information see, for example, in [Arno38], [Buch09], [CoGu96], [DeKo13], [DeKo18], [Dick05], [Gaus01], [Gelf98], [Kost82], [Lege79], [Moze09], [Ore48], [Wiki20].

Exercises

1. Is the congruence $10! \equiv 7!(mod\ 1000)$ true?

2. Is the congruence $\binom{14}{7} \equiv \binom{10}{5}(mod\ 5710)$ true?

3. Find $rest(f(86), 11)$, if $f(x) = 15x^3 - 33x^2 + 7$.

4. Find the smallest 4-digital positive integer, congruent to 23 modulo 101; modulo 100; modulo 99.

5. Prove that $9^{2n+1} + 8^{n+2} \equiv 0(mod\ 73)$ for any $n \geq 0$.

6. For which n it holds:

    (a) $51 \cdot 52 \cdot ... \cdot 300 \equiv 0(mod\ 11^n)$;

    (b) $51 \cdot 52 \cdot ... \cdot 300 \equiv 0(mod\ 15^n)$;

    (c) $31 \cdot 32 \cdot ... \cdot 400 \equiv 0(mod\ 7^n)$;

    (d) $31 \cdot 32 \cdot ... \cdot 400 \equiv 0(mod\ 33^n)$?

7. Check, that:

    (a) $[73]_5 = [-92]_5$;   (c) $[3!]_8 = [-2!]_8$;

    (b) $[99]_6 = [-87]_6$;   (d) $[12!]_9 = [15!]_9$.

8. Find the result of the calculation:

    (a) $[2]_{12} \cdot [9]_{12} + 2 \cdot [5]_{12}$;

    (b) $3 \cdot [4]_{14} \cdot [4]_{14} - 7 \cdot [9]_{14}$;

    (c) $34 \cdot [4]_{17} \div [2]_{17} + 5 \cdot ([4]_{17})^2$;

    (d) $2 \cdot ([5]_{23})^3 - [18]_{23} - 69 \cdot [5]_{23} \div [3]_{23}$.

1.3 Arithmetic functions

A complex-valued function $f(n)$ is called *arithmetic function*, if the value $f(n)$ is determined for any positive integer n.

Examples of arithmetic functions

1.3.1. There are many examples of arithmetic functions, among them:

- *floor function* $\lfloor x \rfloor$, i.e., the greatest integer, less than or equal a given real number x;

- *Möbius function* $\mu(n)$, defined as follows: $\mu(1) = 1$, $\mu(p_1 \cdot \ldots \cdot p_k) = (-1)^k$ for a square-free number $n = p_1 \cdot \ldots \cdot p_k$, and $\mu(n) = 0$, otherwise;

- *divisor function* $\sigma_k(n) = \sum_{d|n} d^k$, where d runs through all positive divisors of a positive integer n; in particular, for $k = 0$ we obtain the *number of divisors' function* $\tau(n) = \sum_{d|n} 1$, while for $k = 1$ we obtain the *sum of divisors' function* $\sigma(n) = \sum_{d|n} d$.

- *Euler's totient function* $\phi(n) = |\{x \in \mathbb{N} : x \leq n, gcd(x, n) = 1\}|$;

- *prime counting function* $\pi(x) = \sum_{p \leq x} 1$.

Multiplicative functions

1.3.2. An arithmetic function f is called *multiplicative*, if $f(1) = 1$, and

$$f(m \cdot n) = f(m) \cdot f(n)$$

for any coprime positive integers n and m.

An multiplicative function f is called *completely multiplicative*, if the condition $f(m \cdot n) = f(m) \cdot f(n)$ holds for any positive integers m and n.

For example, the function $f(n) = n^\alpha$ is completely multiplicative for any real number α. The divisor function, the Möbius function, the Euler's toient function are multiplicative, but not completely multiplicative. The prime counting function $\pi(x)$ is not a multiplicative function.

Properties of multiplicative functions

1. The product of two multiplicative functions is a multiplicative function.

2. If f is a multiplicative function, then, for a given $n = p_1^{\alpha_1} \cdot \ldots \cdot p_k^{\alpha_k}$, one has

$$\sum_{d|n} f(d) = \prod_{i=1}^{k} (1 + f(p_i) + f(p_i^2) + \cdots + f(p_i^{\alpha_k})).$$

3. If f is a multiplicative function, then $h(n) = \sum_{d|n} f(d)$ is a multiplicative function.

The proof of these properties and some additional information see, for example, in [Buch09], [DeKo13], [Chan70], [Ingh32], [Step01].

Euler's totient function

1.3.3. Given a positive integer n, the *Euler's totient function* $\phi(n)$ is defined as the number of positive integers less than or equal to n and coprime to n:

$$\phi(n) = |\{x \in \mathbb{N} : x \le n, gcd(x, n) = 1\}|.$$

For example, $\phi(8) = 4$, since exactly four numbers, 1, 3, 5 and 7, less than or equal to 8, are coprime to 8.

The function ϕ is usually called the *Euler's totient* (or *Euler totient*) after L. Euler, who studied it. The totient function is also called *Euler's phi function* or simply *phi function*, since the letter ϕ is so commonly used for it.

Properties of Euler's totient function

1. $\phi(p) = p - 1$ for any prime p.

2. $\phi(p^\alpha) = p^\alpha - p^{\alpha-1}$ for any prime p and any positive integer α.

3. The Euler's totient function is multiplicative, i.e., $\phi(m \cdot n) = \phi(m) \cdot \phi(n)$ for any coprime positive integers m and n.

4. $\phi(p_1^{\alpha_1} \cdot p_2^{\alpha_2} \cdot ... \cdot p_s^{\alpha_s}) = (p_1^{\alpha_1} - p_1^{\alpha_1-1})(p_2^{\alpha_2} - p_2^{\alpha_2-1}) \cdot ... \cdot (p_s^{\alpha_s} - p_s^{\alpha_s-1})$ for any primes $p_1 < p_2 < \cdots < p_s$ and any positive integers $\alpha_1, \alpha_2, \ldots, \alpha_s$.

5. $\sum_{d|n} \phi(d) = n$, where the sum extends over all positive divisors d of n (*Gauss's identity*).

The Euler's totient function is important mainly because it gives the size of the multiplicative group of integers modulo n. More precisely, $\phi(n)$ is the order of the group of units of the ring $(Z/nZ, +, \cdot)$.

The proof of properties of the Euler's totient function and some additional information see, for example, in [Buch09], [DeKo13].

Fermat's little theorem and Euler's theorem

1.3.4. Now we consider two important theorems: the Fermat's little theorem, and its generalization, the *Euler's theorem.*

Theorem (Fermat's little theorem). *If p is a prime number, then, for any integer a,*

$$a^p \equiv a(mod \ p).$$

It means, that if you start with a number, initialized to 1, and repeatedly multiply, for a total of p multiplications, $p \in P$, that number by a, $a \in \mathbb{Z}$, and then subtract a from the resulting number, the final result will be divisible by p.

A variant of this theorem is stated in the following form: *if p is a prime, and a is an integer coprime to p, then*

$$a^{p-1} \equiv 1(mod \ p).$$

In other words, if p is a prime number and a is any integer that does not have p as a factor, then a raised to the $(p-1)$-th power will leave a remainder of 1 when divided by p.

Fermat's little theorem is the basis for the *Fermat primality test* (see Chapter 2, section 2.3).

1.3.5. The *Euler's theorem* (also known as the *Fermat-Euler theorem* or *Euler's totient theorem*) is a generalization of Fermat's little theorem. L. Euler published a proof of this theorem in 1736.

Theorem (Euler's theorem). *If n is a positive integer, and a is an integer coprime to n, then*

$$a^{\phi(n)} \equiv 1(mod \ n).$$

Here $\phi(n)$ is the Euler's totient function.

The Euler's theorem may be used to easily reduce large powers modulo n.

1.3.6. The Euler's theorem is further generalized by the *Carmichael's theorem*.

Theorem (Carmichael's theorem). *If n is a positive integer, and a is an integer coprime to n, then*

$$a^{\lambda(n)} \equiv 1(mod\ n).$$

Here $\lambda(n)$ is the *Carmichael function*, defined recursively:

- for prime $p \geq 3$ and positive integer α, $\lambda(p^\alpha) = \phi(p^\alpha)$;

- for positive integer $\alpha \geq 3$, $\lambda(2^\alpha) = 2^{\alpha-2}$, while $\lambda(2) = 1$, and $\lambda(4) = 2$;

- at last, $\lambda(p_1^{\alpha_1} \cdot \ldots \cdot p_k^{\alpha_k}) = lcm(\lambda(p_1^{\alpha_1}), \ldots, \lambda(p_k^{\alpha_k}))$.

The proofs of these theorems and some additional information see, for example, in [Buch09], [DeKo13], [DeKo18], [Dick05], [IrRo90], [Lege79], [Moze09], [Ore48], [Sier64], [Step01], [Stra16], [Wiki20].

Exercises

1. Prove, that $n! = p_1^{\alpha_1} \cdot \ldots \cdot p_k^{\alpha_k}$, where p_i are all primes less than or equal to n, and $\alpha_i = \lfloor \frac{n}{p_i} \rfloor + \lfloor \frac{n}{p_i^2} \rfloor + \lfloor \frac{n}{p_i^3} \rfloor + \cdots$.

2. Find:

   (a) $\mu(\sigma(14))$;   (b) $\mu(\tau(10)!)$;   (c) $\mu(\phi(15))$;   (d) $\mu(\frac{10!}{5!5!})$.

3. Find the sum:

   (a) $\sum\limits_{d|n} \frac{\mu(d)}{d}$;

   (b) $\sum\limits_{d|n} \frac{\mu(d)}{\tau(d)}$;

   (c) $\sum\limits_{d|n} \mu(d)\tau^3(d)$;

   (d) $\sum\limits_{d|n} \mu\left(\frac{n}{d}\right)(-7)^{\nu(d)}$;

   (e) $\sum\limits_{d|n} \mu(d)3^{\nu(d)}$;

   (f) $\sum\limits_{d|n} \mu\left(\frac{n}{d}\right)(-3)^{\nu(d)}$;

   (g) $\sum\limits_{d|n} \frac{\mu^2(d)}{\phi(d)}$;

   (h) $\sum\limits_{d|n} \mu\left(\frac{n}{d}\right)\frac{d}{\phi(d)}$;

   (i) $\sum\limits_{d|n} \mu(d)2^{\nu(d)}$;

   (j) $\sum\limits_{d|n} \mu\left(\frac{n}{d}\right)(-1)^{\nu(d)}$.

4. Prove that:

   (a) $\tau(1) + \tau(2) + \cdots + \tau(n) = \lfloor \frac{n}{1} \rfloor + \lfloor \frac{n}{2} \rfloor + \cdots + \lfloor \frac{n}{n} \rfloor$;

   (b) $\sigma(1) + \sigma(2) + \cdots + \sigma(n) = 1 \cdot \lfloor \frac{n}{1} \rfloor + 2 \cdot \lfloor \frac{n}{2} \rfloor + \cdots + n \cdot \lfloor \frac{n}{n} \rfloor$.

5. Prove that $\phi(p_1^{\alpha_1} \cdot p_2^{\alpha_2} \cdot \dots \cdot p_s^{\alpha_s}) = p_1^{\alpha_1 - 1} \cdot p_2^{\alpha_2 - 1} \cdot \dots \cdot p_s^{\alpha_s - 1}(p_1 - 1) \cdot (p_2 - 1) \cdot \dots \cdot (p_s - 1)$.

6. Prove the Gauss *identity* $\sum_{d\mid n} \phi(d) = n$.

7. Find all integers x such that: $\phi(x) = 2$; $\phi(x) = 4$; $\phi(x) = 6$; $\phi(x) = 11$.

8. Find the remainder of the division:

   (a) 5^{14} by 7;   (d) 15^{175} by 11;   (g) 3^{20} by 28;

   (b) 24^{16} by 7;   (e) 3^{100} by 16;   (h) 31^{200} by 28.

   (c) 5^{100} by 11;   (f) 37^{100} by 16;

9. Is it true that for $f(x) = 292x^{181} - 121x^{133} + 252x^{122} - 171x^{121} - 133x^{62} + 3$ it holds:

   (a) $f(24) \equiv -2 (mod\ 13)$;   (c) $f(-55) \equiv -4 (mod\ 13)$;

   (b) $f(-24) \equiv 2 (mod\ 11)$;   (d) $f(55) \equiv 4 (mod\ 11)$?

1.4   Solution of congruences

1.4.1. Let $f(x) = a_m x^m + \cdots + a_1 x + a_0$ be an polynomial with integer coefficients, and $a_m \not\equiv 0 (mod\ n)$. If, for some $x_0 \in \mathbb{Z}$, $f(x_0) \equiv 0 (mod\ n)$, then for any $y \equiv x_0 (mod\ n)$ one has also that $f(y) \equiv 0 (mod\ n)$, and we say, that the residue class $[x_0]_n$, i.e., the class $x \equiv x_0 (mod\ n)$, is a solution of the congruence $f(x) \equiv 0 (mod\ n)$, which is called a *congruence of order m modulo n*.

Linear congruences

1.4.2. The *linear congruence theorem* states that *the linear congruence* $ax + b \equiv 0 (mod\ n)$ *has exactly one solution, if* $gcd(a, n) = 1$, *exactly d solutions, if* $gcd(a, n) = d$, $d\mid n$, *and no solutions, otherwise.*

Theorem (linear congruence theorem).

- *If $gcd(a, n) = 1$, the linear congruence $ax + b \equiv 0(mod\ n)$ has exactly one solution $x \equiv x_0(mod\ n)$;*
- *if $gcd(a, n) = d$, $d|n$, the linear congruence $ax + b \equiv 0(mod\ n)$ has exactly d solutions $x \equiv x_0 + k \cdot \frac{n}{d}(mod\ n)$, $k = 0, \dots, d - 1$, where $x \equiv x_0(mod\ \frac{n}{d})$ is the only solution of the linear congruence $\frac{a}{d}x + \frac{b}{d} \equiv 0(mod\ \frac{n}{d})$;*
- *if $gcd(a, n) = d$, $d \nmid n$, the linear congruence $ax + b \equiv 0(mod\ n)$ has no solutions.*

The only solution $x \equiv x_0(mod\ n)$ of a linear congruence $ax + b \equiv 0(mod\ n)$, $gcd(a, n) = 1$, can be found:

- by consideration of all representatives $0, \dots, n - 1$ of the residue classes modulo n;

- by the formula $x_0 \equiv -b \cdot a^{\phi(n)-1}(mod\ n)$;

- by consideration of congruences $x_0 \equiv \frac{-b+kn}{a}(mod\ n)$, $k = 0, 1, \dots, a - 1$, etc.

For example, a linear congruence

$$11x - 5 \equiv 0(mod\ 7)$$

has the only solution (a residue class modulo 7), as $gcd(11, 7) = 1$. It is equivalent to the congruence $4x \equiv -2(mod\ 7)$.

Checking the numbers $0, 1, 2, \dots, 6$, we obtain that $4 \cdot 3 \equiv -2(mod\ 7)$, i.e., the only solution of the congruence $4x \equiv -2(mod\ 7)$ is the residue class $x \equiv 3(mod\ 7)$.

On the other hand, using the formula

$$4x \equiv -2 + 7k(mod\ 7)$$

for $k = 0, 1, 2 \dots$, we obtain

$$4x \equiv -2(mod\ 7),\ 4x \equiv 5(mod\ 7),\ 4x \equiv 12(mod\ 7).$$

Dividing two sides of the last congruence by 4, we obtain, that $x \equiv 3(mod\ 7)$.

At last, using the Euler's totient function and the fact, that $4^{\phi(7)-1} = 4^5$, we get the following congruence:

$$x \equiv -2 \cdot 4^5 (mod\ 7).$$

As

$$4^5 \equiv (-3)^5 \equiv 9 \cdot (-27) \equiv 2 \cdot 1 \equiv 2(mod\ 7),$$

we get, that $x \equiv -2 \cdot 2 \equiv -4 \equiv 3(mod\ 7)$.

The proof of this theorem and some additional information see, for example, in [Buch09], [DeKo13].

Chinese remainder theorem

1.4.3. The *Chinese remainder theorem* states that *the system of linear congruences with pairwise coprime modules n_i has exactly one solution, a residue class modulo* $N = lcm(n_1, \ldots, n_k) = n_1 \cdot \ldots \cdot n_k$.

Theorem (Chinese remainder theorem). *The system of linear congruences*

$$\begin{cases} x \equiv c_1(mod\ n_1) \\ \ldots \\ x \equiv c_k(mod\ n_k) \end{cases}$$

with pairwise coprime modules $n_1, \ldots, n_k$ has exactly one solution, i.e., a residue class modulo

$$N = lcm(n_1, \ldots, n_k) = n_1 \cdot \ldots \cdot n_k,$$

which has the form

$$x \equiv \frac{N}{n_1}c_1 y_1 + \cdots + \frac{N}{n_k}c_k y_k (mod\ N),$$

where y_i is a solution of the linear congruence $\frac{N}{n_i}y \equiv 1(mod\ n_i)$.

For example, consider a system of linear congruences

$$\begin{cases} x & \equiv & 4 & (mod\ 5) \\ x & \equiv & 1 & (mod\ 12) \\ x & \equiv & 0 & (mod\ 7). \end{cases}$$

It is easy to see, that

$$lcm(5, 12, 7) = lcm(5, 2^2 \cdot 3, 7) = 2^2 \cdot 3 \cdot 5 \cdot 7 = 420.$$

As $x \equiv 1(mod\ 12)$, we get $x = 12t + 1$, where $t \in \mathbb{Z}$.

Using the representation $x = 12t + 1$ in the congruence $x \equiv 0(mod\ 7)$, we get a new congruence $12t + 1 \equiv 0(mod\ 7)$ with argument t. Easy to see, that $-2t \equiv -1(mod\ 7)$, or $-2t \equiv 6(mod\ 7)$, or $t \equiv -3(mod\ 7)$. So, $t = 7t_1 - 3$, where $t_1 \in \mathbb{Z}$, i.e., $x = 12(7t_1 - 3) + 1 = 84t_1 - 35$, $t_1 \in \mathbb{Z}$.

Using the representation $x = 84t_1 - 35$ in the congruence $x \equiv 4(mod\ 5)$, we get $84t_1 - 35 \equiv 4(mod\ 5)$ with argument t_1. Easy to see, that $4t_1 \equiv 4(mod\ 5)$, or $t_1 \equiv 1(mod\ 5)$. So, $t_1 = 5t_2 + 1$, $t_2 \in \mathbb{Z}$, i.e., $x = 84(5t_2 + 1) - 35 = 420t_2 + 49$, $t_2 \in \mathbb{Z}$.

We obtain, that the only solution of our system is the residue class $x \equiv 49(mod\ 420)$.

The original form of the theorem, contained in a III-rd-century book *Sun-tzu Suan-ching* of a Chinese mathematician Sun Tzu and later (1247) republished by Qin Jiushao, is a statement about simultaneous congruences; the general form of the Chinese remainder theorem can be formulated for rings and (two-sided) ideals.

The proof of the theorem and some additional information see, for example, in [Buch09], [DeKo13], [Dick05], [Kost82], [Ore48].

Congruences of order m

1.4.4. Here we consider a general algorithm of a solution of any congruence of order m, $m > 1$, composite modulo n.

I. Given a prime p, the congruence

$$f(x) \equiv 0(mod\ p)$$

of order m modulo p is equivalent to some congruence $x^s + \cdots + b_1 x + b_0 \equiv 0(mod\ p)$ of order $s \leq p - 1$ modulo n, and has at most s solutions.

One can find them by consideration of all representatives $0, \ldots, p - 1$ of the residue classes modulo p.

II. Given a prime p and a positive integer $\alpha > 1$, consider the congruence

$$f(x) \equiv 0 (mod \; p^\alpha)$$

of order m modulo p^α. We can use the following algorithm for finding all solutions of this congruence:

- Consider the congruence $f(x) \equiv 0 (mod \; p)$, find all its solutions; let $x \equiv x_1 (mod \; p)$ be one of these solutions, i.e., $f(x_1) \equiv 0 (mod \; p)$, and $x = x_1 + pt_1$, $t_1 \in \mathbb{Z}$.

- Consider the linear congruence $\frac{f(x_1)}{p} + f'(x_1)t_1 \equiv 0 (mod \; p)$, find all its solutions; let $t_1 \equiv t' (mod \; p)$ be one of these solutions, i.e., $t_1 = t' + pt_2$, and $x = x_2 + p^2 t_2$.

- Consider the linear congruence $\frac{f(x_2)}{p^2} + f'(x_2)t_2 \equiv 0 (mod \; p)$, find all its solutions; let $t_2 \equiv t'' (mod \; p)$ be one of these solutions, i.e., $t_2 = t'' + pt_3$, and $x = x_3 + p^3 t_3$.

- Consider the linear congruence $\frac{f(x_{\alpha-1})}{p^{\alpha-1}} + f'(x_{\alpha-1})t_{\alpha-1} \equiv 0 (mod \; p)$, find all its solutions; let $t_{\alpha-1} \equiv t''^{\cdots'} (mod \; p)$ be one of these solutions, i.e., $t_{\alpha-1} = t''^{\cdots'} + pt_\alpha$, and $x = x_\alpha + p^\alpha t_\alpha$.

Therefore, $x \equiv x_\alpha (mod \; p^\alpha)$ is one of the solutions of the congruence above, and all its solutions can be found on this way.

III. At last, for a given composite number $n = p_1^{\alpha_1} \cdot \ldots \cdot p_k^{\alpha_k}$, all the solutions of the congruence

$$f(x) \equiv 0 (mod \; n)$$

of order m modulo n can be found using the following property: $f(x) \equiv 0 (mod \; n)$ if and only if

$$\begin{cases} f(x) & \equiv & 0 (mod \; p_1^{\alpha_1}) \\ & \cdots & \\ f(x) & \equiv & 0 (mod \; p_k^{\alpha_k}). \end{cases}$$

1.4.5. For example, consider a congruence

$$x^5 - x^4 + 6x^2 + 15x + 45 \equiv 0 (mod \; 675).$$

Let $f(x) = x^5 - x^4 + 6x^2 + 15x + 45$. As $675 = 3^3 \cdot 5^2$, we have a congruence

$$f(x) \equiv 0 (mod\ 3^3 \cdot 5^2).$$

It is equivalent to the system

$$\begin{cases} f(x) & \equiv & 0 (mod\ 3^3) \\ f(x) & \equiv & 0 (mod\ 5^2). \end{cases}$$

A. Consider the congruence $f(x) \equiv 0 (mod\ 3^3)$.

In fact, start with the congruence $f(x) \equiv 0 (mod\ 3)$. Obviously, it is equivalent to the congruence $x^5 - x^4 \equiv 0 (mod\ 3)$.

It is easy to see, that $x \equiv 0 (mod\ 3)$ gives a solution of this congruence.

If $x \not\equiv 0 (mod\ 3)$, then $(x, 3) = 1$, and $x^2 \equiv 1 (mod\ 3)$. For such x we get the congruence $x - 1 \equiv 0 (mod\ 3)$, i.e., a solution $x \equiv 1 (mod\ 3)$.

Therefore, we get all (two) solutions of the congruence $f(x) \equiv 0 (mod\ 3)$: the classes $x \equiv 0 (mod\ 3)$, and $x \equiv 1 (mod\ 3)$.

A1. Consider $x \equiv 0 (mod\ 3)$ — one of the solutions of the congruence $f(x) \equiv 0 (mod\ 3)$. Obviously, $f(0) \equiv 0 (mod\ 3)$, and $x = 3t_1 + 0$, where $t_1 \in \mathbb{Z}$.

Consider the congruence $f(x) \equiv 0 (mod\ 3^2)$, where $x = 3t_1 + 0$, $t_1 \in \mathbb{Z}$. Let $\frac{f(0)}{3} + f'(0) \cdot t_1 \equiv 0 (mod\ 3)$. In fact, $f(0) = 45$. As $f'(x) = 5x^4 - 4x^3 + 12x + 15$, then $f'(0) = 15$. So, we get the congruence $\frac{45}{3} + 15 \cdot t_1 \equiv 0 (mod\ 3)$, or the congruence $15 + 15 \cdot t_1 \equiv 0 (mod\ 3)$. As $15 \equiv 0 (mod\ 3)$, we get three solutions: $t_1 \equiv 0 (mod\ 3)$, $t_1 \equiv 1 (mod\ 3)$, and $t_1 \equiv -1 (mod\ 3)$.

In the first case, $t_1 = 3t_2$, where $t_2 \in \mathbb{Z}$, and $x = 3t_1 + 0 = 3(3t_2) + 0 = 3^2 \cdot t_2 + 0$. In the second case, $t_1 = 3t_2 + 1$, where $t_2 \in \mathbb{Z}$, and $x = 3t_1 + 0 = 3(3t_2 + 1) + 0 = 3^2 \cdot t_2 + 3$. In the last case, $t_1 = 3t_2 - 1$, where $t_2 \in \mathbb{Z}$, and $x = 3t_1 + 0 = 3(3t_2 - 1) + 0 = 3^2 \cdot t_2 - 3$.

A.1.1. Let $f(x) \equiv 0 (mod\ 3^3)$, where $x = 3^2 t_2 + 0$, $t_2 \in \mathbb{Z}$. Consider the congruence $\frac{f(0)}{3^2} + f'(0) \cdot t_2 \equiv 0 (mod\ 3)$. As $f(0) = 45$ and $f'(0) = 15$, then we have the congruence $\frac{45}{9} + 15 \cdot t_2 \equiv 0 (mod\ 3)$, or the congruence $5 + 15 \cdot t_2 \equiv 0 (mod\ 3)$, which has no solutions.

A.1.2. Let $f(x) \equiv 0 (mod\ 3^3)$, where $x = 3^2 t_2 + 3$, $t_2 \in \mathbb{Z}$. Consider the congruence $\frac{f(3)}{3^2} + f'(3), t_2 \equiv 0 (mod\ 3)$. As $f(3) = 306$ and

$f'(3) = 348$, we have the congruence $\frac{306}{9} + 348 \cdot t_2 \equiv 0 (mod\ 3)$, or the congruence $34 + 348 \cdot t_2 \equiv 0 (mod\ 3)$, which has no solutions.

A.1.3. Let $f(x) \equiv 0 (mod\ 3^3)$, where $x = 3^2 t_2 - 3$, $t_2 \in \mathbb{Z}$. Consider the congruence $\frac{f(-3)}{3^2} + f'(-3) \cdot t_2 \equiv 0 (mod\ 3)$. As $f(-3) = 270$, and $f'(-3) = 492$, we have the congruence $\frac{270}{9} + 492 \cdot t_2 \equiv 0 (mod\ 3)$, or the congruence $30 + 492 \cdot t_2 \equiv 0 (mod\ 3)$, which has three solutions: $t_2 \equiv 0 (mod\ 3)$, $t_2 \equiv 1 (mod\ 3)$, and $t_2 \equiv -1 (mod\ 3)$.

In the first case, $t_2 = 3t_3$, where $t_3 \in \mathbb{Z}$, and $x = 3^2 t_2 - 3 = 3^2 (3t_3) - 3 = 3^3 \cdot t_3 - 3$.

In the second case, $t_2 = 3t_3 + 1$, where $t_3 \in \mathbb{Z}$, and $x = 3^2 t_2 - 3 = 3^2 (3t_3 + 1) - 3 = 3^3 \cdot t_3 + 6$.

In the third case, $t_2 = 3t_3 - 1$, where $t_3 \in \mathbb{Z}$, and $x = 3^2 t_2 - 3 = 3^2 (3t_3 - 1) - 3 = 3^2 \cdot t_3 - 12$.

So, we have all solutions of the congruence $f(x) \equiv 0 (mod\ 3^3)$, for which $x \equiv 0 (mod\ 3)$: they are the classes $x \equiv -3 (mod\ 3^3)$, $x \equiv -6 (mod\ 3^3)$, and $x \equiv -12 (mod\ 3^3)$.

A2. Consider $x \equiv 1 (mod\ 3)$ — the other solution of the congruence $f(x) \equiv 0 (mod\ 3)$. Obviously, $f(1) \equiv 0 (mod\ 3)$, and $x = 3t_1 + 1$, where $t_1 \in \mathbb{Z}$.

Let $f(x) \equiv 0 (mod\ 3^2)$, where $x = 3t_1 + 1$, $t_1 \in \mathbb{Z}$. Consider the congruence $\frac{f(1)}{3} + f'(1) \cdot t_1 \equiv 0 (mod\ 3)$. As $f(1) = 66$, and $f'(1) = 28$, we get the congruence $\frac{66}{3} + 28 \cdot t_1 \equiv 0 (mod\ 3)$, or the congruence $22 + 28 \cdot t_1 \equiv 0 (mod\ 3)$. As $22 \equiv 1 (mod\ 3)$, and $28 \equiv 1 (mod\ 3)$, we have, that $1 + t_1 \equiv 0 (mod\ 3)$, and therefore $t_1 \equiv -1 (mod\ 3)$. So, $t_1 = 3t_2 - 1$, where $t_2 \in \mathbb{Z}$, and $x = 3t_1 + 1 = 3(3t_2 - 1) + 1 = 3^2 \cdot t_2 - 2$.

Let $f(x) \equiv 0 (mod\ 3^3)$, where $x = 3^2 t_2 - 2$, $t_2 \in \mathbb{Z}$. Consider the congruence $\frac{f(-2)}{3^2} + f'(-2) \cdot t_2 \equiv 0 (mod\ 3)$. As $f(-2) = -9$, and $f'(-2) = 103$, we have the congruence $\frac{-9}{9} + 103 \cdot t_2 \equiv 0 (mod\ 3)$, or the congruence $-1 + 103 \cdot t_2 \equiv 0 (mod\ 3)$. As $103 \equiv 1 (mod\ 3)$, we get that $-1 + t_2 \equiv 0 (mod\ 3)$, and therefore $t_2 \equiv 1 (mod\ 3)$. So, $t_2 = 3t_3 + 1$, where $t_3 \in \mathbb{Z}$, and $x = 3^2 t_2 - 2 = 3^2 (3t_3 + 1) - 2 = 3^3 \cdot t_3 + 7$.

Therefore, one of the solution of the congruence $f(x) \equiv 0 (mod\ 3^3)$ is a class $x \equiv 7 (mod\ 3^3)$.

So, we got all solution of the congruence $f(x) \equiv 0 (mod\ 3^3)$: the classes $x \equiv -3 (mod\ 3^3)$, $x \equiv -6 (mod\ 3^3)$, $x \equiv -12 (mod\ 3^3)$, and $x \equiv -7 (mod\ 3^3)$.

B. Find all solutions of the congruence $f(x) \equiv 0 (mod\ 5^2)$.

For the first step, consider the congruence $f(x) \equiv 0 (mod\ 5)$, i.e., the congruence $x^5 - x^4 + x^2 \equiv 0 (mod\ 5)$.

Easy to see, that $x \equiv 0 (mod\ 5)$ is a solution of this congruence.

If $x \not\equiv 0 (mod\ 5)$, then $(x, 5) = 1$, and $x^4 \equiv 1 (mod\ 5)$. For such x we get the congruence $x^2 + x - 1 \equiv 0 (mod\ 5)$, which has the only solution $x \equiv 2 (mod\ 5)$.

So, the congruence $f(x) \equiv 0 (mod\ 5)$ has exactly two solutions: $x \equiv 0 (mod\ 5)$, and $x \equiv 2 (mod\ 5)$.

B1. Consider $x \equiv 0 (mod\ 5)$ — one of the solutions of the congruence $f(x) \equiv 0 (mod\ 5)$. Obviously, $f(0) \equiv 0 (mod\ 5)$, and $x = 5t_1 + 0$, where $t_1 \in \mathbb{Z}$.

Let $f(x) \equiv 0 (mod\ 5^2)$, where $x = 5t_1 + 0$, $t_1 \in \mathbb{Z}$. Consider the congruence $\frac{f(0)}{5} + f'(0) \cdot t_1 \equiv 0 (mod\ 5)$. As $f(0) = 45$, and $f'(0) = 15$, we get the congruence $\frac{45}{5} + 15 \cdot t_1 \equiv 0 (mod\ 5)$, or the congruence $9 + 15 \cdot t_1 \equiv 0 (mod\ 5)$, which has no solutions.

B2. Consider $x \equiv 2 (mod\ 5)$ — the other solution of the congruence $f(x) \equiv 0 (mod\ 5)$. Obviously, $f(2) \equiv 0 (mod\ 5)$, and $x = 5t_1 + 2$, where $t_1 \in \mathbb{Z}$.

Let $f(x) \equiv 0 (mod\ 5^2)$, where $x = 5t_1 + 2$, $t_1 \in \mathbb{Z}$. Consider the congruence $\frac{f(2)}{5} + f'(2) \cdot t_1 \equiv 0 (mod\ 5)$. As $f(2) = 115$, and $f'(2) = 87$, we have the congruence $\frac{115}{5} + 87 \cdot t_1 \equiv 0 (mod\ 5)$, or the congruence $23 + 87 \cdot t_1 \equiv 0 (mod\ 5)$. Therefore, we get $t_1 \equiv 1 (mod\ 5)$. So, $t_1 = 5t_2 + 1$, where $t_2 \in \mathbb{Z}$, and $x = 5t_1 + 2 = 5(5t_2 + 1) + 2 = 5^2 \cdot t_2 + 7$.

Therefore, we have proven, that the only solution of the congruence $f(x) \equiv 0 (mod\ 5^2)$ is the class $x \equiv 7 (mod\ 5^2)$.

C. Consider now the system of congruences

$$\begin{cases} x & \equiv & a(mod\ 5^2) \\ x & \equiv & b(mod\ 3^3) \end{cases},$$

where $a = 7$, and $b \in \{-3, 6, 15, 7\}$.

If $x \equiv a(mod\ 25)$, then $x = 25t + a$, where $t \in \mathbb{Z}$. In this case we have $25t + a \equiv b(mod\ 27)$, and hence, $-2t \equiv b - a(mod\ 27)$, $-26t \equiv 13(b - a)(mod\ 27)$, $t \equiv 13(b - a)(mod\ 27)$. So, $t = 27t_1 + 13(b - a)$, where $t_1 \in \mathbb{Z}$, and $x = 25t + a = 25(27t_1 + 13(b - a)) + a = 675t_1 + 325b - 324a$, where $t_1 \in \mathbb{Z}$.

We have proved, that all solutions of the congruence

$$x^5 - x^4 + 6x^2 = 15x + 45 \equiv 0 (mod\ 675)$$

are given by the following residue classes modulo 675:

$$x \equiv 325b - 324a (mod\ 675), \quad \text{where} \quad a = 7, \quad \text{and} \quad b \in \{-3, 6, 15, 7\}.$$

So, we get four solutions: $x \equiv 7 (mod\ 675)$, $x \equiv 132 (mod\ 675)$, $x \equiv 357 (mod\ 675)$, and $x \equiv 582 (mod\ 675)$.

The full explanation of the algorithm above and many examples you can see in [Buch09], [DeKo13]. Some additional information see, for example, in [Arno38], [BiBa70], [Dick05], [Gaus01], [IrRo90], [Lege79], [Moze09], [Ore48], [Stra16], [Wiki20].

Exercises

1. Solve the congruence:

(a) $3x \equiv 1 (mod\ 7)$;

(b) $100x \equiv 21 (mod\ 23)$;

(c) $42x \equiv 33 (mod\ 90)$;

(d) $20x \equiv 12 (mod\ 48)$;

(e) $20x - 50 \equiv 0 (mod\ 35)$;

(f) $78x \equiv 102 (mod\ 273)$;

(g) $315x \equiv -10 (mod\ 275)$;

(h) $76x \equiv 232 (mod\ 220)$;

(i) $45x \equiv 75 (mod\ 100)$.

2. Solve ths system of congruences:

(a) $\begin{cases} x \equiv 1 (mod\ 5) \\ x \equiv 2 (mod\ 6) \ ; \\ x \equiv 3 (mod\ 7) \end{cases}$

(b) $\begin{cases} 3x \equiv 7 (mod\ 10) \\ 2x \equiv 5 (mod\ 15) \ ; \\ 7x \equiv 5 (mod\ 12) \end{cases}$

(c) $\begin{cases} 5x + 7 \equiv 0 (mod\ 12) \\ 3x \equiv 7 (mod\ 8) \end{cases} ;$

(d) $\begin{cases} 4x \equiv 3 (mod\ 7) \\ 5x \equiv 4 (mod\ 11) \ ; \\ 11x \equiv 8 (mod\ 13) \end{cases}$

(e) $\begin{cases} 18x \equiv 226(mod\ 10) \\ 30x \equiv 232(mod\ 24) \end{cases}$;

(h) $\begin{cases} 3x \equiv 5(mod\ 2) \\ x \equiv -3(mod\ 5) \ ; \\ 4x \equiv 7(mod\ 9) \end{cases}$

(f) $\begin{cases} 3x \equiv 1(mod\ 10) \\ 4x \equiv 3(mod\ 5) \ ; \\ 2x \equiv 7(mod\ 9) \end{cases}$

(i) $\begin{cases} 20x \equiv -10(mod\ 15) \\ 2x \equiv -12(mod\ 10) \end{cases}$.

(g) $\begin{cases} 5x \equiv 1(mod\ 12) \\ 5x \equiv 2(mod\ 8) \ ; \\ 7x \equiv 3(mod\ 11) \end{cases}$

3. Solve the congruence:

   (a) $x^3 + x^2 - 2x + 1 \equiv 0(mod\ 5)$;

   (b) $133x^5 - 148x^4 + 85x^3 - 98x^2 + x + 6 \equiv 0(mod\ 7)$;

   (c) $232x^{484} + 852x^{252} - 124x^{202} - x^{200} + 78 \equiv 0(mod\ 11)$;

   (d) $292x^{181} - 121x^{133} + 252x^{122} - 171x^{121} - 133x^{62} + 5 \equiv 0(mod\ 13)$;

   (e) $357x^{427} - 811x^{403} - 127x^{311} + 45 \equiv 0(mod\ 7)$;

   (f) $883x^{693} - 106x^{484} + 59x^{241} + 87x^{233} + 84 \equiv 0(mod\ 5)$;

   (g) $4015x^{10892} + 605x^{9999} + 365x^{1002} + 888x^{1001} - 24 \equiv 0(mod\ 11)$.

4. Solve the congruence:

   (a) $31x^4 + 57x^3 + 96x + 191 \equiv 0(mod\ 675)$;

   (b) $3x^4 - 8x^3 + 8x^2 - 3x + 3 \equiv 0(mod\ 225)$;

   (c) $x^3 + 6x + 7 \equiv 0(mod\ 675)$;

   (d) $x^3 + 2x^2 + 2x + 4 \equiv 0(mod\ 675)$;

   (e) $x^3 - x^2 + x + 3 \equiv 0(mod\ 108)$;

   (f) $x^3 + 5x^2 + 9x + 9 \equiv 0(mod\ 108)$.

1.5 Quadratic residues, Legendre symbol and Jacobi symbol

Given an odd prime p, an integer a coprime to p is called a *quadratic residue* modulo p, if

$$x_0^2 \equiv a(mod\ p)$$

for some integer x_0. Otherwise, a is called quadratic nonresidue.

Legendre symbol

1.5.1. For any integer a coprime to an odd prime p, the *Legendre symbol* (named after Adrien-Marie Legendre) $(\frac{a}{p})$ is defined by $(\frac{a}{p}) = 1$, if a is a quadratic residue modulo p, and by $(\frac{a}{p}) = -1$, if a is a quadratic nonresidue modulo p. If $p|a$, one has $(\frac{a}{p}) = 0$.

Properties of Legendre symbol

1. $(\frac{a}{p}) \equiv a^{\frac{p-1}{2}}(mod\ p)$ (*Euler's criterion*).

2. If $a \equiv b(mod\ p)$, then $(\frac{a}{p}) = (\frac{b}{p})$.

3. $(\frac{ab}{p}) = (\frac{a}{p})(\frac{b}{p})$.

4. $(\frac{a^2}{p}) = 1$; in particular, $(\frac{1}{p}) = 1$.

5. $(\frac{-1}{p}) = (-1)^{\frac{p-1}{2}}$, i.e., $(\frac{-1}{p}) = 1$, if $p \equiv 1(mod\ 4)$, and $(\frac{-1}{p}) = -1$, if $p \equiv -1(mod\ 4)$.

6. $(\frac{2}{p}) = (-1)^{\frac{p^2-1}{8}}$, i.e., $(\frac{2}{p}) = 1$ if $p \equiv \pm 1(mod\ 8)$, and $(\frac{2}{p}) = -1$ if $p \equiv \pm 3(mod\ 8)$.

7. For distinct odd primes p and q,

$$\left(\frac{p}{q}\right) = (-1)^{\frac{p-1}{2} \cdot \frac{q-1}{2}} \left(\frac{q}{p}\right),$$

i.e., $(\frac{p}{q}) = (\frac{q}{p})$, if $p \equiv 1(mod\ 4)$ or $q \equiv 1(mod\ 4)$, and $(\frac{p}{q}) = -(\frac{q}{p})$, if $p \equiv -1(mod\ 4)$ and $q \equiv -1(mod\ 4)$ (*law of quadratic reciprocity*).

For example,

$$\left(\frac{-125}{47}\right) = \left(\frac{-5 \cdot 25}{47}\right) = \left(\frac{-5}{47}\right) \cdot \left(\frac{5^2}{47}\right) = \left(\frac{-5}{47}\right) = \left(\frac{-1}{47}\right)\left(\frac{5}{47}\right)$$

$$= (-1) \cdot \left(\frac{5}{47}\right) = (-1) \cdot \left(\frac{47}{5}\right) = (-1) \cdot \left(\frac{2}{5}\right) = (-1) \cdot (-1) = 1.$$

It means, that -125 is a quadratic residue modulo 47.

Jacobi symbol

1.5.2. The *Jacobi symbol* is a generalization of the Legendre symbol. (Another generalization is the *Kronecker symbol*. Additionally, the Legendre symbol is a *Dirichlet character*.)

Introduced by C.G.J. Jacobi in 1837, it is of theoretical interest in modular Arithmetic and other branches of Number Theory, but its main use is in Computational Number Theory, especially in primality testing and integer factorization.

For any integer a and any positive odd integer n, the *Jacobi symbol* $\left(\frac{a}{n}\right)$ is defined as the product of the Legendre symbols corresponding to the prime factors of n: for $n = p_1^{\alpha_1} \cdot \ldots \cdot p_k^{\alpha_k}$, it holds

$$\left(\frac{a}{n}\right) = \left(\frac{a}{p_1}\right)^{\alpha_1} \cdot \ldots \cdot \left(\frac{a}{p_k}\right)^{\alpha_k}.$$

Like the Legendre symbol, if $\left(\frac{a}{n}\right) = -1$, then a is a quadratic nonresidue modulo n.

But, unlike the Legendre symbol, if $\left(\frac{a}{n}\right) = 1$, then a may or may not be a quadratic residue modulo n.

Properties of Jacobi symbol

1. If $a \equiv b(mod\ n)$, then $\left(\frac{a}{n}\right) = \left(\frac{b}{n}\right)$.

2. $\left(\frac{ab}{n}\right) = \left(\frac{a}{n}\right)\left(\frac{b}{n}\right)$.

3. $\left(\frac{a}{mn}\right) = \left(\frac{a}{m}\right)\left(\frac{a}{n}\right)$.

4. $\left(\frac{a^2}{n}\right) = 1$; in particular, $\left(\frac{1}{n}\right) = 1$.

5. $\left(\frac{-1}{n}\right) = (-1)^{\frac{n-1}{2}}$, i.e., $\left(\frac{-1}{n}\right) = 1$, if $n \equiv 1(mod\ 4)$, and $\left(\frac{-1}{n}\right) = -1$, if $n \equiv -1(mod\ 4)$.

6. $\left(\frac{2}{n}\right) = (-1)^{\frac{n^2-1}{8}}$, i.e., $\left(\frac{2}{n}\right) = 1$, if $n \equiv \pm 1 (mod\ 8)$, and $\left(\frac{2}{n}\right) = -1$, if $n \equiv \pm 3 (mod\ 8)$.

7. For distinct odd positive coprime integers n and m,

$$\left(\frac{m}{n}\right) = (-1)^{\frac{n-1}{2} \cdot \frac{m-1}{2}} \left(\frac{n}{m}\right),$$

i.e., $\left(\frac{m}{n}\right) = \left(\frac{n}{m}\right)$, if $n \equiv 1\ (mod\ 4)$ or $m \equiv 1\ (mod\ 4)$, and $\left(\frac{m}{n}\right) = - \left(\frac{n}{m}\right)$, if $n \equiv -1\ (mod\ 4)$ and $m \equiv -1\ (mod\ 4)$ (*law of quadratic reciprocity*).

For example,

$$\left(\frac{-122}{45}\right) = \left(\frac{13}{45}\right) = \left(\frac{45}{13}\right) = \left(\frac{6}{13}\right) = \left(\frac{2}{13}\right)\left(\frac{3}{13}\right)$$

$$= (-1) \cdot \left(\frac{13}{3}\right) = (-1) \cdot \left(\frac{1}{3}\right) = (-1) \cdot 1 = -1.$$

It means, that -122 is a quadratic nonresidue composite odd modulo 45.

The proofs of the properties of Legendre symbol and Jacobi symbol, as well as some additional information see, for example, in [Buch09], [Gaus01], [DeKo13], [IrRo90]. Historical aspects one can find in [Dick05], [Lege79], [Ore48], [Wiki20].

Exercises

1. Check, that $\left(\frac{-125}{47}\right) = 1$.

2. Find:

(a) $\left(\frac{102}{17}\right)$; (c) $\left(\frac{125}{47}\right)$; (e) $\left(\frac{-5000}{103}\right)$; (g) $\left(\frac{204}{311}\right)$;

(b) $\left(\frac{-88}{23}\right)$; (d) $\left(\frac{5000}{101}\right)$; (f) $\left(\frac{-1116}{73}\right)$; (h) $\left(\frac{219}{383}\right)$.

3. How many solutions has the congruence:

(a) $x^2 - 200 \equiv 0 (mod\ 79)$; (c) $x^2 \equiv 555 (mod\ 101)$;

(b) $x^2 \equiv 56 (mod\ 87)$; (d) $x^2 \equiv 15 (mod\ 209)$;

(e) $x^2 + 27 \equiv 0(mod\ 91)$;   (h) $x^2 \equiv 500(mod\ 1777)$;

(f) $x^2 \equiv 304(mod\ 299)$;   (i) $x^2 - 270 \equiv (mod\ 2803)$.

(g) $x^2 - 990 \equiv 0(mod\ 1787)$;

4. Find the number of solutions of the congruence:

   (a) $2x^2 + 7x + 5 \equiv 0(mod\ 37)$;

   (b) $3x^2 + 5x + 7 \equiv 0(mod\ 87)$;

   (c) $2x^2 - 3x + 4 \equiv 0(mod\ 151)$;

   (d) $3x^2 - x + 7 \equiv 0(mod\ 151)$;

   (e) $5x^2 + 2x - 5 \equiv 0(mod\ 71)$;

   (f) $4x^2 - 2x + 9 \equiv 0(mod\ 137)$;

   (g) $5x^2 + 3x + 25 \equiv 0(mod\ 167)$;

   (h) $76x^2 - 77x + 36 \equiv 0(mod\ 227)$.

5. Find:

   (a) $(\frac{1001}{61})$; (d) $(\frac{342}{677})$; (g) $(\frac{438}{593})$; (j) $(\frac{438}{593})$; (m) $(\frac{1001}{61})$;

   (b) $(\frac{2741}{97})$; (e) $(\frac{514}{327})$; (h) $(\frac{232}{367})$; (k) $(\frac{342}{677})$; (n) $(\frac{514}{727})$;

   (c) $(\frac{342}{667})$; (f) $(\frac{514}{727})$; (i) $(\frac{157}{379})$; (l) $(\frac{2741}{97})$; (o) $(\frac{-88}{263})$.

6. Find all primes p, such that 3 is a quadratic nonresidue modulo p.

7. Find all prime divisors of the quadratic forms $2y^2 + 10$; $3x^2 + 15$.

1.6 Multiplicative orders, primitive roots and indexes

Multiplicative order

1.6.1. Given a positive integer n and an integer a coprime to n, the *multiplicative order* $ord_n\ a$ *of* a *modulo* n is defined as the smallest positive integer γ with

$$a^\gamma \equiv 1(mod \ n).$$

For example, $ord_5 \ 4 = 2$, as $4^2 \equiv 1(mod \ 5)$, but $4^1 \not\equiv 1(mod \ 5)$.

Similarly, $ord_5 \ 3 = 4$, as $3^4 \equiv 1(mod \ 5)$, and $3^1 \not\equiv 1(mod \ 5)$, $3^2 \not\equiv 1(mod \ 5)$, $3^3 \not\equiv 1(mod \ 5)$.

At last, $ord_9 \ 2 = 6$, as $2^6 \equiv 1(mod \ 9)$, and $2^1 \not\equiv 1(mod \ 9)$, $2^2 \not\equiv 1(mod \ 9)$, $\ldots$, $2^5 \not\equiv 1(mod \ 9)$.

Properties of multiplicative order

1. If $a \equiv b(mod \ n)$, then $ord_n \ a = ord_n \ b$.

2. $ord_n \ a | \phi(n)$; in particular, $1 \leq ord_n \ a \leq \phi(n)$.

3. $a^\delta \equiv 1(mod \ n)$ if and only if $ord_n \ a | \delta$.

4. $a^\delta \equiv a^\eta(mod \ n)$ if and only if $\delta \equiv \eta(mod \ ord_n \ a)$.

5. The numbers $a^0, a^1, a^2, \ldots, a^{ord_n \ a - 1}$ represent different residue classes modulo n.

6. $ord_{p_1^{\alpha_1} \cdot \ldots \cdot p_s^{\alpha_s}} \ a = lcm(ord_{p_1^{\alpha_1}} \ a, \ldots, ord_{p_s^{\alpha_s}} \ a)$.

7. $ord_m(a^n) = \frac{ord_m \ a}{gcd(n, ord_m \ a)}$.

8. If $ord_m \ a = \gamma_1 \cdot \gamma_2$, then $ord_m \ a^{\gamma_1} = \gamma_2$.

9. If $ord_{p^k} \ a = \gamma$, then $ord_{p^{k+1}} \ a \in \{\gamma, \gamma \cdot p\}$; if $ord_{p^k} \ a = \gamma$, $ord_{p^{k+1}} \ a = \gamma \cdot p$, and $p^k > 2$, then $ord_{p^{k+2}} \ a = \gamma \cdot p^2$.

For full proofs and examples see [Buch09], [DeKo13].

Primitive roots modulo n

1.6.2. If $ord_n \ g = \phi(n)$, g is called *primitive root* modulo n. In this case the numbers

$$g^0, \ldots, g^{\phi(n)-1}$$

give all representatives of residue classes modulo n, coprime to n.

For example, $\phi(5) = 4$. As $ord_5\,3 = 4$, the number 3 is a primitive root modulo 5; as $ord_5\,4 = 2$, the number 4 is not a primitive root modulo 5.

Similarly, $\phi(9) = 6$; as $ord_9\,2 = 6$, the number 2 is a primitive root modulo 9; as $ord_9\,7 = 3$ ($7^3 \equiv 1(mod\,9)$, but $7^1 \not\equiv 1(mod\,5)$, $7^2 \not\equiv 1(mod\,9)$), the number 7 is not a primitive root modulo 5.

Primitive roots exist only for $n = 2, 4$, p^α, $2p^\alpha$, where p is any odd prime number, and α is any positive integer.

For full proofs and examples see [Buch09], [DeKo13], [Vino03].

Indexes

1.6.3. If g is a primitive root modulo n, then, for any a coprime with n, we have $a \equiv g^\beta(mod\ n)$, where $\beta \in [0, \phi(n) - 1]$. The number β is called *index* (or *discrete logarithm*) *of a modulo n with base g.* In this case we write $\beta = ind\,a = ind_g\,a$.

For example, the number 3 is a primitive root modulo 5; in this case, the index of the number 4 modulo 5 with base 3 is 2, as $4 \equiv 3^2(mod\,5)$. Moreover, as $3^0 \equiv 1(mod\,5)$, $3^1 \equiv 3(mod\,5)$, $3^2 \equiv 4(mod\,5)$ and $3^3 \equiv 2(mod\,5)$, we get $ind\,1 = 0$, $ind\,2 = 3$, $ind\,3 = 1$, and $ind\,4 = 2$.

Properties of indexes

1. If $a \equiv b(mod\ n)$, then $ind\,a \equiv ind\,b(mod\ \phi(n))$.

2. $ind\,a \cdot b \equiv ind\,a + ind\,b(mod\ \phi(n))$.

3. $ind\,a^k \equiv k \cdot ind\,a(mod\ \phi(n))$ for any non-negative integer k.

4. $ord_n\,a = \frac{\phi(n)}{gcd(ind\,a,\phi(n))}$.

Discrete logarithms are quickly computable in a few special cases. However, no efficient method is known for computing them in general. Several important algorithms in Public-key Cryptography base their security on the assumption that the discrete logarithm problem over carefully chosen groups has no efficient solution.

The proofs of the properties of multiplicative orders, primitive roots and indexes, as well as some additional important information see, for example, in [Arno38], [BiBa70], [Buch09], [DeKo13], [DeKo18], [Dick05], [Gaus01], [Gelf98], [IrRo90], [Knut68], [Kost82], [Lege79], [Moze09], [Ore48], [Stra16], [Vino03], [Wiki20].

Exercises

1. Find:

   (a) $ord_{35424}\, 17$;   (f) $ord_{48608}\, 125$;   (k) $ord_{1150}\, 5^{50}$;

   (b) $ord_{334368}\, (-17)$;   (g) $ord_{28768}\, 81$;   (l) $ord_{22009}\, 3$;

   (c) $ord_{6075}\, 8$;   (h) $ord_{203056}\, 125$;   (m) $ord_{2000}\, 343$;

   (d) $ord_{864}\, 625$;   (i) $ord_{20000}\, 81$;   (n) $ord_{63072}\, 49$;

   (e) $ord_{46575}\, 64$;   (j) $ord_{99225}\, 16$;   (o) $ord_{253}\, 3^{55n+2}$.

2. Find all primitive roots modulo:

   (a) 12;   (c) 50;   (e) 98;   (g) 250;   (i) 625;

   (b) 14;   (d) 81;   (f) 242;   (h) 338;   (j) 1250.

3. Find all residue classes:

   (a) $[x]_{13}$, for which $ord_{13}\, x = 4$;   (c) $[x]_{19}$, for which $ord_{19}\, x = 3$;

   (b) $[x]_{17}$, for which $ord_{17}\, x = 8$;   (d) $[x]_{43}$, for which $ord_{43}\, x = 6$.

4. Find all indexes 11 modulo 5; modulo 7; modulo 17.

5. Find the length of the period of the representation $\frac{a}{b} = 0,$ $(q_1...q_k...)_g$, if:

   (a) $\frac{a}{b} = \frac{221}{30000000}$, $g = 7$;   (c) $\frac{a}{b} = \frac{28}{99 \cdot 10^{12}}$, $g = 3$;

   (b) $\frac{a}{b} = \frac{225}{70000}$, $g = 4$;   (d) $\frac{a}{b} = \frac{405}{242 \cdot 30^{10}}$, $g = 8$.

6. Find $rest(37^{32^{90}}, 11)$, using properties of indexes.

7. How many solutions has the congruence:

(a) $x^{12} \equiv 1(mod\ 77)$;   (g) $x^{15} \equiv 1(mod\ 143)$;

(b) $x^{12} \equiv 1(mod\ 91)$;   (h) $x^{60} \equiv 79(mod\ 97)$;

(c) $x^{12} \equiv 1(mod\ 143)$;   (i) $x^{18} \equiv 1(mod\ 77)$;

(d) $3x^{12} \equiv 31(mod\ 41)$;   (j) $x^{18} \equiv 1(mod\ 91)$;

(e) $x^{15} \equiv 1(mod\ 451)$;   (k) $x^{18} \equiv (mod\ 143)$;

(f) $x^{15} \equiv 1(mod\ 287)$;   (l) $x^{55} \equiv 17(mod\ 97)$?

8. Find:

(a) $ord_{2^7 \cdot 11^5 \cdot 47}\ 7$;   (d) $ord_{2^5 \cdot 11^3 \cdot 41^4}\ 3$;   (g) $ord_{2^5 \cdot 3^3 \cdot 83}\ (-7)$;

(b) $ord_{2^7 \cdot 13^3 \cdot 29}\ 11$;   (e) $ord_{2^7 \cdot 13^3 \cdot 53}\ 7$;   (h) $ord_{2^3 \cdot 3^2 \cdot 83}\ 49$.

(c) $ord_{2^6 \cdot 7^3 \cdot 31}\ 3$;   (f) $ord_{2^5 \cdot 3^3 \cdot 89}\ 125$;

1.7 Continued fractions

1.7.1. A *continued fraction* $[a_0, \ldots, a_n, \ldots]$ is defined as $a_0 + \dfrac{1}{a_1 + \dfrac{1}{\cdots \dfrac{1}{a_n + \cdots}}}$, where a_0 is some integer, and all a_n, $n \in \mathbb{N}$, are positive integers (the last number, if it exists, $\neq 1$).

The rational numbers $\delta_k = [a_0, \ldots, a_k] = \frac{P_k}{Q_k}$, $k = 0, \ldots, n, \ldots$, are called *convergents* of the continued fraction $[a_0, \ldots, a_n, \ldots]$.

1.7.2. For example, consider a fraction $\frac{173}{281}$, and use for the numbers 173 and 281 the Euclidean algorithm:

$$173 = 281 \cdot 0 + 173; \quad 281 = 173 \cdot 1 + 108; \quad 173 = 108 \cdot 1 + 65;$$

$$108 = 65 \cdot 1 + 43; \quad 65 = 43 \cdot 1 + 22; \quad 43 = 22 \cdot 1 + 21;$$

$$22 = 21 \cdot 1 + 1; \quad 21 = 1 \cdot 21 + 0.$$

Easy to see, that

$$\frac{173}{281} = 0 + \frac{1}{\frac{281}{173}}, \quad \frac{281}{173} = 1 + \frac{1}{\frac{173}{108}}, \ldots,$$

$$\frac{65}{22} = 1 + \frac{1}{\frac{43}{21}}, \quad \frac{43}{21} = 1 + \frac{1}{\frac{22}{21}}, \quad \frac{22}{21} = 1 + \frac{1}{21}.$$

Therefore,

$$\frac{173}{281} = 0 + \cfrac{1}{1 + \cfrac{1}{\cdots + \frac{1}{21}}} = [0, 1, 1, 1, 1, 1, 1, 21].$$

The convergents of the continued fraction $[0, 1, 1, 1, 1, 1, 1, 21]$ are:
$\delta_0 = [0] = 0;\ \delta_1 = [0, 1] = 0 + \frac{1}{1} = 1;\ \delta_2 = [0, 1, 1] = 0 + \frac{1}{1 + \frac{1}{1}} = 0 + \frac{1}{2} = \frac{1}{2};\ \delta_3 = [0, 1, 1, 1] = \frac{2}{3};\ \delta_4 = [0, 1, 1, 1, 1] = \frac{3}{5};\ \delta_5 = [0, 1, 1, 1, 1, 1] = \frac{5}{8};\ \delta_6 = [0, 1, 1, 1, 1, 1, 1] = \frac{8}{13};\ \delta_7 = [0, 1, 1, 1, 1, 1, 1, 21] = \frac{173}{281}.$

For the decomposition of $\frac{1-3\sqrt{5}}{2}$ we will use a generalization of the Euclidean algorithm.

As $6 < \sqrt{45} < 7$, we obtain for $\frac{1-3\sqrt{5}}{2} = \frac{1-\sqrt{45}}{2}$ the following estimation: $-3 < \frac{1-3\sqrt{5}}{2} < -2.5$, i.e., $a_0 = \lfloor \frac{1-3\sqrt{5}}{2} \rfloor = -3$. Then $\alpha_0 = \frac{1-3\sqrt{5}}{2} = -3 + \frac{1}{\alpha_1}$, where $\frac{1}{\alpha_1} = \frac{1-3\sqrt{5}}{2} - (-3) = \frac{7-3\sqrt{5}}{2}$.

Therefore, $\alpha_1 = \frac{2}{7-3\sqrt{5}} = \frac{2(7+3\sqrt{5})}{(7-3\sqrt{5})(7+3\sqrt{5})} = \frac{2(7+3\sqrt{5})}{49-45} = \frac{2(7+3\sqrt{5})}{4} = \frac{(7+3\sqrt{5})}{2}$.

As $6.5 < \frac{(7+3\sqrt{5})}{2} < 7$, we get $a_1 = \lfloor \frac{(7+3\sqrt{5})}{2} \rfloor = 6$, and $\alpha_1 = \frac{7+3\sqrt{5}}{2} = 6 + \frac{1}{\alpha_2}$, where $\frac{1}{\alpha_2} = \frac{7+3\sqrt{5}}{2} - 6 = \frac{-5+3\sqrt{5}}{2}$.

Therefore, $\alpha_2 = \frac{2}{-5+3\sqrt{5}} = \frac{2(-5-3\sqrt{5})}{(-5+3\sqrt{5})(-5-3\sqrt{5})} = \frac{2(-5-3\sqrt{5})}{25-45} = \frac{2(5+3\sqrt{5})}{20} = \frac{(5+3\sqrt{5})}{10}$.

As $1.1 < \frac{(5+3\sqrt{5})}{10} < 1.2$, then $a_2 = \lfloor \frac{(5+3\sqrt{5})}{10} \rfloor = 1$, and $\alpha_2 = \frac{5+3\sqrt{5}}{10} = 1 + \frac{1}{\alpha_3}$, where $\frac{1}{\alpha_3} = \frac{5+3\sqrt{5}}{10} - 1 = \frac{-5+3\sqrt{5}}{10}$.

Therefore, $\alpha_3 = \frac{10}{-5+3\sqrt{5}} = \frac{10(-5-3\sqrt{5})}{(-5+3\sqrt{5})(-5-3\sqrt{5})} = \frac{10(-5-3\sqrt{5})}{25-45} = \frac{10(5+3\sqrt{5})}{20} = \frac{(5+3\sqrt{5})}{2}$.

As $5.5 < \frac{(5+3\sqrt{5})}{2} < 6$, then $a_3 = \lfloor \frac{(5+3\sqrt{5})}{2} \rfloor = 5$, and $\alpha_3 = \frac{5+3\sqrt{5}}{2} = 5 + \frac{1}{\alpha_4}$, where $\frac{1}{\alpha_4} = \frac{5+3\sqrt{5}}{2} - 5 = \frac{-5+3\sqrt{5}}{2}$.

So, $\frac{1}{\alpha_4} = \frac{1}{\alpha_2}$, i.e., $\alpha_4 = \alpha_2$. Therefore, $\alpha_4 = \frac{5+3\sqrt{5}}{10} = 1 + \frac{1}{\alpha_5}$, where $\frac{1}{\alpha_5} = \frac{-5+3\sqrt{5}}{10}$. In particular, $a_4 = a_2$, and $\frac{1}{\alpha_5} = \frac{1}{\alpha_3}$, i.e., $\alpha_5 = \alpha_3$. Similarly, $a_5 = a_3$, and $\frac{1}{\alpha_6} = \frac{1}{\alpha_4}$, i.e., $\alpha_6 = \alpha_4$; on the same way, $a_6 = a_4$, and $\frac{1}{\alpha_7} = \frac{1}{\alpha_5}$, i.e., $\alpha_7 = \alpha_5$, etc.

So, we get the decomposition

$$\frac{1 - 3\sqrt{5}}{2} = [a_0, a_1, a_2, a_3, a_4, \ldots]$$
$$= [-3, 6, 1, 5, 1, 5, 1, 5, \ldots] = [-3, 6, (1, 5)].$$

The convergents of the continued fraction $[-3, 6, 1, 5, 1, 5, 1, 5, \ldots]$ are: $\delta_0 = [-3] = -3$; $\delta_1 = [-3, 6] = -3 + \frac{1}{6} = -\frac{17}{6}$; $\delta_2 = [-3, 6, 1] = -3 + \frac{1}{6 + \frac{1}{1}} = -3 + \frac{1}{7} = -\frac{20}{7}$, $\delta_3 = [-3, 6, 1, 5] = -\frac{117}{41}$, $\delta_4 = [-3, 6, 1, 5, 1] = -\frac{137}{48}$, etc.

1.7.3. Continued fractions are motivated by a desire to have a "mathematically pure" representation for the real numbers.

Properties of continued fractions

1. If $\delta_k = \frac{P_k}{Q_k}$, then $P_0 = a_0$, $P_1 = a_1 a_0 + 1$, and $P_n = a_n P_{n-1} + P_{n-2}$ for any $n \geq 2$.

2. If $\delta_k = \frac{P_k}{Q_k}$, then $Q_0 = 1$, $Q_1 = a_1$, and $Q_n = a_n Q_{n-1} + Q_{n-2}$ for any $n \geq 2$.

3. $gcd(P_n, Q_n) = 1$.

4. $1 = Q_0 \leq Q_1 < Q_2 < \ldots$.

5. If $P_0 > 1$, then $P_1 > P_2 > P_3 > \ldots$.

6. $\delta_n - \delta_{n-1} = \frac{(-1)^{n+1}}{Q_n Q_{n-1}}$.

7. $\delta_0 < \delta_2 < \cdots < \delta_3 < \delta_1$.

8. Every finite continued fraction is rational, and every rational number can be represented in precisely one way as a finite continued fraction.

9. Every infinite continued fraction is irrational, and every irrational number can be represented in precisely one way as an infinite continued fraction.

10. A continued fraction is periodic if and only if it is the continued fraction representation of a quadratic irrational, that is, a real solution to a quadratic equation with integer coefficients.

11. $[a_0, \ldots, a_n, \ldots] = \frac{a_n P_{n-1} + P_{n-2}}{a_n Q_{n-1} + Q_{n-2}}$, where $a_n = [a_n, \ldots]$.

12. If $\alpha = [a_0, \ldots, a_n, \ldots]$, then $|\alpha - \delta_n| \leq \frac{1}{Q_n Q_{n+1}}$; in particular, for any $n > 1$, δ_n is the best rational approximation for α.

The proof of these properties and some additional information see, for example, in [Buch09], [Dave99], [DeKo13], [Khin97]. The history of the question one can find in [Dick05], [Ore48]. Some number-theoretical applications are represented in [DeDe12].

Exercises

1. Find the decomposition into a continuous fraction:

   (a) $\frac{312}{175}$; (b) $-\frac{19}{15}$; (c) $\frac{72}{103}$; (d) $\frac{3885}{2306}$; (e) $-\frac{1000}{3333}$; (f) $\frac{27899}{36823}$.

2. Find the decomposition into a continuous fraction:

   (a) $\frac{1+3\sqrt{2}}{2}$; (d) $\frac{1-3\sqrt{5}}{2}$; (g) $\frac{2-2\sqrt{6}}{5}$; (j) $\frac{6-5\sqrt{2}}{7}$;

   (b) $\frac{1-3\sqrt{2}}{2}$; (e) $\frac{1+2\sqrt{6}}{5}$; (h) $\frac{2+2\sqrt{6}}{5}$; (k) $\frac{-15-6\sqrt{2}}{8}$;

   (c) $\frac{1+3\sqrt{5}}{2}$; (f) $\frac{1+3\sqrt{3}}{2}$; (i) $\frac{6+2\sqrt{2}}{7}$; (l) $\frac{6\sqrt{2}-15}{8}$.

3. Find the decomposition into a continuous fraction:

   (a) $\frac{\sqrt{2210}-13}{13}$; (b) $\frac{169+\sqrt{63005}}{158}$; (c) $\frac{1170+2\sqrt{93637}}{232}$.

4. Find the rational number, represented as:

   (a) $[1, 2, 3, 4, 6]$; (d) $[-1, 2, 3, 10]$; (g) $[1, 2, 3, 4, 6]$;

   (b) $[-2, 111, 2, 1, 3]$; (e) $[-2, 1, 3, 7]$; (h) $[0, 8, 1, 6, 2, 2]$.

   (c) $[-2, 3, 3, 10]$; (f) $[-3, 1, 3, 9, 5]$;

5. Find the irrational number, represented as:

(a) $= [1, (2)]$;   (h) $= [8, (16)]$;   (o) $= [4, (2, 8)]$;

(b) $= [2, (4)]$;   (i) $= [2, (1, 4)]$;   (p) $[3, (10, 5)]$;

(c) $= [3, (6)]$;   (j) $= [2, (2, 4)]$;   (q) $[3, (1, 5)]$;

(d) $= [4, (8)]$;   (k) $= [3, (1, 6)]$;   (r) $[3, (4, 1)]$;

(e) $= [5, (10)]$;   (l) $= [3, (2, 6)]$;   (s) $[5, (2, 10)]$;

(f) $= [6, (12)]$;   (m) $= [3, (3, 6)]$;   (t) $[-3, 6, (1, 5)]$.

(g) $= [7, (14)]$;   (n) $= [4, (1, 8)]$;

Chapter 1: References

[IrRo90], [Arno38], [BiBa70], [BaCo87], [Buch09], [CoRo96], [Dave99], [Dede63], [DeKo13], [Edva77], [Gelf98], [HaWr79], [Lagr70], [Lege30], [LiNi96], [MSC96], [Moze09], [Nage51], [Ore48], [RaTo57], [Smit84], [Hons91], [Plut78], [Stra16], [Stru87], [Vino03].

Chapter 2

Prime numbers

2.1 History of the question

2.1.1. The Theory of prime numbers forms an important part of classical Number Theory. But prime numbers are more than just mathematical oddities: they have become extremely important in applications, especially in Cryptography, because it is difficult to factor a large composite number into primes. In fact, for general numbers that are few hundred decimal digits long, factorization is nearly impossible; this gave rise to the RSA cryptosystem.

2.1.2. The prime numbers have long and reach history.

Already the *Rhind Mathematical Papyrus*, from around 1550 BC, has Egyptian fraction expansions of different forms for prime and composite numbers.

However, the earliest surviving records of the explicit study of prime numbers come from ancient Greek Mathematics.

The mathematicians of Pythagoras's school (500 BC–300 BC) were interested in numbers for their mystical and numerological properties. They understood the idea of primality and were interested in perfect and amicable numbers.

By the time Euclid's *Elements* appeared in about 300 BC, several important results about primes had been proved.

In Book IX of *Elements*, Euclid proved that *there are infinitely many prime numbers*. This is one of the first proofs known which uses the method of contradiction to establish a result.

Euclid also gave a proof of the fundamental theorem of Arithmetic: *every integer can be written as a product of primes in an essentially unique way.*

Moreover, Euclid showed that *the number $2^{n-1}(2^n - 1)$ is a perfect number, if the number $2^n - 1$ is a prime.* (Much later, in 1747, L. Euler was able to show that all even perfect numbers are of this form.) It is still not known whether there are any odd perfect number.

In about 200 BC, Eratosthenes devised an algorithm for calculating primes. This Greek invention, the *sieve of Eratosthenes*, is still used to construct lists of primes.

There is then a long gap in the history of prime numbers during what is usually called the Dark Ages.

Around 1000 AD, an Islamic mathematician Ibn al-Haytham (Alhazen) found *Wilson's theorem*, characterizing the prime numbers as the numbers n that evenly divide $(n-1)! + 1$. He also conjectured that all even perfect numbers come from Euclid's construction using Mersenne primes, but was unable to prove it.

Another Islamic mathematician, Ibn al-Banna' al-Marrakushi, observed that the sieve of Eratosthenes can be sped up by testing only the divisors up to the square root of the largest number to be tested.

Fibonacci brought the innovations from Islamic Mathematics back to Europe. His book *Liber Abaci* (1202) was the first to describe *trial division* for testing primality, again using divisors only up to the square root.

The next important developments were made by French mathematician Pierre de Fermat at the beginning of the XVII-th century.

He proved a speculation of A. Girard that *every prime number of the form $4n + 1$ can be written in an unique way as the sum of two squares* and was able to show how any number could be written as a sum of four squares. He devised a new method of factorising large numbers which he demonstrated by factorising the number $2027651281 = 44021 \cdot 46061$.

He stated (without proof) the *Fermat's little theorem* (later proved by G.V. Leibniz and L. Euler). The theorem states, that *if p is a prime then for any integer a we have $a^p \equiv a (mod\ p)$*.

This proves one half of what has been called the *Chinese hypothesis* which dates from about 2000 years earlier, that *an integer n is prime if and only if the number $2^n - 2$ is divisible by n*. The other half of this is false, since, for example, $2^{341} - 2$ is divisible by 341 even though $341 = 31 \cdot 11$ is composite.

Fermat's little theorem is the basis for many other results in Number Theory and is the basis for methods of checking whether numbers are prime which are still in use on today's electronic computers.

P. Fermat corresponded with other mathematicians of his day and in particular with the French monk Marin Mersenne. In one of his letters to M. Mersenne he conjectured that *the numbers $2^n + 1$ were always prime if n is a power of 2*. He had verified this for $n = 1, 2, 4, 8$ and 16, and he knew that if n were not a power of 2, the result failed.

Numbers of this form are called *Fermat numbers* and it was not until more than 100 years later that L. Euler showed that the next case $2^{32} + 1 = 4294967297$ is divisible by 641 and so is not prime.

Number of the form $2^n - 1$ also attracted attention because it is easy to show that if unless n is prime these numbers must be composite. These are often called *Mersenne numbers* because M. Mersenne studied them.

Not all numbers of the form $2^n - 1$ with n prime are primes. For example, $2^{11} - 1 = 2047 = 23 \cdot 89$ is composite, though this was first noted as late as 1536.

For many years numbers of this form provided the largest known primes. As of 2020, a total of 51 Mersenne primes have been found.

Christian Goldbach formulated *Goldbach's conjecture*, that *every even number is the sum of two primes*, in 1742, in a letter to L. Euler.

L. Euler's work had a great impact on Number Theory in general, and on Theory of prime numbers, in particular.

L. Euler proved Alhazen's conjecture (now the *Euclid–Euler theorem*) that *all even perfect numbers can be constructed from Mersenne primes*.

He extended Fermat's little theorem and introduced the *Euler's totient function* $\phi(n)$. As was mentioned above, he factorised the 5-th Fermat number $2^{32} + 1$, he found 60 pairs of the amicable numbers, and he stated (but was unable to prove) what became known as the *law of quadratic reciprocity*.

He was the first to realise that the Number Theory could be studied using the tools of Analysis and in so-doing founded the subject of Analytic Number Theory.

He was able to show that not only the so-called Harmonic series $\sum_{n=1}^{\infty} \frac{1}{n}$ is divergent, but the series

$$\frac{1}{2} + \frac{1}{3} + \frac{1}{5} + \frac{1}{7} + \frac{1}{11} + \cdots$$

formed by summing the reciprocals of the prime numbers, is also divergent.

The sum to n terms of the Harmonic series grows roughly like $\log n$, while the latter series diverges even more slowly like $\log \log n$. This means, for example, that summing the reciprocals of all the primes that have been listed, even by the most powerful computers, only gives a sum of about 4, but the series still diverges to ∞.

At first sight the primes seem to be distributed among the integers in rather a strange way. For example, in the 100 numbers immediately before 10 000 000 there are 9 primes, while in the 100 numbers after there are only 2 primes. However, on a large scale, the way in which the primes are distributed is very regular.

A.-M. Legendre and K.F. Gauss did extensive calculations of the density of primes.

Both Legendre and Gauss came to the conclusion that *for large n the density of primes near n is about* $\frac{1}{\log n}$.

Legendre gave an estimate for the prime counting function $\pi(n)$ (the number of primes $p \leq n$) of the form $\pi(n) = \frac{n}{\log n - 1.08366}$, while Gauss's estimate is in terms of the logarithmic integral: $\pi(n) = \int_2^n \frac{1}{\log t} dt$.

The statement that *the number of primes up to x is asymptotic to* $\frac{x}{\log x}$ (i.e., that the density of primes is $\frac{1}{\log x}$) is known as the *Prime number theorem*. Attempts to prove it continued throughout the XIX-th century with notable progress being made by P. Chebyshev and B. Riemann. The result was eventually proved (using powerful methods in Complex Analysis) by J. Hadamard and Ch.J. de la Vallée Poussin in 1896. There are still many open questions (some of them dating back hundreds of years) relating to prime numbers.

Another important XIX-th century result was *Dirichlet's theorem on arithmetic progressions*, which states, that *certain arithmetic progressions contain infinitely many primes*.

2.1.3. Many mathematicians have worked on primality tests for numbers larger than those where trial division is practicably applicable. Methods that are restricted to specific number forms include Pépin's test for Fermat numbers (1877), Proth's theorem (1878), the Lucas–Lehmer primality test (1856), and the generalized Lucas primality test.

Since 1951 all the largest known primes have been found using these tests on computers. The search for ever larger primes has generated interest outside mathematical circles, through the Great Internet Mersenne Prime Search and other distributed computing projects.

2.1.4. The idea that prime numbers had few applications outside of pure Mathematics was shattered in the 1970's when the *Public-key Cryptography* and the *RSA cryptosystem* were invented, using prime numbers as their basis.

The increased practical importance of computerized primality testing and integer factorization led to the development of improved methods capable of handling large numbers of unrestricted form.

For additional information see, for example, [Bras88], [Dick05], [Ore48], [Salo90], [Sier64], [Sing00], [Stru87].

2.2 Elementary properties of prime numbers

Prime and composite numbers

2.2.1. A positive integer number p which has exactly two natural divisors, is called a *prime number*, or simply a *prime*.

The set of prime numbers is denoted by P. So,

$$P = \{2, 3, 5, 7, 11, 13, 17, 19, 23, 29, \dots \}.$$

A positive integer number n which has more than two natural divisors, is called a *composite number.*

The set of composite numbers is denote by S. So,

$$S = \{4, 6, 8, 9, 10, 12, 14, 15, 16, 18, \dots \}.$$

As the unity has exactly one positive integer divisor, i.e., is neither prime, nor composite, we obtain, that

$$\mathbb{N} = P \cup S \cup \{1\}.$$

Infiniteness of the set of prime numbers

2.2.2. About two thousand years ago Euclid (circa 300 BC) showed that *there are infinitely many prime numbers.*

Theorem (Euclid). *There are infinitely many prime numbers.*

□ The Euclid's proof is thus.

Suppose that there are only finite number of primes, and

$$P = \{p_1, p_2, \dots, p_k\}$$

are all the primes. Let

$$E = p_1 \cdot p_2 \cdot \dots \cdot p_k + 1$$

and let p be a prime dividing E. Then p cannot be any of $p_1, p_2, \dots, p_k$, otherwise p would divide the difference $E - p_1 \cdot p_2 \cdot \dots \cdot p_k = 1$, which is impossible. So, this prime p is still another prime, and $p_1, p_2, \dots, p_k$ would not be all the primes. □

L. Euler gave an analytic proof of this theorem.

□ In fact, if $P = \{p_1, p_2, \dots, p_k\}$ are all the primes, consider the rational number

$$X = \prod_{i=1}^{k} \left(1 - \frac{1}{p_i}\right)^{-1}.$$

As the value $(1-\frac{1}{p_i})^{-1}$ is the sum $1+\frac{1}{p_i}+\frac{1}{p_i^2}+\cdots$ of infinite geometric progression $1, \frac{1}{p_i}, \frac{1}{p_i^2}, \ldots$, we obtain that

$$\prod_{i=1}^{k}\left(1-\frac{1}{p_i}\right)^{-1} = \prod_{i=1}^{k}\left(1+\frac{1}{p_i}+\frac{1}{p_i^2}+\cdots\right)$$

$$= \sum_{\alpha_i \geq 0} \frac{1}{p_1^{\alpha_1} \cdot \ldots \cdot p_k^{\alpha_k}} = \sum_{n=1}^{\infty} \frac{1}{n}.$$

But $\sum_{n=1}^{\infty} \frac{1}{n}$ is the harmonic series, which is divergent, a contradiction. $\square$

Euler had shown also, that
- *the sum $\sum_p \frac{1}{p}$ of the reciprocals of the primes is divergent.*

More exactly, there exists an asymptotic formula

$$\sum_{p \leq x} \frac{1}{p} = \log x + o(\log \log x).$$

The proof of this asymptotic formula and many additional information see, for example, in [Buch09], [DeKo13], [Ingh32], [Kara83], [Titc87].

Fundamental theorem of Arithmetics

2.2.3. The *fundamental theorem of Arithmetics* shows that the set P of primes forms a multiplicative basis of the set of all positive integers.

Theorem (Fundamental theorem of Arithmetics). *Any natural number $n > 1$ can be represented in one and only one way as a product of prime numbers:*

$$n = p_1^{\alpha_1} \cdot p_2^{\alpha_2} \cdot \ldots \cdot p_s^{\alpha_s}, \ p_i \in P, \ p_1' < p_2 < \cdots < p_s, \ \alpha_i \in \mathbb{N}.$$

This representation is called the *canonical form of factorization.*

For example, $60 = 2^2 \cdot 3^1 \cdot 5^1$; $100 = 2^2 \cdot 5^2$; $101 = 101^1$.

The proof of this theorem and some additional information see, for example, in [Buch09], [DeKo13], [Dick05], [Sier64], [Ingh32].

Exercises

1. Prove, that any positive integer $n > 1$ has a prime divisor.

2. Prove, that if $n \in S$, then $n = a \cdot b$, $1 < a \le b < n$.

3. Prove, that any composite number n has a prime divisor $p \le \sqrt{n}$.

4. Prove, that for any prime number p and for any integers a, b we have the following property: $p|a \cdot b$ if and only if $p|a$ or $p|b$.

5. Prove, that $p_n \le 2^{2^n}$ for any positive integer n, where p_n is the n-th prime number.

6. Prove, that $p_n \le 2^{2^{n-1}}$ for $n > 1$.

7. Prove, that, for a given $n \ge 3$, between n and $n!$ there exists at least one prime number; using this fact, prove, that there exist infinitely many prime numbers.

8. Prove, that if p is an odd prime, then it has the form $4k - 1$, or the form $4k + 1$.

9. Prove, that if p is a prime, and $p > 3$, then it has the form $6k - 1$, or the form $6k + 1$.

10. Prove, that there exist infinitely many prime numbers of the form $4k - 1$.

11. Prove, that there exist infinitely many prime numbers of the form $6k - 1$.

12. Find all prime twins, i.e., pairs of primes $(p, p+2)$, less then 100.

2.3   How to recognize whether a natural number is a prime?

The problem of distinguishing prime n from composite numbers is known to be one of the most important and useful in Arithmetic.

— Gauss, 1801.

Sieve of Eratosthenes

2.3.1. An easy method of finding consecutive prime numbers was given by a Greek mathematician Eratosthenes (circa 276 BC – circa 194 BC).

Consider the sequence

$$2, 3, 4, 5, 6, 7, 8, ..., n, ...$$

Then, since 2 is the first prime number p_1, we remove from this sequence all the numbers greater than p_1 and divisible by p_1, i.e., starting with 2, each second number of the table.

The first of the reminding numbers is $3 = p_2$. Now we remove all the numbers greater than p_2 and divisible by p_2, i.e., starting with 3, each third number of the table.

The first of the reminding numbers is $5 = p_3$, and so on.

Thus, we obtain $p_1 = 2$, $p_2 = 3$, $p_3 = 5$, $p_4 = 7$, $p_5 = 11$, ..., $p_{1000} = 7917$, $p_{6000000} = 104395301$, ... Here p_n means the n-th prime number.

Modern sieves include the *Legendre sieve*, *Brun sieve*, the *Selberg sieve*, and the *Large sieve*. One of the original purposes of the Sieve Theory was to try to prove conjectures in Number Theory such as the *twin prime conjecture*. While the original broad aims of Sieve Theory still are largely unachieved, there has been some partial successes, especially in combination with other number-theoretical tools.

Some additional information see, for example, in [Buch09], [Hool76], [Moto83], [DeKo13], [Dick05], [Moto83], [Ore48].

Simplest primality tests

2.3.2. A *primality test* is a test to determine whether or not a given positive integer n is prime, as opposed to actually decomposing the number into its constituent prime factors, which is known as *prime (integer) factorization*.

The simplest primality test is as follows:

• *given an positive integer n, we see if any integer m from 2 to $n - 1$ divides n; if n is divisible by any m, then n is composite, otherwise it is prime.*

2.3.3. Rather than testing all m up to $n - 1$, we need only test m up to $\sqrt{n}$:

- *if n is composite, then it has a prime divisor $p \leq \sqrt{n}$.*

□ In fact, if n is composite, there exists a non-trivial positive divisor a of n, i.e., n can be represented as

$$n = a \cdot b, \ 1 < a \leq b < n.$$

As $a > 1$, it has a prime divisor p, which is, obviously, a prime divisor of n. As $p, a \in \mathbb{N}$ and $p|a$, then $p \leq a$, and, hence, $p \leq b$, i.e., $p^2 \leq a \cdot b$. So, $p^2 \leq n$, and $p \leq \sqrt{n}$. □

Therefore, we get the simplest practical primality test, called *trial division*, which is commonly used also for factorization of small integers:

- *given an positive integer n, we see if any prime p from 2 to $\sqrt{n}$ divides n; if n is divisible by any such p, then n is composite, otherwise it is prime.*

2.3.4. A good way to speed up these methods (and all the others mentioned below) is to pre-compute and store a list of all primes up to a certain bound (such a list can be computed with the sieve of Eratosthenes). Then, before testing n for primality with a serious method, one first checks whether n is divisible by any prime from the list.

Primality tests come in two varieties: *deterministic* and *probabilistic*.

Deterministic tests determine with absolute certainty whether a number is prime. Examples of deterministic tests include the Lucas test, the Lucas-Lehmer test, the Pépin's test, the elliptic curve primality proving, etc.

Most popular primality tests are *probabilistic tests*. Examples of probabilistic test include the Fermat test, the Miller–Rabin test, the Solovay–Strassen test, etc. These tests use, apart from the tested number n, some other numbers a which are chosen at random from some sample space; the usual randomized primality tests never report a prime number as composite, but it is possible for a composite number to be reported as prime. For almost all probabilistic tests the probability of error can be reduced by repeating the test with several independently chosen a's.

Some additional information see, for example, in [Buch09], [DeKo13], [DeKo18], [Wiki20].

Fermat primaliry test; Poulet and Carmichael numbers

2.3.5. The simplest probabilistic primality test is the *Fermat primality test*.

It is only a heuristic test; some composite numbers (*Carmichael numbers*) will be declared probably prime no matter what *witness a* is chosen.

Nevertheless, it is sometimes used if a rapid screening of numbers is needed, for instance in the key generation phase of the RSA Public-key cryptographical algorithms.

The Miller-Rabin primality test and the Solovay-Strassen primality test are more sophisticated variants of a probabilistic primality test.

The Fermat primality test uses the *Fermat's little theorem*.

Theorem (Fermat's little theorem). *If p is a prime and a is an integer coprime to p, then*

$$a^{p-1} \equiv 1 (mod\ p).$$

□ Let proof this theorem by induction.
For $a = 1$, one has $1^p \equiv 1 (mod\ p)$.
Going from a to $a + 1$, we obtain

$$(a + 1)^p \equiv a^p + \binom{p}{1} a^{p-1} + \cdots + \binom{p}{p-1} a + 1 \equiv a + 1 (mod\ p),$$

as the binomial coefficients $\binom{p}{k} \equiv 0 (mod\ p)$ for $k = 1, ..., p - 1$. □

In fact, if we want to test if n is prime, then we can pick random a in the interval $[2, n - 1]$ and see if the congruence

$$a^{n-1} \equiv 1 (mod\ n)$$

holds. If the congruence does not hold for a value of a, then n is composite. If the equality does hold for many values of a, then we can say that n is a *probably prime*, or a *pseudoprime*.

2.3.6. More exactly, a composite number x is called a *Fermat pseudoprime* to base a if the number a is coprime to x, and it holds

$$a^{x-1} \equiv 1 (mod\ x).$$

A *Poulet number* (or *Sarrus number*) is a Fermat pseudoprime to base 2.

The smallest Poulet number is 341. It is not a prime, since it holds

$$341 = 11 \cdot 31,$$

but it satisfies the Fermat's little theorem:

$$2^{340} \equiv 1 (mod\ 341).$$

The first few Poulet numbers are 341, 561, 645, 1105, 1387, 1729, 1905, 2047, 2465, 2701, ... (sequence A001567 in the OEIS).

2.3.7. A number x that is a Fermat pseudoprime for all values of a that are coprime to x is called a *Carmichael number*.

It is proven (Korselt, 1899), that

• *a positive composite integer n is a Carmichael number if and only if n is square-free, and for all prime divisors p of n, it is true that $p - 1$ divides $n - 1$.*

A. Korselt was the first who observed these properties, but he could not find an example.

In 1910, R. Carmichael found the first and smallest such number, 561.

That 561 is a Carmichael number can be seen with the Korselt's theorem. Indeed,

$$561 = 3 \cdot 11 \cdot 17$$

is squarefree, and $2|560, 10|560, 16|560$.

The next few Carmichael numbers are 1105, 1729, 2465, 2821, 6601, 8911, 10585, 15841, 29341, ... (sequence A002997 in the OEIS).

In fact, $1105 = 5 \cdot 13 \cdot 17$, $1729 = 7 \cdot 13 \cdot 19$, $2465 = 5 \cdot 17 \cdot 29$, $2821 = 7 \cdot 13 \cdot 31$, $6601 = 7 \cdot 23 \cdot 41$, $8911 = 7 \cdot 19 \cdot 67$, etc.

We can see, that the first Carmichael numbers are very special: they are obtained as a product of three distinct primes. The corresponding property was proven by J. Chernick (1939):

• *the number* $(6k + 1)(12k + 1)(18k + 1)$ *is a Carmichael number if its three factors are all prime.*

The first Carmichael number with 4 prime factors is 41041:

$$41041 = 7 \cdot 11 \cdot 13 \cdot 41.$$

The first Carmichael number with 5 prime factors is 825265:

$$825265 = 5 \cdot 7 \cdot 17 \cdot 19 \cdot 73.$$

(For more infirmation see sequence A006931 in the OEIS.)

It is proven (Alford, Granville and Pomerance, 1994), that

• *there exist infinitely many Carmichael numbers.*

Specifically, for sufficiently large n, there are at least $n^{2/7}$ Carmichael numbers between 1 and n.

This makes tests based on Fermat's little Theorem slightly risky compared to others such as the Solovay-Strassen primality test.

Still, as numbers become larger, Carmichael numbers become very rare. For example, there are 1401644 Carmichael numbers between 1 and 10^{18} (approximately one in 700 billion numbers.)

The proofs of these properties and some additional information see, for example, in [AGP94], [Buch09], [Cher39], [Deza17], [DeKo18], [Uspe76], [Vile14].

Solovay-Strassen primality test

2.3.8. The *Solovay-Strassen primality test*, developed by R.M. Solovay and V. Strassen, is a probabilistic test to determine if a number is composite or probably prime.

It has been largely superseded by the Miller-Rabin primality test, but has great historical importance in showing the practical feasibility of the RSA cryptosystem.

L. Euler proved that *for a prime number p and any* $a \in [1, p-1]$, *it holds*

$$\left(\frac{a}{p}\right) \equiv a^{\frac{p-1}{2}} (mod \ p),$$

where $(\frac{a}{p})$ *is the Legendre symbol.* It is so called *Euler's criterion.*

□ In fact, by the Fermat's little theorem, $a^{p-1} \equiv 1(mod \ p)$, and, hence, $a^{p-1} - 1 \equiv 0(mod \ p)$. In other words,

$$\left(a^{\frac{p-1}{2}}\right)^2 - 1^2 \equiv \left(a^{\frac{p-1}{2}} - 1\right)\left(a^{\frac{p-1}{2}} + 1\right) \equiv 0(mod \ p).$$

As p is a prime, we get

$$a^{\frac{p-1}{2}} - 1 \equiv 0(mod \ p), \quad \text{or} \quad a^{\frac{p-1}{2}} + 1 \equiv 0(mod \ p).$$

Therefore, for any a coprime with an odd prime p, it holds

$$a^{\frac{p-1}{2}} \equiv 1(mod \ p), \quad \text{or} \quad a^{\frac{p-1}{2}} \equiv -1(mod \ p).$$

If a is a quadratic residue modulo p, there exists an integer x_0, such that $x_0^2 \equiv a(mod \ p)$. In this case, using the Fermat's little theorem, we have

$$a^{\frac{p-1}{2}} \equiv x_0^{p-1} \equiv 1(mod \ p).$$

As for a quadratic residue a modulo p its Legendre symbol $(\frac{a}{p})$ is equal to 1, we get the congruence

$$a^{\frac{p-1}{2}} \equiv \left(\frac{a}{p}\right)(mod \ p).$$

For a quadratic nonresidue a modulo p in holds $(\frac{a}{p}) = -1$, and the proof follows by similar way. □

The Jacobi symbol $(\frac{a}{n})$, defined by

$$\left(\frac{a}{n}\right) = \left(\frac{a}{p_1}\right)^{\alpha_1} \cdot \ldots \cdot \left(\frac{a}{p_k}\right)^{\alpha_k} \quad \text{for odd } n = p_1^{\alpha_1} \cdot \ldots \cdot p_k^{\alpha_k},$$

is a generalisation of the Legendre symbol. The Jacobi symbol can be computed in time $O((\log n)^2)$ using the law of quadratic reciprocity.

Thus, if we have a value n and want to determine if it is prime, we can check many random values of $a \in [2, n-1]$ and make sure that the congruence

$$\left(\frac{a}{n}\right) \equiv a^{\frac{n-1}{2}} \,(mod\ n)$$

holds. If it does not hold for some a, we know, that n must not be prime.

2.3.9. Much like with the Fermat primality test, however, there are liars.

A value a is known as an *Euler liar* if the congruence $\left(\frac{a}{n}\right) \equiv a^{\frac{n-1}{2}} \,(mod\ n)$ holds, but n is composite.

An *Euler witness* is a value of a such that the equality does not hold when n is composite; that is to say that a is a witness for the compositeness of n.

Unlike the Fermat primality test, for every composite n at least half of all $a \in [1, p-1]$ are Euler witnesses. Therefore, there are no such values that are guaranteed to be liars all the time, like Carmichael numbers are for Fermat test.

So, the test declares a composite n as probably prime with a probability at most 2^{-k}, where k is the number of different values of a we test.

The algorithm of calculation of Jacobi symbol, as well as some additional information see in Chapter 1, section 1.3, and, for example, in [Buch09], [DeKo13], [DeKo18], [IrRo90], [Knut68].

Miller-Rabin primality test

2.3.10. Just like with the Fermat and Solovay-Strassen tests, with the *Miller-Rabin test* we will rely on an equality or a set of equalities that hold true for prime values, and then see whether or not they hold for a number that we want to test for primality.

The original version of the Miller-Rabin primality test, due to G.L. Miller, was deterministic, but it relies on the unproven *generalized Riemann hypothesis*; M.O. Rabin modified it to obtain an unconditional probabilistic algorithm.

First, consider a behaviour of square roots of unity in the finite field $(Z/pZ, +, \cdot)$, where p is a prime. Certainly, 1 and -1 always yield 1 when squared modulo p; call these *trivial square roots of 1*. As p is a prime, there are no non-trivial square roots of 1. It is, for example, a special case of the result that in a field, a polynomial has no more zeros than its degree; so, the polynomial $x^2 - 1$ has the only roots ± 1.

Now, let p be an odd prime, then we can write $p-1$ as $2^k \cdot d$, where k is an integer, and d is odd. Then, *one of the following congruences must be true for any integer a coprime to p:*

$$a^d \equiv 1 (mod\ p), \quad or \quad a^{2^r d} \equiv -1 (mod\ p) \quad for\ some \quad r \in [0, k-1].$$

$\square$ In fact, for odd prime p and for any integer a coprime to p we get by the Fermat's little theorem the congruence $a^{p-1} \equiv 1 (mod\ p)$. Then, using the decomposition $p - 1 = 2^k \cdot d$, we obtain

$$a^{p-1} - 1 \equiv a^{2^k \cdot d} - 1 \equiv (a^{2^{k-1} \cdot d})^2 - 1^2 \equiv (a^{2^{k-1} \cdot d} + 1)(a^{2^{k-1} \cdot d} - 1)$$

$$\equiv (a^{2^{k-1} \cdot d} + 1)(a^{2^{k-2} \cdot d} + 1)(a^{2^{k-2} \cdot d} - 1)$$

$$\equiv (a^{2^{k-1} \cdot d} + 1)(a^{2^{k-2} \cdot d} + 1) \cdot \ldots \cdot (a^{2 \cdot d} + 1)(a^{2 \cdot d} - 1)$$

$$\equiv (a^{2^{k-1} \cdot d} + 1)(a^{2^{k-2} \cdot d} + 1) \cdot \ldots \cdot (a^{2 \cdot d} + 1)(a^d + 1)(a^d - 1)$$

$$\equiv 0 (mod\ p).$$

As p is a prime, then

$$a^{2^{k-1} \cdot d} + 1 \equiv 0 (mod\ p), \quad or \quad a^{2^{k-2} \cdot d} + 1 \equiv 0 (mod\ p), \quad \ldots$$

$$\ldots \quad or \quad a^{2 \cdot d} + 1 \equiv 0 (mod\ p), \quad or \quad a^d + 1 \equiv 0 (mod\ p),$$

$$or \quad a^d - 1 \equiv 0 (mod\ p).$$

Therefore,

$$a^{2^{k-1} \cdot d} \equiv -1 (mod\ p), \quad or \quad a^{2^{k-2} \cdot d} \equiv -1 (mod\ p), \quad \ldots$$

$$\ldots \quad or \quad a^{2 \cdot d} \equiv -1 (mod\ p), \quad or \quad a^d \equiv -1 (mod\ p),$$

$$or \quad a^d \equiv 1 (mod\ p). \quad \square$$

The Miller-Rabin primality test is based on the contrapositive of the above claim. That is, for a given odd number n, for which

$n - 1 = 2^k \cdot d$, if we can find an a such that

$$a^d \not\equiv 1(mod\ n), \quad \text{and} \quad a^{2^r d} \not\equiv -1(mod\ n) \quad \text{for all} \quad r \in [0, n-1],$$

then a is a *strong witness* for the compositeness of n. Otherwise n can be prime, but can also be composite.

2.3.11. If n is a composite number, such that

$$a^d \equiv 1(mod\ n), \quad \text{or} \quad a^{2^r d} \equiv -1(mod\ n) \quad \text{for some} \quad r \in [0, n-1],$$

the base a is called a *strong liar*, and n is a *strong probable prime* to base a.

The smallest odd composite strong probable primes (SPRP's) are the following:

- $2047 = 23 \cdot 89$ is a 2-SPRP;
- $121 = 11 \cdot 11$ is a 3-SPRP;
- $781 = 11 \cdot 71$ is a 5-SPRP;
- $25 = 5 \cdot 5$ is a 7-SPRP.

For every odd composite n, there are many witnesses a. In fact, for a given composite number n, at least $\frac{3}{4}$ of numbers $a \in [2, n-1]$ are witnesses of compositeness of n.

However, no simple way of generating such an a is known. The solution is to make the test probabilistic: we choose a randomly, and check whether or not it is a witness for the compositeness of n. If n is composite, most of the numbers a are witnesses; thus, the test will detect n as composite with high probability: the test declares a composite n as probably prime with a probability at most 4^{-k}, where k is the number of different values of a we test.

The examples of primality testing and some additional information see, for example, in [Buch09], [DeKo13], [DeKo18].

Criterions of prime numbers

However, there exist several classical criterions of primality of a given positive integer n.

2.3.12. A well-known criterion of primality is given by the *Wilson's theorem* (also known as *Al-Haytham's theorem*).

Theorem (Wilson's theorem). *A positive integer p is a prime number if and only if*

$$(p-1)! \equiv -1 (mod\ p).$$

Consider two proofs of this theorem.

☐ The first proof uses the fact that if p is an odd prime, then the set of residue classes $Z/pZ^* = \{[1]_p, [2]_p, ..., [p-1]_p\}$ forms a group under multiplication modulo p.

This means that for each element $[a]_p \in Z/pZ^*$, there is an unique inverse element $[b]_p \in Z/pZ^*$ such that $ab \equiv 1(mod\ p)$.

If $a \equiv b(mod\ p)$, then $a^2 \equiv 1(mod\ p)$, which forces

$$a^2 - 1 = (a+1)(a-1) \equiv 0(mod\ p),$$

and since p is prime, this forces

$$a \equiv 1, \quad or \quad a \equiv -1(mod\ p),$$

i.e., $a = [1]_p$, or $a = [p-1]_p$.

In other words, $[1]_p$ and $[p-1]_p$ are each their own inverse, but every other element has a distinct inverse, and so if we collect the elements from Z/pZ^* pairwise in this fashion and multiply them all together, we get the product -1.

For example, if $p = 11$, we have

$$10! \equiv (1 \cdot 10)(2 \cdot 6)(3 \cdot 4)(5 \cdot 9)(6 \cdot 8) \equiv -1(mod\ 11).$$

If $p = 2$, the result is trivial to check.

For a converse, suppose the congruence

$$(n-1)! \equiv -1(mod\ n)$$

holds for a composite n, and note that then n has a proper divisor d with $1 < d < n$.

Clearly, d divides $(n-1)!$. But by the congruence above, d also divides $(n-1)! + 1$, so that d divides 1, a contradiction. ☐

The second proof is thus.

☐ Suppose p is an odd prime. Consider the polynomial

$$g(x) = (x-1)(x-2) \cdot ... \cdot (x-(p-1)).$$

If $f(x)$ is a non-zero polynomial of degree d over a field F, then $f(x)$ has at most d roots over F. Now, with $g(x)$ as above, consider the polynomial

$$f(x) = g(x) - (x^{p-1} - 1).$$

Since the leading coefficients cancel, we see that $f(x)$ is a polynomial of degree at most $p - 2$. Reducing modulo p, we see that $f(x)$ has at most $p - 2$ roots modulo p.

But by the Fermat's little theorem, each of the elements 1, 2, ..., $p - 1$ is a root of $f(x)$. This is impossible, unless $f(x)$ is identically zero modulo p, i.e., unless each coefficient of $f(x)$ is divisible by p. But since p is odd, the constant term of $f(x)$ is just $(p-1)! + 1$, and the result follows. $\square$

This theorem was first discovered by Ibn al-Haytham (also known as Alhazen), but it is named after J. Wilson (a student of an English Mathematician E. Waring) who rediscovered it more than 700 years later. E. Waring announced the theorem in 1770, although neither he nor J. Wilson could prove it. J.-L. Lagrange gave the first proof in 1773. There is evidence that G.W. Leibniz was also aware of the result a century earlier, but he never published it.

Unfortunately, the Wilson's theorem is useless as a primality test, since computing $(n - 1)!$ is difficult for large n (see, for example, Chapter 5, section 5.6).

2.3.13. There is also a generalization of the Wilson's theorem, due to C.F. Gauss:

• *for a given positive integer m, it holds*

$$\prod_{1 \le a < m, gcd(a,m)=1} a \equiv -1 (mod\ m), \quad for\ \ m = 4, p^\alpha, 2p^\alpha,$$
$$where\ p \in P \backslash \{2\}, \alpha \in \mathbb{N};$$
$$\prod_{1 \le a < m, gcd(a,m)=1} a \equiv 1 (mod\ m), \quad otherwise.$$

2.3.14. It is easy to check, that

• *an odd positive integer p greater than 1 is a prime number if and only if it can be represented in only one way as the difference of two squares of positive integers:*

$$p = x^2 - y^2, \ x, y \in \mathbb{N}.$$

□ In fact, for given odd positive integer $n > 1$, if $n = a \cdot b$, $a \le b, a, b \in \mathbb{N}$, then

$$n = \left(\frac{a+b}{2}\right)^2 - \left(\frac{b-a}{2}\right)^2, \quad \frac{a+b}{2}, \frac{b-a}{2} \in \mathbb{N};$$

moreover, any representation of n in the form $n = x^2 - y^2$, $x, y \in \mathbb{N}$, gives decomposition $n = a \cdot b$, $a \le b, a, b \in \mathbb{N}$, where $a = x - y$, $b = x + y$, and, hence, $x = \frac{a+b}{2}, y = \frac{b-a}{2}$.

If p is a prime, we have the only decomposition $p = 1 \cdot p$; so, we get the only representation of p as the difference of two squares of positive integers:

$$p = \left(\frac{1+p}{2}\right)^2 - \left(\frac{p-1}{2}\right)^2, \quad \frac{1+p}{2}, \frac{p-1}{2} \in \mathbb{N}.$$

In n is a composite number, it has at least two decompositions,

$$n = 1 \cdot n, \quad \text{and} \quad n = a \cdot b, \ 1 < a \le b < n,$$

so, we get at least two representations of n as the difference of two squares. □

This theorem is a base of the *Fermat's factorization method.*

For a representation $n = x^2 - y^2$, we have $x^2 > n$ and, obviously, $y^2 = x^2 - n$.

So, using Fermat's factorization method, one tries various values of x, starting from $x = \lceil \sqrt{n} \rceil$, hoping that $x^2 - n = y^2$, a square; in this case, $n = (x + y)(x - y)$. For an odd composite number n, we find such square for some $x < \frac{n+1}{2}$.

In its simplest form, Fermat's method might be even slower than trial division (worst case). Nonetheless, the combination of trial division and Fermat's is more effective than either.

2.3.15. The following criterion of primality for a given positive integer n is more effective:

• *a positive integer $n > 2$ is a prime number if and only if there exists some positive integer a, $1 < a < n$, such that*

$$a^{n-1} \equiv 1 (mod \ n), \quad \text{and} \quad a^\gamma \not\equiv 1 (mod \ n)$$

for all positive integers $\gamma | (n - 1)$.

□ In fact, by multiplicative order's properties, we obtain from the condition $a^{n-1} \equiv 1(mod\ n)$, that $ord_n\ a | (n-1)$, and from the second condition it follows, that $ord_n\ a = n - 1$. As $ord_n\ a | \phi(n)$, then $n - 1 | \phi(n)$; as $\phi(n) < n$ for $n \neq 1$, then $\phi(n) = n - 1$, i.e., $n \in P$.

Conversely, if n is a prime number, then there exists a primitive root g modulo n. It is a generator of the group Z/nZ^*. Such a generator has order $|Z/nZ^*| = n - 1$ and both equivalences will hold for any such primitive root:

$$g^{n-1} \equiv 1(mod\ n), \quad \text{and} \quad g^\gamma \not\equiv 1(mod\ n)$$

for all positive integers $\gamma | (n-1)$. □

Of course, there is no need to check all positive integer exponents $\gamma | (n-1)$. It is enough to check only γ of the form $\frac{n-1}{q}$ for all prime divisors of n. So, we get the following criterion of primality:

• *a positive integer $n > 2$ is a prime number if and only if there exists some positive integer a, $1 < a < n$, such that*

$$a^{n-1} \equiv 1(mod\ n), \quad \text{and} \quad a^{\frac{n-1}{q}} \not\equiv 1(mod\ n)$$

for any prime factor q of n.

It is a result of a French mathematician Francois Édouard Anatole Lucas (1842–1891).

In Computational Number Theory, the corresponded test is called the *Lucas test*.

It is deterministic primality test, which can be applied to any positive integer $n > 2$. However, it requires that all prime factors of $n - 1$ are known.

The Lucas test relies on the fact that the multiplicative order of a number g modulo n is $n - 1$ for a prime n, when g is a primitive root modulo n. If we can show g is primitive root modulo n, we can show n is prime.

For example, take $n = 71$. Then $n - 1 = 70$, and the prime factors of 70 are $2, 5$ and 7. We randomly select an $a = 17$, $a < n$. Now we compute

$$17^{70} \equiv 1(mod\ 71).$$

It is known that for any integer a coprime to n it holds

$$a^{n-1} \equiv 1(mod\ n) \quad \text{if and only if} \quad ord_n(a)|(n-1).$$

Therefore, the multiplicative order of 17 modulo 71 is not necessarily 70 because some factor of 70 may also work. So check 70 divided by its prime factors:

$$17^{35} \equiv 70 \not\equiv 1(mod\ 71), \quad 17^{14} \equiv 25 \not\equiv 1(mod\ 71), \quad 17^{10} \equiv 1(mod\ 71).$$

Unfortunately, we get that $17^{10} \equiv 1(mod\ 71)$. So we still don't know if 71 is prime or not.

Let try another random a, this time choosing $a = 11$. Now we compute

$$11^{70} \equiv 1(mod\ 71).$$

Again, this does not show that the multiplicative order of 11 modulo 71 is 70 because some factor of 70 may also work. So check 70 divided by its prime factors:

$$11^{35} \equiv 70 \not\equiv 1(mod\ 71), \quad 11^{14} \equiv 54 \not\equiv 1(mod\ 71),$$

$$11^{10} \equiv 32 \not\equiv 1(mod\ 71).$$

So, the multiplicative order of 11 modulo 71 is 70, and thus 71 is prime.

To carry out these modular exponentiations, one could use a fast exponentiation algorithm like binary or addition-chain exponentiation.

Note that if there exists an $a < n$ such that the first equivalence fails, we get that a is a Fermat witness for the compositeness of n.

2.3.16. The *Pocklington–Lehmer primality test* is a deterministic primality test (devised by H.C. Pocklington and D.H. Lehmer), which uses a partial factorization of $n - 1$ to prove that an integer n is a prime number.

It produces a *primality certificate* to be found with less effort than the Lucas primality test, which requires the full factorization of $n-1$.

The basic version of the test relies on the *Pocklington criterion* which is formulated as follows.

Theorem (Pocklington criterion). *A positive integer $n > 1$ is a prime number if and only if there exist numbers a and p such that:*

a) $a^{n-1} \equiv 1(mod\ n)$;

b) p *is prime,* $p|(n-1)$, *and* $p > \sqrt{n} - 1$;

c) $gcd(a^{\frac{n-1}{p}} - 1, n) = 1$.

For the proof of this criterion, see Chapter 5, section 5.5.

Note, that the first condition $a^{n-1} \equiv 1(mod\ n)$ is simply the Fermat primality test.

If we find any value of a, not divisible by n, such that this congruence is false, we may immediately conclude that n is not prime. For example, let $n = 15$. With $a = 2$, we find that

$$a^{n-1} \equiv 2^{14} \equiv 4 \not\equiv 1(mod\ 15).$$

This is enough to prove that $n = 15$ is not a prime number.

The proofs of all theorems, as well as many examples and some additional information see, for example, in [Bras88], [Buch09], [DeKo13], [DeKo18], [Dick05], [Kobl87], [Kost82], [LiNi96], [Wiki20].

Exercises

1. Find all primes on the interval $[1, 100]$; $[1, 300]$.

2. Check the primality of the number 7; 17; 27; 37; 47, 57; 67; 77, 87; 97.

3. Prove, that the last digit of any odd prime number is 1, 3, 7, 9.

4. Prove, that $n \in P$ if and only if $\sum_{m=1}^{n} \lfloor \frac{n}{m} \rfloor - \lfloor \frac{n-1}{m} \rfloor = 2$.

5. Prove, that $p \equiv -1(mod\ 4)$ is a prime if and only if p can be represented, and in only one way, as a difference of two squares of integers.

6. Prove, that $p \equiv 1(mod\ 4)$ is a prime if and only if p can be represented, and in only one way, as a sum of two squares of integers.

7. Check, that 561, 645, 1105, 1387 are Poulet numbers.

8. Check, that 1105, 1729, 2465, 6601 are Carmichael numbers.

9. Prove, that the number $(6k+1)(12k+1)(18k+1)$ is a Carmichael number if its three factors are all primes.

10. Prove the Gauss formula above.

2.4   Formulas of primes

2.4.1. In Mathematics, a *formula for primes* is a formula generating the prime numbers, exactly and without exception. No such formula is known.

There are many examples of formulas generating important sets of integers.

Thus, the n-th *triangle number* is defined as $S_3(n) = S_3(n-1)+n$, $S_3(1) = 1$, and we get the closed form expression

$$S_3(n) = \frac{n(n+1)}{2}.$$

The n-th *m-gonal number* is defined as $S_m(n) = S_m(n-1)+(m-2)n$, $S_m(1) = 1$, and we get the closed form expression

$$S_3(n) = \frac{n((m-2)n - m + 4)}{2}.$$

The n-th *Fibonacci number* is defined as $u_n = u_{n-1} + u_{n-2}$, $u_1 = u_2 = 1$, and we get the closed form expression

$$u_n = \frac{\alpha^n - \beta^n}{\sqrt{5}},$$

where $\alpha = \frac{1+\sqrt{5}}{2}$, $\beta = \frac{1-\sqrt{5}}{2}$ (*Binet's formula*).

All *Pythagorean triples* (x, y, z) can be obtained as

$$x = k(n^2 - m^2), \ y = k(2mn), \ z = k(n^2 + m^2),$$

where $k, n, m \in \mathbb{N}$, $n > m$, and n, m are coprime and not both odd.

In fact, to obtain prime numbers, it is natural to ask for functions, defined for all positive integer n, which are computable in practice and produce all or some prime numbers. For example, one of the following conditions should be satisfied:

- $f(n) = p_n$ for all $n \geq 1$;

- $f(n)$ is always a prime number, and if $m \neq n$, then $f(n) \neq f(m)$;

- the set of prime numbers is equal to the set of positive values assumed by the function $f(n)$.

Prime-generating polynomials

2.4.2. It is known (Goldbach, 1752) that no non-constant polynomial $f(x)$ with integer coefficients can give primes for all integer values of x.

Theorem (Goldbach's theorem). *There is no non-constant polynomial $f(x)$ with integer coefficients, which gives a prime for any integer value of x.*

The proof is simple.

□ Suppose such a polynomial existed. Then $f(1)$ would evaluate to a prime p, so $f(1) \equiv 0 (mod\ p)$. But for any k, $f(1 + kp) \equiv 0 (mod\ p)$ also, so $f(1+kp)$ cannot also be prime (as it would be divisible by p) unless it were p itself, but the only way $f(1 + kp) = f(1)$ for all k is if the polynomial function is constant.□

A.-M. Legendre showed that

- *there is no rational algebraic function which always gives primes.*

Using more Algebraic Number Theory, one can show an even stronger result:

- *no non-constant polynomial $f(n)$ exists that evaluates to a prime number for almost all integers n.*

It was shown also (Reiner, 1943), that

- *if f_i, g_i, $i = 1, ..., n$, are polynomials with positive coefficients, then $\sum_{i=1}^{n} f_i(x)^{g_i(x)}$ cannot gives only primes for $x \in \mathbb{N}$.*

2.4.3. However, there exist polynomials which can give many prime numbers.

So, the *Euler polynomial* (Euler, 1772)

$$f(x) = x^2 + x + 41$$

gives a prime for all non-negative integers less than 40. The primes for $n = 0, 1, 2, 3...$ are 41, 43, 47, 53, 61, 71, ... The differences between the terms are 2, 4, 6, 8, 10, ... For $n = 40$, it produces a square number, 1681, which is equal to 41^2, the smallest composite number for this formula. In fact, if 41 divides n, it divides $f(n)$, too.

The quadratic polynomial $x^2 - x + 41$ (Legendre, 1798) gives the same prime numbers for $x = 1, ..., 40$, and gives composite numbers for 41 and 42.

The polynomial

$$g(x) = x^2 - 79x + 1601$$

gives primes for $x = 0, ..., 79$, corresponding to the 40 primes given by the above formula, taken twice each.

Moreover, any polynomial $x^2 + x + q$, $q = 2, 3, 5, 11, 17$, gives primes for all $x = 0, ..., q - 2$.

It is known, based on the *Dirichlet's theorem on the existence of primes in a given arithmetic progression* (Dirichlet, 1837), that

• *linear polynomial function $f(x) = ax + b$ produce infinitely many primes as long as a and b are relatively prime.*

Note that no such function will assume prime values for all values of n.

In particular, there are infinitely many primes of the form $4x - 1$, $4x + 1$, $6x - 1$, $6x + 1$.

It is not known, however, whether there exists a polynomial of degree greater than one that assumes an infinite number of values that are primes. In particular, it is not known, whether there are infinitely many primes of the form $ax^2 + bx + c$, and especially, of the form $x^2 + 1$.

Howewer, it was proven by J.P.G.L. Dirichlet, that, for integers a, b, c, such that $gcd(a, 2b, c) = 1$ and $b^2 = ac$, the polynomial in two variables $ax^2 + bxy + cy^2$ represents infinitely many primes. In particular, there are infinitely many primes of the form $x^2 + y^2$, and of the form $x^2 + y^2 + 1$.

The proofs of these facts and some additional information see, for example, in [Buch09], [ChFa07], [DeKo13] [DiDe63], [HaWr79], [Lege79], [Step01].

Prime formulas of the first kind

2.4.4. In this subsection we consider some formulas, which give the n-th prime number as a function of the number n. These formulas have little practical value: most primality tests are far more efficient. Moreover, they usually involve constants that cannot be determined without previous knowledge of the primes themselves.

W. Sierpiński, 1952, suggested define a constant A as follows:

$$A = \sum_{n=1}^{\infty} p_n 10^{-2^n} = 0.02030005000000070\ldots$$

Then, using the floor function $\lfloor x \rfloor$ (the greatest integer less than or equal to x) we have this exact folrmula for n-th prime number:

$$p_n = \left\lfloor 10^{2^n} \cdot A \right\rfloor - 10^{2^{n-1}} \left\lfloor 10^{2^{n-1}} \cdot A \right\rfloor.$$

☐ In fact,

$$10^{2^n} A = \sum_{k=1}^{n-1} p_k 10^{2^n - 2^k} + p_n + \sum_{k>n} p_k 10^{2^n - 2^k},$$

and we obtain that

$$\left\lfloor 10^{2^n} A \right\rfloor = \sum_{k=1}^{n-1} p_k 10^{2^n - 2^k} + p_n + \left\lfloor \sum_{k>n} p_k 10^{2^n - 2^k} \right\rfloor.$$

As $p_n \le 2^{2^{n-1}}$ for $n > 1$, we have that

$$p_{n+1} 10^{2^n - 2^{n+1}} \le (0.2)^{2^n}, \quad \text{and} \quad p_{n+t} 10^{2^n - 2^{n+t}} \le (0.2)^{2^{n+(t-1)}}.$$

So,

$$\sum_{k>n} p_k 10^{2^n - 2^k} \le \sum_{t \ge 1} (0.2)^{2^{n+(t-1)}} \le \sum_{t \ge 1} (0.2)^{2^n + (t-1)} = (0.2)^{2^n + 1} < 1,$$

i.e., $\left\lfloor \sum_{k>n} p_k 10^{2^n - 2^k} \right\rfloor = 0$, and $\left\lfloor 10^{2^n} A \right\rfloor = \sum_{k=1}^{n-1} p_k 10^{2^n - 2^k} + p_n$.

On the same way, we obtain that

$$\left\lfloor 10^{2^{n-1}} A \right\rfloor = \sum_{k=1}^{n-2} p_k 10^{2^{n-1} - 2^k} + p_{n-1},$$

and, therefore,

$$10^{2^{n-1}} \left\lfloor 10^{2^{n-1}} \cdot A \right\rfloor = \sum_{k=1}^{n-2} p_k 10^{2^n - 2^k} + 10^{2^{n-1}} p_{n-1}.$$

As $10^{2^n - 2^{n-1}} = 10^{2^{n-1}}$, we obtain the result. $\square$

2.4.5. G.H. Hardy and E.M. Wright, 1979, gave a variant of the formula above. Let r be an integer greater than one, and define a constant B as follows:

$$B = \sum_{n=1}^{\infty} p_n r^{-n^2}.$$

Then we obtain the n-th prime from the formula

$$p_n = \left\lfloor r^{n^2} B \right\rfloor - r^{2n-1} \left\lfloor r^{(n-1)^2} B \right\rfloor.$$

2.4.6. By using the Wilson's theorem, an other formula to test, whether any number x is a prime, can be established (Willans, 1964). In fact, consider a function

$$f(x) = \left\lfloor \cos^2 \pi \frac{(x-1)! + 1}{x} \right\rfloor.$$

The function gives the value 1, if x is prime, and 0, if x is composite; $f(1) = 1$.

Then prime counting function $\pi(x)$, i.e., the number of primes less than or equal to x, can be written now as

$$\pi(x) = -1 + \sum_{j=1}^{x} f(j).$$

The n-th prime number p_n can then be written as

$$p_n = 1 + \sum_{m=1}^{2^n} \left\lfloor \left(\frac{n}{1 + \pi(m)} \right)^{\frac{1}{n}} \right\rfloor,$$

or, equivalently, as

$$p_n = 1 + \sum_{m=1}^{2^n} \left\lfloor \left(\frac{1}{n} \sum_{x=1}^{m} \left\lfloor \cos^2 \pi \frac{(x-1)! + 1}{x} \right\rfloor \right)^{-\frac{1}{n}} \right\rfloor.$$

2.4.7. Alternatively, we can write $\pi(m) = \sum_{j=2}^{m} \left\lfloor \frac{(j-1)!+1}{j} - \left\lfloor \frac{(j-1)!}{j} \right\rfloor \right\rfloor$, and then write

$$p_n = 1 + \sum_{m=1}^{2^n} \left\lfloor \left(\frac{n}{1+\pi(m)} \right)^{\frac{1}{n}} \right\rfloor.$$

Consider the construction of the last formula.

□ It is known, that $p_n \leq 2^n$ for any $n \in \mathbb{N}$. As $\pi(p_n) = n$, we obtain that $\pi(m) \geq n$ for any $m \geq p_n$, and, therefore,

$$\frac{n}{1+\pi(m)} \leq \frac{n}{n+1} < 1, \quad \text{and} \quad \left\lfloor \left(\frac{n}{1+\pi(m)} \right)^{1/n} \right\rfloor = 0.$$

On the other hand, $\pi(m) < n$ for any $1 \leq m < p_n$, and, therefore,

$$1 \leq \frac{n}{1+\pi(m)} \leq n < p_n \leq 2^n, \quad \text{i.e.,} \quad 1 \leq \left(\frac{n}{1+\pi(m)} \right)^{1/n} < 2,$$

and $\left\lfloor \left(\frac{n}{1+\pi(m)} \right)^{1/n} \right\rfloor = 1$.

So,

$$1 + \sum_{m=1}^{2^n} \left\lfloor \left(\frac{n}{1+\pi(m)} \right)^{\frac{1}{n}} \right\rfloor = 1 + \sum_{m=1}^{p_n-1} 1 = 1 + (p_n - 1) = p_n. \ \square$$

2.4.8. Another approach (Ruiz, 2000) does not use factorials and the Wilson's theorem, but also heavily employs the floor function: first define

$$\pi(k) = k - 1 + \sum_{j=2}^{k} \left\lfloor \frac{2}{j} \left(1 + \sum_{s=1}^{\lfloor \sqrt{j} \rfloor} \left(\left\lfloor \frac{j-1}{s} \right\rfloor - \left\lfloor \frac{j}{s} \right\rfloor \right) \right) \right\rfloor,$$

and then

$$p_n = 1 + \sum_{k=1}^{2(\lfloor n \log n + 1 \rfloor)} \left(1 - \left\lfloor \frac{\pi(k)}{n} \right\rfloor \right).$$

2.4.9. One more formula of such kind was obtained in 1971:

$$p_n = \left\lfloor 1 - \frac{1}{\log 2} \log \left(-\frac{1}{2} + \sum_{d|P_{n-1}} \frac{\mu(d)}{2^d - 1} \right) \right\rfloor,$$

where $\mu(n)$ is the Möbius function, and $P_{n-1} = p_1 \cdot \ldots \cdot p_{n-1}$.

Consider the proof of the last formula.

□ It is sufficiently to show, that p_n is the only one integer, such that

$$1 < 2^{p_n} \left(-\frac{1}{2} + \sum_{d|P_{n-1}} \frac{\mu(d)}{2^d - 1} \right) \leq 2.$$

In fact, if $p_n \leq \alpha < p_{n+1}$, then

$$2^{p_n} \leq 2^\alpha < 2^{p_{n+1}}, \quad \text{i.e.,} \quad 2^{p_n} \cdot 2^{\log_2 \left(-\frac{1}{2} + \sum_{d|P_{n-1}} \frac{\mu(d)}{2^d-1} \right)} \leq 2,$$

and

$$2^{p_n} \cdot 2^{\log_2 \left(-\frac{1}{2} + \sum_{d|P_{n-1}} \frac{\mu(d)}{2^d-1} \right)} > 1,$$

$$\text{i.e.,} \quad 1 < 2^{p_n} \left(-\frac{1}{2} + \sum_{d|P_{n-1}} \frac{\mu(d)}{2^d - 1} \right) \leq 2.$$

Let $S = \sum_{d|P_{n-1}} \frac{\mu(d)}{2^d-1}$.

Then (using the property $\sum_{d|n} \mu(d) = 1$ for $n = 1$, and $\sum_{d|n} \mu(d) = 0$, otherwise) we have:

$$(2^{P_{n-1}} - 1)S = \sum_{d|P_{n-1}} \mu(d) \frac{2^{P_{n-1}} - 1}{2^d - 1}$$

$$= \sum_{d|P_{n-1}} \mu(d)(1 + 2^d + \cdots + 2^{P_{n-1}-d})$$

$$= \sum_{0 \leq t \leq P_{n-1}} \left(\sum_{d|gcd(t,P_{n-1})} \mu(d) \right) 2^t$$

$$= \sum_{0 < t < P_{n-1}, gcd(t,P_{n-1})=1} 2^t = 2^{P_{n-1}}$$

$$+ \sum_{0 < t < P_{n-1}-1, gcd(t,P_{n-1})=1} 2^t.$$

Then

$$2 \left(2^{P_{n-1}} - 1\right) \left(-\frac{1}{2} + S\right)$$

$$= 2^{P_{n-1}} + \sum_{0<t<P_{n-1}-1, gcd(t,P_{n-1})=1} 2^{t+1} - 2^{P_{n-1}} + 1$$

$$= 1 + \sum_{0<t<P_{n-1}-1, gcd(t,P_{n-1})=1} 2^{t+1} = 1$$

$$+ \sum_{0<t\leq P_{n-1}-p_n, gcd(t,P_{n-1})=1} 2^{t+1}.$$

Really, if $2 \leq j < p_n$, then there exists $q \in P$, such that $q|j$. But then $q < p_n$, i.e., $q|P_{n-1}$, and, hence, $q|P_{n-1} - j$. Therefore, if $0 < t < P_n - 1$, and $gcd(t, P_{n-1}) = 1$, then $0 < t \leq P_{n-1} - p_n$. Let $\sum_{0<t\leq P_{n-1}-p_n, gcd(t,P_{n-1})=1} 2^{t+1} = \sum$. Then

1. $1 + \sum \leq 1 + 2 + 2^2 + \cdots + 2^{P_{n-1}-p_n+1} = 2^{P_{n-1}-p_n+1}$;

2. $1 + \sum \geq 2^{P_{n-1}-p_n+1}$, as for $t = P_{n-1} - p_n$ we have $gcd(t, P_{n-1}) = 1$: if $q|t$, $q|P_{n-1}$, then $q|p_n$, a contradiction. Therefore,

$$\frac{2^{P_{n-1}-p_n+1}}{2^{P_{n-1}+1}} < -\frac{1}{2} + S = \frac{1 + \sum}{2(2^{P_{n-1}} - 1)} < \frac{2^{P_{n-1}+2} - 1}{2^{P_{n-1}+1} - 2} < \frac{2^{P_{n-1}-p_n+2}}{2^{P_{n-1}+1}},$$

and $1 < 2^{p_n}\left(-\frac{1}{2} + S\right) < 2$.

Some additional information see, for example, in [Deza17], [DeKo13], [Dick05], [HaWr79], [Sier64], [Step01].

Primes formulas of the second kind

2.4.10. In this subsection we consider some formulas, which give distinct prime numbers for any value of argument.

The first such formula known was established in 1947 by W.H. Mills, who proved that there exists a real number θ such that

$$\lfloor \theta^{3^n} \rfloor$$

is a prime number for all positive integers n. The smallest such θ has a value of around 1.3063... and is known as the *Mills' constant*.

This formula has no practical value, because very little is known about this constant (not even whether it is rational), and there is no known way of calculating the constant without finding primes in the first place.

2.4.11. Similarly (Wright, 1951), the function

$$g(n) = \left\lfloor 2^{2^{\cdots^{\omega}}} \right\rfloor$$

(a string of n exponents) gives a prime for every $n \geq 1$.

Here ω is a number which is roughly equal to 1.9287800... .

So, $g(1) = 3$, $g(2) = 13$, $g(3) = 16381$, and $g(4)$ has more than 5000 digits.

The fact that θ and ω are known only approximately and the numbers grow very fast, making these formulas no more than curiosities.

The following function yields all the primes, and only primes, for non-negative integers n:

$$f(n) = 2 + rest(2n!, n+1).$$

This formula is based on the Wilson's theorem; the number two is generated many times and all other primes are generated exactly once by this function. In fact, a prime p is generated for $n = p - 1$, and 2 is generated, otherwise; that is, 2 is generated when $n + 1$ is composite.

Some additional information see, for example, in [Buch09], [DeKo13], [Sier64], [HaWr79], [Step01].

Prime formulas of the third kind

2.4.12. In this subsection we consider the formulas, which generate the set of primes as the set of their positive values.

It was proved by Y. Matiyasevich in 1971, that

• *there is a polynomial in several variables with integer coefficients such that the set of primes is equal to the set of positive values of this polynomial obtained as the variables run through all non-negative integers.*

The first such polynomial discovered was the following polynomial in 26 variables of degree 25 (Sato, Wada and Wiens, 1976):

$(k+2)\{1-[wz+h+j-q]^2-[(gk+2g+k+1)(h+j)+h-z]^2-$
$[2n+p+q+z-e]^2-[16(k+1)^3(k+2)(n+1)^2+1-f^2]^2-[e^3(e+$
$2)(a+1)^2+1-o^2]^2-[(a^2-1)y^2+1-x^2]^2-[16r^2y^4(a^2-1)+1-$
$u^2]^2-[((a+u^2(u^2-a))^2-1)(n+4dy)^2+1-(x+cu)^2]^2-[n+l+$
$v-y]^2-[(a^2-1)l^2+1-m^2]^2-[ai+k+1-l-i]^2-[p+l(a-n-$
$1)+b(2an+2a-n^2-2n-2)-m]^2-[q+y(a-p-1)+s(2ap+$
$2a-p^2-2p-2)-x]^2-[z+pl(a-p)+t(2ap-p^2-1)-pm]^2\}.$

Notice that this polynomial factors. Look at the special form of the second part: it is one minus a sum of squares, so the only way for it to be positive is for each of the squared terms to be zero. Can you find a values $(a, b, c, d, \dots, z)$ (all non-negative) for which the polynomial above is positive?

The record for the lowest degree of such a polynomial is 5 (with 42 variables), and the record for fewest variables is 10 (with degree about $1.6 \cdot 10^{45}$).

In the view of the polynomial above, we can say, that a set of Diophantine equations in 26 variables can be used to obtain primes. In fact, a given number $k+2$ is prime if and only if the following system of Diophantine equations has a solution in the natural numbers:

$$
\begin{cases}
0 &= wz+h+j-q, \\
0 &= (gk+2g+k+1)(h+j)+h-z, \\
0 &= 16(k+1)^3(k+2)(n+1)^2+1-f^2, \\
0 &= 2n+p+q+z-e, \\
0 &= e^3(e+2)(a+1)^2+1-o^2, \\
0 &= (a^2-1)y^2+1-x^2, \\
0 &= 16r^2y^4(a^2-1)+1-u^2, \\
0 &= n+l+v-y, \\
0 &= (a^2-1)l^2+1-m^2, \\
0 &= ai+k+1-l-i, \\
0 &= ((a+u^2(u^2-a))^2-1)(n+4dy)^2+1-(x+cu)^2, \\
0 &= p+l(a-n-1)+b(2an+2a-n^2-2n-2)-m, \\
0 &= q+y(a-p-1)+s(2ap+2p-p^2-2p-2)-x, \\
0 &= z+pl(a-p)+t(2ap-p^2-1)-pm.
\end{cases}
$$

A general theorem of Matiyasevich says that the range of every recursively enumerable function may be given by the solutions to a set of Diophantine equations in 9 variables, but in general of very high degree. This implies the existence of a set of Diophantine equations in 9 variables that has the same property as the above one. Jones showed that there also exists such a set of equations of degree only 4, but in 58 variables.

Some additional information see, for example, in [Buch09], [DeKo13], [Mati93].

Formulas of primes and Fermat and Mersenne numbers

2.4.13. At first glance, it seems likely that primes could be produced by taking a high power of a fixed integer k and adding or subtracting 1.

In fact, after the simplest algebraic function, a polynomial, it is natural to consider the exponential function, $f(n) = k^n$, or, as the value k^n cannot give infinitely many primes, the variant of it of the form

$$f(n) = k^n \pm b^n.$$

As for any positive integer n one has

$$k^n - b^n = (k - b)\left(k^{n-1} + k^{n-2}b + \cdots + k^2 b^{n-2} + b^{n-1}\right),$$

then $k^n - b^n$ can be prime only in the case $k - b = 1$, or $k = b + 1$. In the simplest case we have, that $b = 1$, i.e.,

$$f(n) = k^n - 1.$$

The same choice of b give us in the case $k^n + b^n$ the function

$$f(n) = k^n + 1.$$

For $k^m + 1$ to be prime, k must be even and the exponent m must be a power of 2. The simplest case $k = 2$ gives us the *Fermat numbers*

$$F_n = 2^{2^n} + 1, \ n = 0, 1, 2, 3, \ldots .$$

The general case gives us the *generalized Fermat numbers*

$$F_n(b) = b^{2^n} + 1, \ 2|b, \ n = 0, 1, 2, 3, \dots .$$

If $k^m - 1$ is a prime, then $k = 2$ and m is a prime, so the problem is reduced to that of *Mersenne numbers*

$$M_p = 2^p - 1.$$

The proof of these properties and some additional information see, for example, in [Buch09], [DeDe12], [Deza17], [Deza18], [DeKo13], [HaWr79], [Lege79], [Sier64].

Exercises

1. Prove, that the function $(-1)^n + 4$ generates prime numbers for any positive integer n.

2. Prove, that if $a^n + 1$ is a prime, then $n = 2^k$.

3. Prove, that if $a^n - 1$ is a prime, then $a = 2$, and $n \in P$.

4. Prove, that any number of the form $4n^4 + 1$, $n > 1$, is composite.

5. Prove, that any number of the form $4n^4 + n^2 + 1$, $n > 1$, is composite.

6. Prove, that any polynomial $x^2 + x + q$, $q = 2, 3, 5, 11, 17$, gives primes for all $x = 0, \dots, q - 2$.

7. Prove, that there are infinitely many primes of the form $x^2 - y^2$.

8. Prove, that there are infinitely many primes of the form $x^2 + y^2$.

9. Prove, that the function $f(x) = \lfloor \cos^2 \pi \frac{(x-1)!+1}{x} \rfloor$ gives the value 1 if x is prime, and 0 if x is composite; moreover, $f(1) = 1$.

10. Check, formulas for p_n, given above.

11. Check, that the function $f(n) = 2 + rest(2n!, n + 1)$ generates a prime p for $n = p - 1$, and generates 2, otherwise.

2.5 Prime numbers in the family of special numbers

Fibonacci and Lucas primes

2.5.1. *Fibonacci numbers* u_n, $n = 0, 1, 2, ...$, are defined by famous recurrence

$$u_{n+2} = u_{n+1} + u_n, \ u_0 = 0, u_1 = 1.$$

The first few Fibonacci numbers are 0, 1, 1, 2, 3, 5, 8, 13, 21, 34, ... (sequence A000045 in the OEIS).

A *Fibonacci prime* is a Fibonacci number that is prime, a type of integer sequence prime.

The first Fibonacci primes are 2, 3, 5, 13, 89, 233, 1597, 28657, 514229, 433494437, ... (sequence A005478 in the OEIS).

It is not known whether there are infinitely many Fibonacci primes. With the indexing starting with $u_1 = u_2 = 1$, the first 34 are u_n for the values $n = 3, 4, 5, 7, 11, 13, 17, 23, 29, 43, 47, 83, 131, 137, 359, 431, 433, 449, 509, 569, 571, 2971, 4723, 5387, 9311, 9677, 14431, 25561, 30757, 35999, 37511, 50833, 81839, 104911$ (see sequence A001605 in the OEIS).

In addition to these proven Fibonacci primes, there have been found probable primes for $n = 130021, 148091, 201107, 397379, 433781, 590041, 593689, 604711, 931517, 1049897, 1285607, 1636007, 1803059, 1968721, 2904353, 3244369, 3340367$.

Except for the case $n = 4$, all Fibonacci primes have a prime index, because if m divides n, then u_m also divides u_n, but not every prime is the index of a Fibonacci prime.

The number of prime factors in the Fibonacci numbers with prime index are 0, 1, 1, 1, 1, 1, 1, 2, 1, 1, 2, 3, 2, 1, 1, 2, 2, 2, 3, 2, 2, 2, 1, 2, 4, 2, 3, 2, 2, 2, 2, 1, 1, 3, ... (sequence A080345 in the OEIS).

The number u_p is prime for 8 of the first 10 primes p; the exceptions are $u_2 = 1$ and $u_{19} = 4181 = 37 \cdot 113$.

However, Fibonacci primes appear to become rarer as the index increases. The number u_p is prime for only 26 of the 1229 primes p below 10000.

As of 2020, the largest known certain Fibonacci prime is u_{104911}, with 21925 digits (Steine and Water, 2015).

The largest known probable Fibonacci prime is $u_{3340367}$ (Lifchitz, 2018).

It was proved by N. MacKinnon, that the only Fibonacci numbers that are also members of the set of twin primes are 3, 5, and 13.

2.5.2. *Lucas numbers* L_n, $n = 0, 1, 2, ...$, are defined by the same recurrence as Fibonacci numbers, but with different starting values:

$$L_{n+2} = L_{n+1} + L_n, \ L_0 = 2, L_1 = 1.$$

The first few Lucas numbers are 2, 1, 3, 4, 7, 11, 18, 29, 47, 76, ... (sequence A000032 in the OEIS).

A *Lucas prime* is a Lucas number that is prime.

The first few Lucas primes are 2, 3, 7, 11, 29, 47, 199, 521, 2207, 3571, ... (sequence A005479 in the OEIS).

The indices of these primes are 0, 2, 4, 5, 7, 8, 11, 13, 16, 17, ... (sequence A001606 in the OEIS).

If L_n is prime, then n is 0, a prime, or a power of 2. L_{2m} is prime for $m = 1, 2, 3$ and 4, and no other known values of m.

The proofs of the propositions above and some additional information see, for example, in [Buch09], [DeKo13], [Hogg69], [Voro61].

Prime numbers in Pascal's triangle

2.5.3. *Pascal's triangle* is a number triangle, the sides of which are formed by 1, and any inner entry is obtained by adding the two entries diagonally above.

It is easy to show, that elements of Pascal's triangle coinside with *binomial coefficients*, i.e., coefficients $\binom{n}{k}$, $n = 0, 1, 2, 3, ...$, $k = 0, 1, 2, ..., n$, of the following decomposition:

$$(1 + x)^n = \binom{n}{0} + \binom{n}{1}x + \binom{n}{2}x^2 + \cdots + \binom{n}{n}x^n.$$

The first few rows of the Pascal's triangle are

$$\binom{0}{0} = 1$$

$$\binom{1}{0} = 1 \quad \binom{1}{1} = 1$$

$$\binom{2}{0} = 1 \quad \binom{2}{1} = 2 \quad \binom{2}{2} = 1$$

$$\binom{3}{0} = 1 \quad \binom{3}{1} = 3 \quad \binom{3}{2} = 3 \quad \binom{3}{3} = 1$$

$$\binom{4}{0} = 1 \quad \binom{4}{1} = 4 \quad \binom{4}{2} = 6 \quad \binom{4}{3} = 4 \quad \binom{4}{4} = 1$$

$$\binom{5}{0} = 1 \quad \binom{5}{1} = 5 \quad \binom{5}{2} = 10 \quad \binom{5}{3} = 10 \quad \binom{5}{4} = 5 \quad \binom{5}{5} = 1$$

$$\binom{6}{0} = 1 \quad \binom{6}{1} = 6 \quad \binom{6}{2} = 15 \quad \binom{6}{3} = 20 \quad \binom{6}{4} = 15 \quad \binom{6}{5} = 6 \quad \binom{6}{6} = 1$$

$$\cdots \quad \cdots \quad \cdots$$

It is known, that $\binom{n}{k} = \frac{n!}{k!(n-k)!}$. Using this fact, it is easy to show, that

• *for a prime p, the binomial coefficient $\binom{p}{k}$ is divisible by p for any $k = 1, 2, ..., p - 1$:*

$$\binom{p}{k} \equiv 0 (mod\ p), \quad p \in P, \quad k = 1, 2, ..., p - 1.$$

□ Consider the formula $\binom{p}{k} = \frac{p!}{k!(n-k)!}$. Since $p \in P$, then p has no factors except 1 and p itself. Every entry in the triangle is an integer, so therefore by definition $(p - k)!$ and $k!$ are factors of $p!$. However, there is no possible way p itself can show up in the denominator, so therefore p (or some multiple of it) must be left in the numerator, making the entire entry a multiple of p. □

This fact allows to obtain a simple, but important property of Pascal's triangle:

• *in the p-th row of Pascal's triangle, where p is a prime number, all the terms except the unities are multiples of p.*

In this case we say, that the *p-th row of the Pascal's triangle is divisible by p.*

For additional information about divisibility of the rows of Pascal's triangle by prime p, we need to use the *Lucas's theorem.*

This theorem, first appeared in 1878 in papers by É. Lucas, expresses the remainder of division of the binomial coefficient $\binom{n}{m}$ by a prime number p in terms of the base p expansions of the integers m and n:

• *for non-negative integers n and m and a prime p, if*

$$n = n_k p^k + n_{k-1} p^{k-1} + \cdots + n_1 p + n_0,$$

$$m = m_k p^k + m_{k-1} p^{k-1} + \cdots + m_1 p + m_0$$

are the base p expansions of n and m, respectively, then

$$\binom{n}{m} \equiv \prod_{i=0}^{k} \binom{n_i}{m_i} (mod\ p).$$

This uses the convention that $\binom{m}{n} = 0$ if $m < n$.

From the Lucas's theorem we get easily that *a binomial coefficient* $\binom{m}{n}$ *is divisible by a prime p if and only if at least one of the base p digits of n is greater than the corresponding digit of m.*

This fact allow to obtain a criterion of divisibility of a row of Pascal's triangle by a prime p:

• *the n-th row of Pascal's triangle is divisible by a prime p if and only if n is a power of p, i.e., $n = p^l$, $l \in \mathbb{N}$.*

□ In fact, if $n = p^l$, $l \in \mathbb{N}$, then $n = 0 + 0 \cdot p + 0 \cdot p^2 + \cdots + 1 \cdot p^l$, and for $0 < m < n$ we have $m = m_0 + m_1 p + \cdots + m_{l-1} p^{l-1} + 0 \cdot p^l$. So,

$$\binom{n}{m} \equiv \binom{0}{m_0} \cdot \binom{0}{m_1} \cdot \binom{0}{m_2} \cdot \dots \cdot \binom{1}{0} (mod\ p).$$

As there is at least one element $m_i \neq 0$ in the set $\{m_0, m_1, ..., m_{l-1}\}$, then $\binom{0}{m_i} = 0$; it means, that $\binom{n}{m}$, $0 < n < m$, is divisible by p.

Conversely, let $n \neq p^l$, $l \in \mathbb{N}$. In this case $n = n_0 + n_1 p + n_2 p^2 + \cdots + n_k p^k$, where $n_k \neq 0$. For $n = n_0$ it holds $n < p$, and the first element $\binom{n}{1}$ of n-th row of Pascal's triangle is equal to n and, hence, is not divisible by p. If $n \geq p$ and $n \neq p^l$, $l \in \mathbb{N}$, then $k \geq 1$, and there exists at least one element n_i in the decomposition $n = n_0 + n_1 p + n_2 p^2 + \cdots + n_k p^k$, such that $n_i \neq n_k$, and $n_i \neq 0$. In this case, consider $\binom{n}{m}$, where $m = n_0 + n_1 p + \cdots + n_{i-1} p^{i-1} + 0 \cdot p^i + n_{i+1} p^{i+1} + \cdots + n_k p^k$. It is easy to see, that

$$\binom{n}{m} \equiv \binom{n_0}{n_0} \cdot \binom{n_1}{n_1} \cdot \binom{n_i}{0} \cdot \dots \cdot \binom{n_k}{n_k} \equiv 1 (mod\ p).$$

So, m-th element $\binom{n}{m}$, $0 < m < n$, of the n-th row of Pascal's triangle is not divisible by p, and, therefore, the n-th row itself is not divisible by p. □

So, we have proven, that for any p^k-th row of Pascal's triangle, where p is a prime number, and k is a positive integer, all the terms of this row except the unities are multiples of p. Moreover, it is known also, that

• *for a n-th row of Pascal's triangle all the terms in this row except the unities are multiples of m if and only if m is a prime number p, and $n = p^k$, $k \in \mathbb{N}$.*

More exactly,

• *the greatest common divisor of all inner terms of n-th row of Pascal's triangle is equal to p, if $n = p^k$, $k \in \mathbb{N}$, and is equal to 1, otherwise.*

If we consider $p = 2$, we get that all inner entries in 2-nd, 4-th, 8-th, ..., 2^k-th, ... rows of Pascal's triangle are even. Easy to see, that there exist exactly two inner odd entries in the $(2^k + 1)$-th row, etc. This leads to the Sierpiński construction in the Pascal's triangle modulo 2.

The proof of Lucas's theorem and some additional information see, for example, in [Bond93], [Uspe76], [Deza18].

Catalan primes

2.5.4. The *Catalan numbers* form a sequence of natural numbers that occur in various counting problems, often involving recursively defined objects.

The n-th Catalan number can be given directly in terms of binomial coefficients by

$$C_n = \frac{1}{n+1}\binom{2n}{n} = \frac{(2n)!}{(n+1)!\,n!} = \prod_{k=2}^{n} \frac{n+k}{k}, \quad n \geq 0.$$

The first Catalan numbers for $n = 0, 1, 2, 3, \ldots$ are 1, 1, 2, 5, 14, 42, 132, 429, 1430, 4862, ... (sequence A000108 in the OEIS).

It is easy to check, that the only Catalan numbers C_n that are odd are those for which $n = 2^k - 1$, all others are even. Moreover,

• *the only prime Catalan numbers are $C_2 = 2$, and $C_3 = 5$.*

□ Really, using the formula

$$C_n = \frac{1}{n+1}\binom{2n}{n} = \frac{1}{n+1} \cdot \frac{(2n)!}{n!\,n!}$$

we get, that if a prime number p divides C_n, therefore $p \le 2n$. On the other hand, it is easy to see, that

$$C_n > 2n + 1, \text{ if } n \ge 4.$$

In fact, for $n = 4$ it holds $C_4 = 14 > 2 \cdot 4 + 1 = 9$. Going from n to $n + 1$, we get that $\frac{2(2n-1)}{n+1} \ge 2$ for all $n \ge 3$, therefore,

$$C_n = \frac{2(2n - 1)}{n + 1} C_n \ge 2(2n - 1) = 4n - 2 > 2n + 1, \quad \text{if } n \ge 4.$$

So, for $n \ge 4$ the number C_n is a composite number. For $n < 4$ we have primes only for $n = 2$ and $n = 3$. $\square$

There are some interesting divisibility properties of Catalan numbers. For example,

• *if a positive integer $n + 2$ is a power of a prime, $n + 2 = p^k$, $k \in \mathbb{N}$, then C_n is divisible by p.*

For example, $2 + 2 = 2^2$, and $C_2 = 2 \equiv 0 (mod\ 2)$; $5 + 2 = 7$, and $C_5 = 42 \equiv 0 (mod\ 7)$; $7 + 2 = 3^2$, and $C_7 = 429 \equiv 0 (mod\ 3)$.

Similarly, we can obtain the following number-theoretical property:

• *if $n = p^k$, $k \in \mathbb{N}$, then $C_n \equiv 2 (mod\ p)$.*

For example, if $2 = 2^1$, then $C_2 = 2 \equiv 2 (mod\ 2)$; if $3 = 3^1$, then $C_3 = 5 \equiv 2 (mod\ 3)$; if $4 = 2^2$, then $C_4 = 14 \equiv 2 (mod\ 2)$; if $5 = 5^1$, then $C_5 = 42 \equiv 2 (mod\ 5)$; if $7 = 7^1$, then $C_7 = 429 \equiv 2 (mod\ 7)$; if $8 = 2^3$, then $C_8 = 1430 \equiv 2 (mod\ 2)$; if $9 = 3^2$, then $C_9 = 4862 \equiv 2 (mod\ 3)$.

For references and some additional information see, for example, [Bern99], [Cata44], [Deza18], [Goul85], [Stan15].

Cullen and Woodall primes

2.5.5. A *Cullen number* is a positive integer of the form

$$Cu(n) = n \cdot 2^n + 1, \quad n \ge 0.$$

Cullen numbers were first studied by R.J. Cullen in 1905.

The first few Cullen numbers are 1, 3, 9, 25, 65, 161, 385, 897, 2049, 4609, ... (sequence A002064 in the OEIS).

A *Cullen prime* is a Cullen number, which is a prime.

The only known Cullen primes are those for $n = 1$, 141, 4713, 5795, 6611, 18496, 32292, 32469, 59656, 90825, 262419, 361275, 481899, 1354828, 6328548 and 6679881 (sequence A005849 in the OEIS).

It was shown (Hooley, 1976) that

• *almost all Cullen numbers are composite.*

H. Suyama showed that it works for any sequence of numbers $n \cdot 2^{n+a} + b$, where a and b are integers, and in particular also for Woodall numbers.

Still, it is conjectured that *there are infinitely many Cullen primes.*

There are several important divisibility properties of Cullen numbers. For example,

• *a Cullen number $Cu(n)$ is divisible by $p = 2n - 1$, if p is a prime number of the form $8k - 3$.*

Furthermore, it follows from the Fermat's little theorem, that

• *an odd prime p divides $Cu(m(k))$ for each $m(k) = (2^k - k) \cdot (p - 1) - k$, $k \in \mathbb{N}$.*

It has also been shown that the prime number p divides $Cu(\frac{p+1}{2})$ when the Legendre symbol $(\frac{2}{p})$ is equal to -1, and that p divides $Cu(\frac{3p-1}{2})$ when the Legendre symbol $(\frac{2}{p})$ is equal to 1. Using the properties of $(\frac{2}{p})$, we get, that

• *a prime number p divides $Cu(\frac{p+1}{2})$ when $p \equiv \pm 3 (mod\ 8)$, and that p divides $C(\frac{3p-1}{2})$ when $p \equiv \pm 1 (mod\ 8)$.*

It is unknown whether there exists a prime number p such that $Cu(p)$ is also prime.

Sometimes, a *generalized Cullen number* is defined to be a number of the form

$$GCu(n) = n \cdot b^n + 1,\ n + 2 > b.$$

If a prime can be written in this form, it is called a *generalized Cullen prime.*

Least n such that $n \cdot b^n + 1$ is prime for $b = 1, 2, 3, \dots$ are 1, 1, 2, 1, 1242, 1, 34, 5, 2, 1, ... (sequence A240234 in the OEIS).

As of 2020, the largest known generalized Cullen prime is $2805222 \cdot 25^{2805222} + 1$. It has 3921539 digits and was discovered by T. Greer, a *PrimeGrid* participant.

2.5.6. A *Woodall number* (or *Riesel numer*, or *Cullen number of the second kind*) is a positive integer of the form

$$W_n = n \cdot 2^n - 1, n \in \mathbb{N}.$$

Woodall numbers were first studied by A.J.C. Cunningham and H.J. Woodall in 1917, inspired by Cullen's earlier study of the similarly defined Cullen numbers.

The first few Woodall numbers are 1, 7, 23, 63, 159, 383, 895, 2047, 4607, 10239, ... (Sloan's A003261).

Woodall numbers that are also prime numbers are called *Woodall primes*; the first few exponents n for which the corresponding Woodall numbers W_n are primes are $n = 2, 3, 6, 30, 75, 81, 115, 123, 249, 362, ...$ (sequence A002234 in the OEIS); the Woodall primes themselves begin with $7, 23, 383, 32212254719, ...$ (sequence A050918 in the OEIS).

As of 2020, the largest known Woodall prime is $17016602 \cdot 2^{17016602} - 1$. It has 5122515 digits and was found by D. Bertolotti in 2018 in the distributed computing project *PrimeGrid*.

It is proven (Keller, Suyama, 1995) that *almost all Woodall numbers are composite*.

Nonetheless, it is also conjectured that *there are infinitely many Woodall primes*.

Like Cullen numbers, Woodall numbers have many divisibility properties.

For example, if p is a prime, then p divides $W_{\frac{p+1}{2}}$ if the Legendre symbol $\left(\frac{2}{p}\right)$ is equal to 1, and p divides $W_{\frac{3p-1}{2}}$ if the Legendre symbol $\left(\frac{2}{p}\right)$ is equal to 1. So, one can say, that

• *a given prime number p divides $W_{\frac{p+1}{2}}$ if $p \equiv \pm 3 (mod\ 8)$, and p divides $W_{\frac{3p-1}{2}}$ if $p \equiv \pm 1 (mod\ 8)$.*

A *generalized Woodall number* is defined to be a number of the form

$$GW(n, b) = n \cdot b^n - 1, \ n + 2 > b.$$

If a prime can be written in this form, it is called a *generalized Woodall prime*.

Least n such that $n \cdot b^n - 1$ is prime for $b = 1, 2, 3, \ldots$ are 3, 2, 1, 1, 8, 1, 2, 1, 10, 2, ... (sequence A240235 in OEIS).

As of 2020, the largest known generalized Woodall prime is

$$GW(17016602, 17016602) = 17016602 \cdot 2^{17016602} - 1.$$

The needed proofs and some additional information see, for example, in [Guy94], [Cull05], [CuWu17], [Hool76], [Kell95].

Bernoulli numbers and Staudt primes

2.5.7. *Bernoulli numbers* B_n, $n \geq 0$, form a sequence of rational numbers which occur frequently in Number Theory and in other domains of Mathematics. The Bernoulli numbers appear in (and can be defined by) the Taylor series expansions of the tangent and hyperbolic tangent functions, in Faulhaber's formula for the sum of m-th powers of the first n positive integers, in the Euler–Maclaurin formula, and in expressions for certain values of the Riemann zeta-function.

Formally, the n-th *Bernoulli number* B_n can be defined by the following recurrence:

$$B_n = -\frac{1}{n+1} \sum_{k=1}^{n} \binom{n+1}{k+1} B_{n-k}, \quad \text{with} \quad B_0 = 0.$$

The first few Bernoulli numbers are 0, $-\frac{1}{2}$, $\frac{1}{6}$, 0, $-\frac{1}{30}$, 0, $\frac{1}{42}$, 0, $-\frac{1}{30}$, 0, ...

The *von Staudt–Clausen theorem* is a result determining the fractional part of Bernoulli numbers using some special set of primes. It was proven independently by K. von Staudt (1840) and T. Clausen (1840). Specifically,

• *if n is a positive integer and we add $\frac{1}{p}$ to the Bernoulli number B_{2n} for every prime p such that $p - 1$ divides $2n$, we obtain an integer, i.e.,*

$$B_{2n} + \sum_{p-1 \mid 2n} \frac{1}{p} \in \mathbb{Z}.$$

In other words, for any Bernoulli number B_{2n} with even index $2n > 0$, there exists a representation

$$B_{2n} = G_{2n} - \frac{1}{p_1} - \frac{1}{p_2} - \frac{1}{p_3} - \dots - \frac{1}{p_s}, \quad G_{2n} \in \mathbb{Z},$$

which uses all primes p_i, such that $p_i - 1 | 2n$.

For $n = 1$, we can see, that $p_1 = 2$, and $p_2 = 3$, as $1|2$, and $2|2$. So,

$$B_2 = G_2 - \frac{1}{2} - \frac{1}{3}, \quad G_2 \in \mathbb{Z}.$$

In fact, $B_2 = \frac{1}{6}$, $\frac{1}{2} + \frac{1}{3} = \frac{5}{6}$, so, $G_2 = \frac{1}{6} + (\frac{1}{2} + \frac{1}{3}) = \frac{1}{6} + \frac{5}{6} = 1$. Therefore,

$$B_2 = 1 - \frac{1}{2} - \frac{1}{3}.$$

For $n = 5$, we have $p_1 = 2$, $p_2 = 3$, and $p_3 = 11$, as $1|10$, $2|10$, and $10|10$. So,

$$B_{10} = G_{10} - \frac{1}{2} - \frac{1}{3} - \frac{1}{11}, \quad G_{10} \in \mathbb{Z}.$$

In fact, $B_{10} = \frac{5}{66}$, and $\frac{1}{2} + \frac{1}{3} + \frac{1}{11} = \frac{61}{66}$, so, $G_{10} = \frac{5}{66} + (\frac{1}{2} + \frac{1}{3} + \frac{1}{11}) = \frac{5}{66} + \frac{61}{66} = 1$. Therefore,

$$B_{10} = 1 - \frac{1}{2} - \frac{1}{3} - \frac{1}{11}.$$

For $n = 8$, we get $p_1 = 2$, $p_2 = 3$, $p_3 = 5$, $p_4 = 17$, as $1|16$, $2|16$, $4|16$, and $16|16$. So,

$$B_{16} = G_{16} - \frac{1}{2} - \frac{1}{3} - \frac{1}{5} - \frac{1}{17}, \quad G_{16} \in \mathbb{Z}.$$

In fact, $B_{16} = -7\frac{47}{510}$, and $\frac{1}{2} + \frac{1}{3} + \frac{1}{5} + \frac{1}{17} = \frac{557}{510}$, so, $G_{16} = -7\frac{47}{510} + (\frac{1}{2} + \frac{1}{3} + \frac{1}{5} + \frac{1}{17}) = -7 - \frac{47}{510} + \frac{557}{510} = -7 + 1 = -6$. Therefore,

$$B_{16} = -6 - \frac{1}{2} - \frac{1}{3} - \frac{1}{5} - \frac{1}{17}.$$

The theorem immediately allows us to characterize the denominators of the non-zero Bernoulli numbers B_{2n} as the product of all primes p such that $p - 1$ divides $2n$; consequently the denominators are square-free and divisible by 6.

These denominators are 6, 30, 42, 30, 66, 2730, 6, 510, 798, 330, ... (sequence A002445 in the OEIS).

The proof of this theorem and some additional information see, for example, in [Abra74], [Apos86], [Clau40], [Deza18], [KGP94], [Stau40].

Motzkin primes

2.5.8. *Motzkin numbers*, named after an Israeli-American mathematician T. Motzkin, have diverse applications in Geometry, Combinatorics and Number Theory.

The n-th Motzkin number MN_n can be defined by the recurrence relation

$$MN_n = MN_{n-1} + \sum_{i=0}^{n-2} MN_i MN_{n-2-i}, \quad MN_0 = MN_1 = 1.$$

It is the number of different ways of drawing non-intersecting chords between n points on a circle.

The first few Motzkin numbers MN_n for $n = 0, 1, 2, \ldots$ form the sequence 1, 1, 2, 4, 9, 21, 51, 127, 323, 835, ... (sequence A001006 in the OEIS).

A *Motzkin prime* is a Motzkin number that is prime. As of 2020, only four such primes are known. They are 2, 127, 15511, 953467954114363 (sequence A092832 in the OEIS).

Some additional information see, for example, in [Bern99], [Deza18], [DoSh77], [Motz48].

Palindromic and permutable primes

2.5.9. A *palindrome* is an positive integer that reads the same forward or backward. Some examples of palindromes are 11, 121, 12321, 1234321, 111, etc.

A *palindromic prime* is a prime which is a palindrome. The first few palindromic primes are 2, 3, 5, 7, 11, 101, 131, 151, 181, 191, ... (sequence A002385 in the OEIS).

As of 2020, the largest known palindromic prime (Batalov, 2014) is $10^{474500} + 999 \cdot 10^{237249} + 1$; it has 474501 digits.

The first non-trivial palindromic prime is 11. In fact, it is the only palindromic prime with even number of digits. Except for 11, all palindromic primes have an odd number of digits, because the divisibility test for 11 tells us that every palindromic number with an even number of digits is a multiple of 11.

It is not known if *there are infinitely many palindromic primes.*

Obviously, the palindromicity of a number depends on the base in which the number is written. For example, the sequence of binary palindromic primes begins (in binary) from 11, 101, 111, 10001, 11111, 1001001, 1101011, 1111111, 100000001, 100111001, ... (sequence A117697 in the OEIS) and contains all *Mersenne primes* and *Fermat primes.*

It is easy to check, that *all binary palindromic primes except binary 11 (decimal 3) have an odd number of digits*; those palindromes with an even number of digits are divisible by 3.

It is known that, *for any base, almost all palindromic numbers are composite*, i.e., the ratio between palindromic composites and all palindromes below n tends to 1.

2.5.10. A *permutable prime* (or *absolute prime*) is a prime which remains prime on every rearrangement (permutation) of the digits. For example, 337 is a permutable because each of 337, 373 and 733 is prime.

In base 10 the only known permutable primes are 2, 3, 5, 7, 13, 17, 37, 79, 113, 199, 337, their permutations, and the repunit primes $R_2 = 11$, R_{19}, R_{23}, R_{317}, and R_{1031} (see sequences A003459 and A258706 in the OEIS).

There is no n-digit permutable prime for $3 < n < 6 \cdot 10^{175}$ which is not a repunit. It is conjectured that *there are no non-repunit permutable primes other than those listed above.*

Obviously, in base 10 permutable primes of two or more digits may not have the digits 2, 4, 6, 8 or 5. So, all permutable primes of two or more digits are composed from the digits 1, 3, 7, 9.

It is proven that *no permutable prime exists which contains three different of the four digits 1, 3, 7, 9*, as well as that *there exists no permutable prime composed of two or more of each of two digits selected from 1, 3, 7, 9.* Looking modulo 7 we also see *they may not have all four of the digits 1, 3, 7, and 9 simultaneously.* In fact, there exist the following theorem:

• *every permutable prime is a near-repdigit, that is, it is a permutation of the integer* $B_n(a,b) = (aaa...aab)$, *where* a *and* b *are distinct digits from the set* $\{1,3,7,9\}$.

In base 2, only repunits can be permutable primes, because any 0 permuted to the ones place results in an even number. Therefore,

• *the base 2 permutable primes are the Mersenne primes.*

The proofs of the properties above and some additional information see, for example, in [BaWe76], [Gees20], [John77], [Malc86], [Ribe96].

Exercises

1. Find first five Fibonacci numbers. Prove, that any prime Fibonacci number has a prime index.

2. Find first five Lucas primes. What one can say about their indexes?

3. Find the first 10 triangle numbers. What we can say about primality of triangle numbers?

4. Find the first 10 pentagonal numbers. What we can say about primality of pentagonal numbers?

5. Find several prime numbers in the Pascal's triangle. Check, that 2-th, 4-th, 8-th row of Pascal's triangle is divisible by 2. Check, that 3-th, 9-th, 27-th row of Pascal's triangle is divisible by 3.

6. Prove, that any prime number cannot to be a perfect number.

7. Prove, that the only prime Catalan numbers are 2 and 5.

8. Check, that the set $\{2,3,5\}$ is the maximal subset of P, such that the following property holds: for any $m \in M$, each Pythagorean triple has a number, divisible by m.

9. Check, that starting with $W_4 = 63$ and $W_5 = 159$, every sixth Woodall number is divisible by 3.

10. Check, that for a positive integer m, the Woodall number W_{2^m} may be prime only if $2^m + m$ is prime.

2.6 Open problems

There are many famous unsolved problems in Mathematics, which are closely connected with prime numbers. Most of them have very simple form and are clean even for amateurs of Mathematics. However, often having an elementary formulation, many of these problems have withstood proof for decades.

Twin primes conjecture

2.6.1. Well-known is the irregularity of the distribution of prime numbers on the number line.

For a given N, *there exists an interval of length N which does not contain prime numbers*: in fact, the number $(N + 1)! + 2$ is divisible by 2, $(N + 2)! + 3$ is divisible by 3, ..., $(N + 1)! + (N + 1)$ is divisible by $(N + 1)$.

So, we have proven the existence of arbitrarily large prime gaps. However, large prime gaps occur much earlier than this argument shows. For example, the first prime gap of length 8 is between the primes 89 and 97, much smaller than $8! = 40320$.

2.6.2. On the other hand, there exist many so called *twin primes*. If you look down the list of primes, you will quite often see two consecutive odd numbers: 3 and 5, 5 and 7, 11 and 13, 17 and 19, etc. We call these pairs $(p, p + 2)$ of primes numbers *twin primes* (this name was coined by P.G.S. Stäckel in 1916).

The first such pairs are $(3, 5)$, $(5, 7)$, $(11, 13)$, $(17, 19)$, $(29, 31)$, $(41, 43)$, $(59, 61)$, $(71, 73)$, $(101, 103)$, $(107, 109)$, ... (sequence A077800 in the OEIS).

A question to which the answer is not known is
- *whether there exist infinitely many twin primes.*

It has been proved by V. Brun in 1919 that the series of the reciprocals of the prime numbers of the pairs of twin primes is finite or convergent.

This famous result expresses the scarcity of twin primes, even if there are infinitely many of them.

The sum B_2 of the series has been calculated by V. Brun in 1919 to six decimal places, and now is called *Brun's constant:*

$$B_2 = \left(\frac{1}{3} + \frac{1}{5}\right) + \left(\frac{1}{5} + \frac{1}{7}\right) + \left(\frac{1}{11} + \frac{1}{13}\right) + \dots = 1.902160\dots\,.$$

More exactly, let $\pi_2(x)$ denote the number of primes p such that $p \leq x$ and $p+2$ is also a prime. V. Brun announced in 1919 that for $x \geq x_0$ one has

$$\pi_2(x) < \frac{100x}{(\log x)^2}.$$

The proof appeared in 1920.

R.P. Brent, 1976, has found that there are 152892 pairs of twin primes less than 10^{11}.

P. Sebah and P. Demichel, 2002, had shown, using all twin primes up to 10^{16}, that $B_2 = 1.902160583104\dots$.

The greatest known pair of twin primes is the pair $2996863034895 \cdot 2^{1290000} \pm 1$ (2016). It has 388342 digits.

2.6.3. Another question to which the answer is not known is
- *whether there exist infinitely many primes p for which p, $p+2$, $p+6$ and $p+8$ are all prime numbers.*

A quadruple $(p, p+2, p+6, p+8)$ of this type is called a *prime quadruplet*. (Obviously, the numbers p, $p+2$, $p+4$, $p > 3$, cannot be all primes, because one from them is divisible by 3.) The first six consecutive quadruplets are obtained for $p = 5, 11, 101, 191, 821, 1481$.

The smallest prime quadruplet of 50 digits (Stevens, 1995) has the form

$$1000\dots058537891, 93, 97, 99.$$

As of 2020, the largest known prime quadruplet (Kaiser, 2019) has 10132 digits and starts with

$$p = 667674063382677 \cdot 2^{33608} - 1.$$

The constant representing the sum of the reciprocals of all prime quadruplets, so called *Brun's constant for prime quadruplets,* is approximately $0.87058\dots$.

2.6.4. There exist also other "relative" primes. So, a *cousin primes* is a pair of prime numbers that differ by four. The first cousin primes are $(3, 7)$, $(7, 11)$, $(13, 17)$, $(19, 23)$, $(37, 41)$, $(43, 47)$, $(67, 71)$, $(79, 83)$, $(97, 101)$, $(103, 107)$, ... (see sequences A023200 and A046132 in the OEIS).

The largest known cousin prime pair has 20008 digits and is $(p, p+4)$ for

$$p = 4111286921397 \cdot 2^{66420} + 1;$$

It is known, that cousin primes have the same asymptotic density as twin primes. An analogy of Brun's constant for twin primes is defined for cousin primes, with the initial term $(3, 7)$ omitted:

$$B_4 = \left(\frac{1}{7} + \frac{1}{11}\right) + \cdots = 1.1970449\ldots.$$

2.6.5. A *sexy prime* is a pair $(p, p+6)$ of prime numbers that differ by six. The name "sexy prime" stems from the Latin word for six, *sex*.

The first sexy primes are $(5, 11)$, $(7, 13)$, $(11, 17)$, $(13, 19)$, $(17, 23)$, $(23, 29)$, $(31, 37)$, $(37, 43)$, $(41, 47)$, $(47, 53)$, ... (see sequences A023201 and A046117 in the OEIS).

The largest known pair of sexy primes (Kaiser, 2019) is $(p, p+6)$ for

$$p = (520461 \cdot 2^{55931} + 1) \cdot (98569639289$$
$$\cdot (520461 \cdot 2^{55931} - 1)^2 - 3) - 1.$$

It has 50539 digits.

2.6.6. Like twin primes, sexy primes can be extended to larger constellations. Triplets of primes $(p, p+6, p+12)$ such that $p+18$ is composite are called *sexy prime triplets*.

The first sexy prime triplets are $(7, 13, 19)$, $(17, 23, 29)$, $(31, 37, 43)$, $(47, 53, 59)$, $(67, 73, 79)$, $(97, 103, 109)$, $(151, 157, 163)$, $(167, 173, 179)$, $(227, 233, 239)$, $(257, 263, 269)$, ... (see sequences A046118, A046119, and A046120 in the OEIS).

The largest known sexy prime triplet (Lamprecht and Luhn, 2019) has 10602 digits; in it,

$$p = 2683143625525 \cdot 2^{35176} + 1.$$

Since every fifth number of the form $6n \pm 1$ is divisible by 5, only one *sexy prime quintuplet exists*, namely, (5, 11, 17, 23, 29), and no larger sequences of sexy primes are possible.

Some additional information see, for example, in [Buch09], [DeKo13], [HaWr79], [Prim20], [Sloa20], [Wiki20].

Prime gaps conjectures

2.6.7. All problems, considered above, belong to a classical type of open problems in the Theory of prime numbers concerns *prime gaps*, i.e., the differences between consecutive primes.

It is conjectured that there are infinitely many pairs of primes with difference 2; this is the *twin prime conjecture*.

Prime gaps can be generalized to prime k-tuples, patterns in the differences between more than two prime numbers. Their infinitude and density are the subject of the *first Hardy–Littlewood conjecture*, which can be motivated by the heuristic that the prime numbers behave similarly to a random sequence of numbers with density given by the Prime number theorem.

The fist Hardy-Littlewood conjecture is concerned with the distribution of prime constellations, including twin primes, in analogy to the Prime number theorem. In particular, if $\pi_2(x)$ denote the number of primes $p \le x$ such that $p + 2$ is also prime, and if the constant C_2 is defined as

$$C_2 = \prod_{\substack{p \text{ prime} \\ p \ge 3}} \left(1 - \frac{1}{(p-1)^2}\right) \approx 0.6601618158...,$$

then a special case of the first Hardy-Littlewood conjecture is that

$$\pi_2(x) \sim 2C_2 \frac{x}{(\ln x)^2} \sim 2C_2 \int_2^x \frac{dt}{(\ln t)^2}.$$

The conjecture can be justified (but not proven) by assuming that $\frac{1}{\log n}$ describes the density function of the prime distribution.

This assumption, which is suggested by the Prime number theorem, implies the twin prime conjecture, as shown in the formula for $\pi_2(x)$ above.

The general first Hardy–Littlewood conjecture deals with prime k-tuples.

2.6.8. The *Polignac's conjecture* states more generally that
- *for every positive integer k, there are infinitely many pairs of consecutive primes that differ by $2k$.*

2.6.9. *Andrica's conjecture, Brocard's conjecture, Legendre's conjecture*, and *Oppermann's conjecture* all suggest that the largest gaps between primes from 1 to n should be at most approximately $\sqrt{n}$, a result that is known to follow from the *Riemann hypothesis*, while the much stronger *Cramér's conjecture* sets the largest gap size at $O((\log n)^2)$.

In fact, the *Bertrand's postulate* states, that
- *for any integer $n > 3$, there always exists at least one prime number p with $n < p < 2n - 2$.*

A less restrictive formulation is: *for every $n > 1$ there is always at least one prime p between n and $2n$.*

This statement was first conjectured in 1845 by J. Bertrand, and was completely proved by P. Chebyshev in 1852; so the postulate is also called the *Bertrand–Chebyshev theorem*.

It can also be stated as a relationship with $\pi(x)$, where $\pi(x)$ is the prime counting function (the number of primes less than or equal to x):

$$\pi(x) - \pi(\tfrac{x}{2}) \geq 1 \quad \text{for all} \quad x \geq 2.$$

The Prime number theorem implies that the number of primes up to x is roughly $\frac{x}{\log x}$, so if we replace x with $2x$ then we see the number of primes up to $2x$ is asymptotically twice the number of primes up to x, as the terms $\log 2x$ and $\log x$ are asymptotically equivalent.

Therefore, the number of primes between n and $2n$ is roughly $\frac{n}{\log n}$, when n is large, and so in particular there are many more primes in this interval than are guaranteed by Bertrand's Postulate. So, Bertrand's postulate is comparatively weaker than the Prime number theorem. But the Prime number theorem is a deep theorem, while

Bertrand's Postulate can be stated more memorably and proved more easily, and also makes precise claims about what happens for small values of n.

The similar and still unsolved *Legendre's conjecture* asks whether
- *for every $n > 1$ there is a prime p, such that $n^2 < p < (n+1)^2$.*

Again we expect that there will be not just one but many primes between n^2 and $(n+1)^2$, but in this case the Prime number theorem doesn't help: the number of primes up to x^2 is asymptotic to $\frac{x^2}{\log x^2}$ while the number of primes up to $(x+1)^2$ is asymptotic to $\frac{(x+1)^2}{\log(x+1)^2}$, which is asymptotic equal to the estimate on primes up to x^2.

So, unlike the previous case of x and $2x$ we don't get a proof of Legendre's conjecture even for all large n.

If Legendre's conjecture is true, the gap between any prime p and the next largest prime would always be at most on the order of $\sqrt{p}$; in other words, the gaps are $O(\sqrt{p})$.

Three stronger conjectures, *Andrica's conjecture*, *Brocard's conjecture* and *Oppermann's conjecture*, imply that the gaps have the same magnitude.

Andrica's conjecture (named after D. Andrica) states that the inequality $\sqrt{p_{n+1}} - \sqrt{p_n} < 1$ holds for all n, where p_n is the n-th prime number.

Brocard's conjecture (named after H. Brocard) states that
- *there are at least four prime numbers between $(p_n)^2$ and $(p_{n+1})^2$, where p_n is the n-th prime number, for every $n \geq 2$.*

Note, that Legendre's conjecture that there is a prime between consecutive integer squares directly implies that there are at least two primes between prime squares for $p_n \geq 3$ since $p_{n+1} - p_n \geq 2$.

Oppermann's conjecture is closely related to but stronger than Legendre's conjecture, Andrica's conjecture, and Brocard's conjecture. It is named after Danish mathematician L. Oppermann, who announced it in 1877.

The conjecture states that,
- *for every integer $x > 1$, there is at least one prime number between $x(x-1)$ and x^2, and at least another prime between x^2 and $x(x+1)$.*

It can also be phrased equivalently as stating that the prime counting function must take unequal values at the endpoints of each range. That is:

$$\pi(x^2 - x) < \pi(x^2) < \pi(x^2 + x) \quad \text{for} \quad x > 1.$$

The end points of these two ranges are a square between two pronic numbers, with each of the pronic numbers being twice a pair triangular number.

H. Cramér conjectured that the gaps are always much smaller, of the order $(\log p)^2$. If Cramér's conjecture is true, Legendre's conjecture would follow for all sufficiently large n.

Cramér also proved that the Riemann hypothesis implies a weaker bound of $O(\sqrt{p}\log p)$ on the size of the largest prime gaps. A counterexample near 10^{18} would require a prime gap fifty million times the size of the average gap.

Note, that it is proven, that
• *for all sufficiently large n, there is a prime between the consecutive cubes n^3 and $(n + 1)^3$.*

R.S. Baker, G. Harman and J. Pintz proved also (2001), that
• *there is a prime in the interval $[x,\ x + O(x^{\frac{21}{40}})]$ for all large x.*

Some additional information see, for example, in [Broc76], [Broc85], [Buch09], [Deza17], [McDa98], [McDa98a].

Goldbach's conjecture

2.6.10. One of the most known open problems in Number Theory is the *Goldbach's conjecture*, which asserts that
• *every even integer n greater than 2 can be written as a sum of two primes.*

As of 2014, this conjecture has been verified for all numbers up to $n = 4 \cdot 10^{18}$.

However, some weaker statements have been proven. For example, the *Vinogradov's theorem* says that
• *every sufficiently large odd integer can be written as a sum of three primes.*

The *Chen's theorem* says that
- *every sufficiently large even number can be expressed as the sum of a prime and a semi-prime (the product of two primes).*

Also, it is proven, that
- *any even integer greater than 10 can be written as the sum of six primes.*

The history of this conjecture was started on 7 June 1742, when the German mathematician C. Goldbach wrote a letter to L. Euler; in the margin of his letter, he proposed the following conjecture: "Every integer greater than 2 can be written as the sum of three primes."

L. Euler replied in a letter, dated 30 June 1742, and reminded C. Goldbach of an earlier conversation, in which Goldbach had conjectured that "every positive even integer can be written as the sum of two primes."

Note, that C. Goldbach was following the now-abandoned convention of considering 1 to be a prime number, so that a sum of units would indeed be a sum of primes. A modern version of the marginal conjecture is:
- *every integer greater than 5 can be written as a sum of three primes.*

And a modern version of Goldbach's older conjecture of which L. Euler reminded him is:
- *every even integer greater than 2 can be written as a sum of two primes.*

It is known as the *strong* (or *even*, or *binary*) *Goldbach's conjecture.*

A weaker form of this statement, known as the *weak* (or *odd*, or *ternary*) Goldbach's conjecture asserts that
- *very odd integer greater than 7 can be written as a sum of three odd primes.*

Note that the weak conjecture would be a corollary of the strong conjecture: if $n - 3$ is a sum of two primes, then n is a sum of three primes. But the converse implication and thus the strong Goldbach conjecture remain unproven.

A *Goldbach number* is defined a positive even integer that can be expressed as a sum of two odd primes. So, another form of the statement of Goldbach's conjecture is that *all even integers greater than 4 are Goldbach numbers.*

The expression of a given even number as a sum of two primes is called a *Goldbach partition* of that number. The following are examples of Goldbach partitions for some even numbers:

$$4 = 2 + 2; \quad 6 = 3 + 3; \quad 8 = 3 + 5; \quad 10 = 3 + 7 = 5 + 5; \quad ...$$

$$... \ 100 = 3 + 97 = 11 + 89 = 17 + 83 = 29 + 71$$

$$= 41 + 59 = 47 + 53.$$

The number of ways in which $2n$ can be written as the sum of two primes (for n starting at 1) is: 0, 1, 1, 1, 2, 1, 2, 2, 2, 2, 3, 3, 3, 2, 3, 2, 4, 4, 2, 3, 4, 3, 4, 5, 4, 3, 5, 3, 4, 6, 3, 5, 6, 2, 5, 6, 5, 5, 7, 4, 5, 8, 5, 4, 9, 4, 5, 7, 3, 6, 8, 5, 6, 8, 6, 7, 10, 6, 6, 12, 4, 5, 10, 3, ... (sequence A045917 in the OEIS).

The proof of these properties and some additional information see, for example, in [Apos86], [Buch09], [Dick05], [Moze09], [Kara83], [Vino03].

Primes in some sequences

2.6.11. In Number Theory, *primes in arithmetic progression* are any sequence of at least three prime numbers that are consecutive terms in an arithmetic progression.

An example is the sequence of primes $(3, 7, 11)$, which is given by $a_n = 3 + 4n$ for $0 \le n \le 2$.

According to the *Green–Tao theorem* (2004),

• *there exist arbitrarily long sequences of primes in arithmetic progression.*

Sometimes the phrase may also be used about primes which belong to an arithmetic progression which also contains composite numbers. For example, it can be used about primes in an arithmetic progression of the form $a \cdot n + b$, where a and b are coprime. According to *Dirichlet's theorem on arithmetic progressions* it contains infinitely many primes, but along with infinitely many composites.

There are many open questions in this branch of the Theory of prime numbers. For example,

• *are there infinitely many sets of 3 consecutive primes in an arithmetic progression?*

Obviously, it is true if we omit the word "consecutive". Other question:

• *is there an arithmetic progression of consecutive primes for any given length?*

For example, the arithmetic progression 251, 257, 263, 269 has length 4, while the largest known example has length 10.

It is known, that $n^2 - n + 41$ is prime for $0 \le n \le 40$ (see Chapter 2, section 2.4). One of the open questions is:

• *are there infinitely many primes of this form?*

The same question can be applied to the polynomial $n^2 - 79n + 1601$, which is prime for $0 \le n \le 79$.

2.6.12. As for factorials $n!$ ($n!$ is the product of all positive integers $k \le n$) and primorials ($n\#$ is the product of all primes $p \le n$), the following open questions exist:

• *are there infinitely many primes of the form $n\# \pm 1$?*

• *are there infinitely many primes of the form $n! \pm 1$?*

The proof of these properties and some additional information see, for example, in [Buch09], [ChFa07], [DeKo13], [Sier64].

Riemann hypothesis

2.6.13. One of the most famous unsolved questions in Mathematics, dating from 1859, is the *Riemann hypothesis*. It was proposed by German mathematician B. Riemann and asks where the zeros of the Riemann zeta function $\zeta(s)$ are located.

It is of great interest in Number Theory because it implies results about the distribution of prime numbers. Many mathematicians consider it to be the most important unsolved problem in pure Mathematics (Bombieri, 2000).

The *Riemann zeta function* $\zeta(s)$ is a function whose argument $s = \sigma + it$ may be any complex number other than 1, and whose values are also complex. In fact, it is an analytic function on the complex numbers, other than 1.

For $s = \sigma + it$, $\sigma > 1$, the Riemann zeta function is defined by the absolutely convergent infinite series

$$\zeta(s) = \sum_{n=1}^{\infty} \frac{1}{n^s} = 1 + \frac{1}{2^s} + \frac{1}{3^s} + \frac{1}{3^s} + \cdots + \frac{1}{n^s} + \cdots .$$

L. Euler already considered this series in the 1730's for real values of s, in conjunction with his solution ($\sum_{n=1}^{\infty} \frac{1}{n^2} = \frac{\pi^2}{6}$) to the *Basel problem*. He also proved that

$$\zeta(s) = \prod_{p \in P} \frac{1}{1 - p^{-s}} = \frac{1}{1 - 2^{-s}} \cdot \frac{1}{1 - 3^{-s}} \cdot \frac{1}{1 - 5^{-s}} \cdot \frac{1}{1 - 7^{-s}} \cdot \frac{1}{1 - 11^{-s}} \cdots,$$

where the infinite product extends over all prime numbers p.

So, for complex numbers $s = \sigma + it$ with real part σ greater than one $\zeta(s)$ equals both an infinite sum over all integers, and an infinite product over all prime numbers:

$$\zeta(s) = \sum_{n=1}^{\infty} \frac{1}{n^s} = \prod_{p \in P} \frac{1}{1 - p^{-s}}.$$

This equality between a sum and a product is called an *Euler product*. The Euler product can be derived from the fundamental Theorem of Arithmetic. It shows the close connection between the zeta function and the set of prime numbers.

The function $\zeta(s)$ has zeros at the negative even integers; that is, $\zeta(s) = 0$ when s is one of $-2, -4, -6, \ldots$. These zeros are called its *trivial zeros*.

However, the negative even integers are not the only values for which the zeta function is equal to zero. The other ones are called *non-trivial zeros*.

All of them belong to so called *critical strip* $0 < \sigma < 1$, consisting of the complex numbers $s = \sigma + it$, where σ is a real number from the interval $(0, 1)$. The *Riemann hypothesis* is concerned with exact location of these non-trivial zeros, and states that

- *the real part σ of every non-trivial zero of the Riemann zeta function is equal to $\frac{1}{2}$.*

Thus, if the hypothesis is correct, all the non-trivial zeros lie on the *critical line* $\sigma = \frac{1}{2}$, consisting of the complex numbers $s = \frac{1}{2} + it$.

The original proof of the Prime number theorem was based on a weak form of this hypothesis, states that there are no zeros with real part equal to 1, although after other more elementary proofs have been found.

In fact, the prime counting function can be expressed by Riemann's explicit formula as a sum in which each term comes from one of the zeros of the zeta function; the main term of this sum is the logarithmic integral, and the remaining terms cause the sum to fluctuate above and below the main term.

In this sense, the zeros control how regularly the prime numbers are distributed. If the Riemann hypothesis is true, these fluctuations will be small, and the asymptotic distribution of primes given by the Prime number theorem will also hold over much shorter intervals (of length about the square root of x for intervals near a number x.

The Riemann hypothesis and some of its generalizations, along with the Goldbach's conjecture and the twin prime conjecture, comprise Hilbert's eighth problem in David Hilbert's list of 23 unsolved problems.

It is also one of the Clay Mathematics Institute's Millennium Prize Problems.

There are several useful books on the Riemann hypothesis; For example, the book [MaSt15] gives mathematical introduction, while [Kara83], [KaVo92], [Ivic85] and [Titc87] are advanced monographs.

Landau's problems

2.6.14. At the 1912 International Congress of Mathematicians, Edmund Landau listed four basic problems about prime numbers. These problems were characterised in his speech as "unattackable at the present state of Mathematics" and are now known as *Landau's problems*. They are as follows.

- *Goldbach's conjecture*: can every even integer greater than 2 be written as the sum of two primes?

- *Twin prime conjecture*: are there infinitely many primes p such that $p + 2$ is prime?

- *Legendre's conjecture*: does there always exist at least one prime between consecutive perfect squares?

- Are there infinitely many primes p such that $p - 1$ is a perfect square? In other words: are there infinitely many primes of the form $n^2 + 1$?

As of November 2020, all four problems are unresolved.

For references, an additional information see, for example, [Guy94], [HaWr79], [John77], [Ingh32], [Ivic85], [Kara83], [Lagr70], [Land09], [Lege30], [Wiki20].

Exercises

1. Find first ten pairs of twin primes.

2. Prove, that for any pair of twin primes (p, q) with $q > p > 3$ it holds $p \equiv -1 (mod\ 6)$, and $q \equiv 1 (mod\ 6)$.

3. Prove, that there are no prime triples of the form $(p, p+2, p+4)$, $p, p+2, p+4 \in P$.

4. Represent any positive integer $4 \leq n \leq 100$ as a sum of at most tree primes. How many such representations exist for each n?

5. Prove, that there are infinity many primes in the arithmetical progression $ax - 1$, where $a = 2, 4, 6$.

6. What we can say about a number of primes in a given geometric progression?

7. Prove, that any square number $S_4(n) = n^2$, $n = 1, 2, 3, \ldots$ cannot to be a prime. Find a prime of the form $S_4(n) - 1$. Find several prime of the form $S_4(n) - 1$.

8. For first 100 positive integers, check, that there exist at least one prime between two consecutive perfect squares.

Chapter 2: References

[Abra74], [AbSt72], [IrRo90], [AGP94], [Andr98], [Arno38], [Apos86], [Avan67], [BaCo87], [Bern99], [Dave47], [BaWe76], [Bond93], [Broc85], [Buch09], [Cata44], [ChFa07], [Clau40], [Cohn64], [Cohn65], [CoGu96], [CoRo96], [Dede63], [DeDe12], [DeKo13], [Deza17], [Deza18], [Dick05], [Diop74], [DoSh77], [ErOb37], [Eule48], [Gard61], [Gard88], [Gees20], [GKP94], [Goul85], [Guy94], [HaWr79], [John77], [Ingh32], [Ivic85], [Kara83], [Lagr70], [Land09], [Lege30], [LiNi96], [LiNi96], [LiNi96], [Luca75], [Mada79], [McDa98], [McDa98a], [MSC96], [Motz48], [Moze09], [Nage51], [Ore48], [Hogg69], [Pasc54], [Plut78], [RaTo57], [SlPl95], [Smit84], [Gard88], [Kobl87], [Kobl87], [Shan93], [Sier64], [Sier03], [Vile14], [Weis99], [Well86], [Wiki20].

Chapter 3

Mersenne numbers

3.1 History of the question

3.1.1. A *Mersenne number* is a positive integer of the form

$$M_n = 2^n - 1, \ n \in \mathbb{N}.$$

The first Mersenne numbers are 1, 3, 7, 15, 31, 63, 127, 255, 511, 1023, 2047, ... (sequence A000225 in OEIS).

For $2^n - 1$ to be prime, it is necessary that n itself be prime. By this reason, many authors require that the exponent n in the definition above be a prime.

Mersenne numbers $2^p - 1$ with prime p form the sequence 3, 7, 31, 127, 2047, 8191, 131071, 524287, 8388607, 536870911, ... (sequence A001348 in OEIS).

In fact, the exponents n which give Mersenne primes are 2, 3, 5, 7, 13, 17, 19, 31, ... (sequence A000043 in the OEIS), and the resulting Mersenne primes are 3, 7, 31, 127, 8191, 131071, 524287, 2147483647, ... (sequence A000668 in the OEIS).

However, we will consider these numbers for any positive integer n.

This class of special numbers is of great interest because the *Mersenne primes* (prime Mersenne numbers) are among the oldest and most studied of all primes.

101

In fact, Mersenne primes are very rare: not all numbers of the form $2^p - 1$ with a prime p are primes; for example,

$$2^{11} - 1 = 2047 = 23 \cdot 89$$

is not a prime number.

It is proven, that for the 2610944 prime numbers p up to 43112609, the corresponding Mersenne numbers $M_p = 2^p - 1$ is prime for only 47 of them. Today we know only 51 Mersenne primes.

The numbers of the form $2^n - 1$ had been investigated long before the XVII-th century, when they got their name.

In fact, they are closely connected with *perfect numbers*.

Many ancient cultures were concerned with the relationship of a number with the sum of its divisors, often giving mystic interpretations. Here we are concerned only with one such relationship.

In Number Theory, a *perfect number* is a positive integer that is equal to the sum of its positive divisors, excluding the number itself, i.e., as the sum of its proper divisors.

For instance, 6 has divisors 1, 2 and 3 (excluding itself), and

$$1 + 2 + 3 = 6;$$

12 has proper divisors 1, 2, 3, 4, and 6, and

$$1 + 2 + 3 + 4 + 6 > 12;$$

4 has proper divisors 1 and 2, and

$$1 + 2 < 4.$$

So, 6 is a *perfect number*; 12 is an *abundant number*; 4 is a *deficient number*.

It is easy to check, that 6 is the first perfect number.

The second perfect number is 28:

$$28 = 1 + 2 + 4 + 7 + 14.$$

The next two are 496 and 8128.

Euclid proved that *the number $2^{p-1}(2^p - 1)$ is an even perfect number whenever $2^p - 1$ is a prime*; in fact, a Mersenne prime. Over a millennium after Euclid, Alhazen (circa 1000 AD) conjectured that *every even perfect number is of the form* $2^{p-1}(2^p - 1)$, *where* $2^p - 1$ *is prime*, but he was not able to prove this result. It was not until the XVIII-th century that L. Euler proved that the formula $2^{p-1}(2^p - 1)$, $2^p - 1 \in P$, will yield all the even perfect numbers. Thus, each Mersenne prime generates one even perfect number, and vice versa. This result is known as the *Euclid–Euler theorem*.

In fact, the first four perfect numbers are generated by the formula $2^{p-1}(2^p - 1)$, with p a prime number, as follows:

for $p = 2$: $2^1(2^2 - 1) = 2 \cdot 3 = 6$;
for $p = 3$: $22(23 - 1) = 4 \cdot 7 = 28$;
for $p = 5$: $2^4(2^5 - 1) = 16 \cdot 31 = 496$;
for $p = 7$: $2^6(2^7 - 1) = 64 \cdot 127 = 8128$.

So the search for Mersenne primes is also the search for even perfect numbers.

These first four perfect numbers, corresponding to $p = 2, 3, 5$ and 7, were the only ones known to early Greek Mathematics. They were given in the *Arithmetic* of Nicomachus of Gerasa.

Philo of Alexandria in his first-century book *On the creation* mentioned perfect numbers, claiming that the world was created in 6 days and the moon orbits in 28 days because 6 and 28 are perfect.

Philo is followed by Origen, and by Didymus the Blind, who added the observation that there are only four perfect numbers that are less than 10000.

The fifth perfect number 33550336 corresponding to $p = 13$ was discovered by Regiomontanus, a mathematician, astrologer and astronomer of the German Renaissance (XV-th century).

In the XVI-th century, the German scientist Scheibel found two more perfect numbers, 8589869056 and 137438691328. They correspond to $p = 17$ and $p = 19$.

Many writers felt that the numbers of the form $2^n - 1$ were prime for all primes n, but in 1536 Hudalricus Regius showed that $2^{11} - 1 = 2047$ was not prime; it is $23 \cdot 89$.

By 1603 Pietro Cataldi had correctly verified that $2^{17} - 1$ and $2^{19} - 1$ were both prime, but then incorrectly stated $2^n - 1$ was also prime for $23, 29, 31$ and 37.

In 1640 Pierre Fermat showed that P. Cataldi was wrong about 23 and 37; then L. Euler in 1732 showed that P. Cataldi was also wrong about 29. Sometime later L. Euler (1750) showed that Cataldi's assertion about 31 was correct.

3.1.2. But finally the numbers of the form $2^n - 1$ were named after Marin Mersenne (1588–1648), a French monk and mathematician, whose works touched a wide variety of fields.

M. Mersenne was born in Sarthe, France. He was educated at the Jesuit College of La Flèche.

Between 1614 and 1618, he taught Theology and Philosophy at Nevers, but he returned to Paris and settled at the convent of L'Annonciade in 1620. There he studied Mathematics and met with René Descartes, Étienne Pascal, Pierre Petit, Gilles de Roberval, Thomas Hobbes, and Nicolas-Claude Fabri de Peiresc. He corresponded with Giovanni Doni, Jacques Alexandre Le Tenneur, Constantijn Huygens, Galileo Galilei, and other scholars in Italy, England and the Dutch Republic.

In 1635 he set up the informal Académie Parisienne (Academia Parisiensis), which had nearly 140 correspondents, including astronomers and philosophers as well as mathematicians. He was not afraid to cause disputes among his learned friends in order to compare their views, notable among which were disputes between R. Descartes and P. de Fermat and J. de Beaugrand. In 1635 Mersenne met with Tommaso Campanella. In 1643–1644 Mersenne also corresponded with the German Socinian Marcin Ruar concerning the Copernican ideas of Pierre Gassendi.

M. Mersenne discussed numbers of the form $2^n - 1$ in his work *Cogita physico mathematics* (1644) and stated conjectures about the number's ocurrence.

One of his most famous conjectures is called the *Mersenne conjecture*:

- $2^n - 1$ *is prime for* $n = 2, 3, 5, 7, 13, 17, 19, 31, 67, 127, 257$, *and is composite for all other integers* $2 \leq n \leq 257$.

It is obvious, that M. Mersenne could not have tested all of these numbers. It took three centuries and several mathematical discoveries (such as the *Lucas-Lehmer test*), before the exponents in Mersenne's conjecture had been completely checked.

It was determined that M. Mersenne had made five errors (three primes, for $n = 61, 89, 107$, omitted, two composites, for $n = 67, 257$, listed), and the correct list is:

$$n = 2, 3, 5, 7, 13, 17, 19, 31, 61, 89, 107 \text{ and } 127.$$

Primality of the number $2147483647 = 2^{31} - 1$ was proved by L. Euler (1750).

Primality of an 19-digit number $2305843009213693951 = 2^{61} - 1$ was proved by Russian mathematician I. Pervushin (1883).

R.E. Powers showed that an 27-digit number $2^{89} - 1$ is prime (1911), as well as an 33-digit number $2^{107} - 1$ (1913).

É. Lukas (1876) showed that $2^{127} - 1 \in P$. This 39-digit Mersenne is the largest prime number open in a precomputer era.

The next result was obtained only in 1952 by R.M. Robinson, who with the help of computers managed to prove primality of Mersenne numbers $2^{521} - 1$ (157 decimal digits), $2^{607} - 1$ (183 decimal digits), $2^{1279} - 1$ (386 decimal digits), $2^{2203} - 1$ (664 decimal digits), and $2^{2281} - 1$ (687 decimal digits).

So, the Mersenne's conjecture was not completely correct, but his name is still attached to the numbers of the form $2^n - 1$.

Where did M. Mersenne get his list? Perhaps we find a hint in this quote. In a letter to Tanner É. Lucas stated that M. Mersenne (1644, 1647) implied that a necessary and sufficient condition that

• $2^p - 1$ *be a prime is that p be a prime of one of the forms* $2^{2n} + 1$, $2^{2n} \pm 3$, $2^{2n+1} - 1$.

Tanner expressed his belief, and noted the sufficient condition would be false if $2^{67} - 1$ is composite.

There is also a missing restriction on the size of the prime because $2^3 - 1$ is prime, but 3 is not one of these forms.

So there seems to be a belief that the exponents p of Mersenne primes have a special form.

If you check the numbers under 257, you will get Mersenne's list (except 3) plus the prime 61.

Sadly, the conditions quoted above are neither necessary nor sufficient. The Mersenne numbers M_p are composite for the following primes p: $257 = 2^8 + 1$, $1021 = 2^{10} - 3$, $67 = 2^6 + 3$, and $8191 = 2^{13} - 1$. So, none of the "sufficient" conditions hold.

Also, M_p is prime for $p = 89$, but 89 cannot be written in any of the listed forms.

3.1.3. P.T. Bateman, J.L. Selfridge and Jr.S.S. Wagstaff made (1989) the *New Mersenne Conjecture*. It is thus:

• *Let p be any odd positive integer. If two of the following conditions hold, then so does the third:*

1. $p = 2^k \pm 1$ or $p = 4^k \pm 3$;

2. $2^p - 1$ is a prime (a *Mersenne prime*);

3. $\frac{2^p+1}{3}$ is a prime (a *Wagstaff prime*).

This conjecture has been verified for all primes p less than 100000, and for all 51 known Mersenne primes. Some feel that "conjecture" is too strong of a word for the above and that perhaps this is even another case of *Guy's law of small numbers*.

3.1.4. The Mersenne numbers give us most of prime number records.

In 1811 P. Barlow wrote in his text *Theory of Numbers* that $2^{30}(2^{31} - 1)$ "... is the greatest perfect number that will be discovered; for as they are merely curious, without being useful, it is not likely that any person will attempt to find one beyond it." Obviously no one in the late 1800's had any idea of the power of modern computers.

After the 23-rd Mersenne prime was found at the University of Illinois (Gilles, 1963), the Mathematics department was so proud that the chair, Dr. Bateman, had their postage meter changed to stamp "$2^{11213} - 1$ is prime" on each envelope. This was used until the *four color theorem* was proved in 1976.

The 25-th Mersenne prime was found by high-school students Laura Nickel and Landon Curt Noll (1978), who, though they had little understanding of the Mathematics involved, used Lucas' simple test on the local university's mainframe.

D. Slowinski, who works for Cray computers, has written a version of the Lucas test that he has convinced many Cray labs around the world to run in their spare time. D. Slowinski does not search for record primes systematically. In fact, looking at the table of Mersenne primes you see he missed the 29-th prime but found the 30-th (1983) and 31-st (1985).

W. Colquitt and L. Welsh worked to fill in the gaps and found the 29-th Mersenne prime in 1988.

Starting in late 1995, E. G. Woltman, gathered up the disparate databases and combined them into one. Then he placed this database, and a free, highly optimized program for search for Mersenne primes onto the web. This began the Great Internet Mersenne Prime Search (GIMPS), which has now found the largest known Mersenne primes, combining the efforts of dozens of experts and thousands of amateurs.

Since 1997, all newly found Mersenne primes have been discovered by the Great Internet Mersenne Prime Search.

In late 1997, S. Kurowski (and others) established PrimeNet to automate the selection of ranges and reporting of results for GIMPS.

In 1999 (Hajratwala, Woltman, Kurowski *et al.*), the first prime with more than 1000000 decimal digits (in fact, more than 2000000) was obtained. It is the 38-th Mersenne prime, $M_{6972593}$; it has 2098960 decimal digits.

Let a *megaprime* be a prime number with at least one million decimal digits, one can say, that the first megaprime, $M_{6972593}$, was found in 1999.

In 2018 (Laroche *et al.*), the last known (in fact, the 51-th known) Mersenne prime, $M_{82589933}$, was obtained. It has 24862048 decimal digits.

So, as of 2020, 51 Mersenne primes are known. In December 2020, a major milestone in the GIMPS project was passed after all exponents below 100 million were checked at least once.

List of known Mersenne primes

	p	Number of digits in M_p	Year	Discoverer
1	2	1	- - - -	- - - -
2	3	1	- - - -	- - - -
3	5	2	- - - -	- - - -
4	7	3	- - - -	- - - -
5	13	4	1456	anonymous
6	17	6	1588	Cataldi
7	19	6	1588	Cataldi
8	31	10	1772	Euler
9	61	19	1883	Pervushin
10	89	27	1911	Powers
11	107	33	1914	Powers
12	127	39	1876	Lucas
13	521	157	1952	Robinson
14	607	183	1952	Robinson
15	1279	386	1952	Robinson
16	2203	664	1952	Robinson
17	2281	687	1952	Robinson
18	3217	969	1957	Riesel
19	4253	1281	1961	Hurwitz
20	4423	1332	1961	Hurwitz
21	9689	2917	1963	Gillies
22	9941	2993	1963	Gillies
23	11213	3376	1963	Gillies
24	19937	6002	1971	Tuckerman
25	21701	6533	1978	Noll and Nickel
26	23209	6987	1979	Noll
27	44497	13395	1979	Nelson and Slowinski
28	86243	25962	1982	Slowinski
29	110503	33265	1988	Colquitt and Welsh
30	132049	39751	1983	Slowinski
31	216091	65050	1985	Slowinski
32	756839	227832	1992	Slowinski, Gage *et al.*
33	859433	258716	1994	Slowinski and Gage
34	1257787	378632	1996	Slowinski and Gage
35	1398269	420921	1996	Armengaud, Woltman *et al.*
36	2976221	895932	1997	Spence, Woltman *et al.*
37	3021377	909526	1998	Clarkson, Woltman, Kurowski *et al.*
38	6972593	2098960	1999	Hajratwala, Woltman, Kurowski *et al.*

(*Continued*)

	p	Number of digits in M_p	Year	Discoverer
39	13466917	4053946	2001	Cameron, Woltman, Kurowski *et al.*
40	20996011	6320430	2003	Shafer, Woltman, Kurowski *et al.*
41	24036583	7235733	2004	Findley, Woltman, Kurowski *et al.*
42	25964951	7816230	2005	Nowak, Woltman, Kurowski *et al.*
43	30402457	9152052	2005	Cooper, Boone, Woltman, Kurowski *et al.*
44	32582657	9808358	2006	Cooper, Boone, Woltman, Kurowski *et al.*
45	37156667	11185272	2008	Elvenich, Woltman, Kurowski *et al.*
46	42643801	12837064	2009	Strindmo, Woltman, Kurowski *et al.*
47	43112609	12978189	2008	Smith, Woltman, Kurowski *et al.*
48 (?)	57885161	17425170	2013	Cooper, Woltman, Kurowski *et al.*
49 (?)	74207281	22338618	2016	Cooper, Woltman, Kurowski, Blosser *et al.*
50 (?)	77232917	23249425	2017	Pace, Woltman, Kurowski, Blosser *et al.*
51 (?)	82589933	24862048	2018	Laroche, Woltman, Blosser *et al.*

In the table above, we put question marks instead of a number for the last of the Mersenne primes because it will not be known if there are other Mersenne's in between these until a check and double check has been completed by GIMPS. See the GIMPS Status Page for more information. Not all smaller exponents have been tested.

For additional information see, for example, [DeKo18], [Dick05], [Ore48], [Salo90], [Sier64], [Prim20], [Wiki20].

3.2 Elementary properties of Mersenne numbers

In this section, we give the large list of simplest properties of Mersenne numbers. All of them can be proven using elementary properties of divisibility of integers.

Elementary conditions of primality of Mersenne numbers

3.2.1. First of all, consider elementary properties of Mersenne numbers, connected with their primality.

It is easy to see, that a number of the form $2^n - 1$ can be a prime only if n is a prime:

- *for a positive integer n, if $2^n - 1 \in P$, then $n \in P$.*

□ In fact, consider $n \in \mathbb{N}$. If $n = 1$, one has $2^2 - 1 = 1$; it is not a prime. If n is composite, one can find positive integers a, b, such that $n = a \cdot b$, and $1 < a \le b < n$. Then

$$2^n - 1 = 2^{ab} - 1 = (2^a)^b - 1 = ((2^a) - 1) \cdot K, \ K \in \mathbb{N}.$$

So, the number $2^n - 1$ has a positive integer divisor $2^a - 1$. As $a < n$, it holds $2^a - 1 < 2^n - 1$. It means, that $2^a - 1$ is a proper divisor of $2^n - 1$; so, $2^n - 1$ is composite. □

As the consequence, we get the following elementary fact:

- *for a positive integer n, if $n \in S$, then $2^n - 1 \in S$.*

So, it is proven, that

- *there are infinite many composite Mersenne numbers.*

However, there exist prime numbers p, such that $2^p - 1$ is composite. The first such number is 11:

$$M_{11} = 2^{11} - 1 = 1027 = 23 \cdot 89.$$

The next are 23, 29, 37, 41, 43, 53, 59, 67, 71, 73, ... (sequence A054723 in the OEIS).

In order to find some prime divisors of Mersenne numbers, note, that *any odd prime number divides at least one Mersenne number:*

- *given prime number $p \ge 3$, it holds $p|M_{p-1}$.*

□ Due to the Fermat's little theorem: as $gcd(2, p) = 1$ for odd prime p, it holds

$$2^{p-1} \equiv 1 (mod \ p), \quad \text{or} \quad M_{p-1} \equiv 0 (mod \ p). \ □$$

Moreover, *any odd prime number divides infinity many Mersenne numbers:*

- *given prime number $p \ge 3$, it holds $p|M_{(p-1)k}$ for any $k \in \mathbb{N}$.*

□ Indeed, for such p it holds $2^{p-1} \equiv 1(mod\ p)$; therefore, for any positive integer k we get $2^{(p-1)k} \equiv 1(mod\ p)$, and $p|M_{(p-1)k}$. □

Last decimal digits of Mersenne numbers

3.2.2. Consider now possible reminders of Mersenne numbers modulo 10. It will give us the information about last digit of the Mersenne number M_n.

In fact, *the last digit of a Fermat number can be any odd digit except 9*:

- *for $n \geq 1$, $n = 4l$, it holds $M_n \equiv 5(mod\ 10)$, i.e., the last digit of M_n is 5;*

- *for $n \geq 1$, $n = 4l + 1$, it holds $M_n \equiv 1(mod\ 10)$, i.e., the last digit of M_n is 1;*

- *for $n \geq 1$, $n = 4l + 2$, it holds $M_n \equiv 3(mod\ 10)$, i.e., the last digit of M_n is 3;*

- *for $n \geq 2$, $n = 4l + 3$, it holds $F_n \equiv 7(mod\ 10)$, i.e., the last digit of M_n is 7.*

□ The reader can easy check this facts, noting that

$$2^{4l+k} - 1 \equiv (2^4)^l \cdot 2^k - 1 \equiv 6^l \cdot 2^k - 1 \equiv 6 \cdot 2^k - 1(mod\ 10);$$

as $6 \cdot 2^0 - 1 \equiv 5(mod\ 10)$, $6 \cdot 2^1 - 1 \equiv 1(mod\ 10)$, $6 \cdot 2^2 - 1 \equiv 3(mod\ 10)$, and $6 \cdot 2^3 - 1 \equiv 7(mod\ 10)$, the result holds. □

Any odd prime number p is congruent to 1 or 3 modulo 4. So, *given odd prime p, the last digit of a Mersenne number M_p is 1 or 7:*
- *given prime number $p \geq 3$, it holds $M_p \equiv 1, 7(mod\ 10)$, i.e., the last digit of M_p is 1 or 7.*

In particular, it is proven, that
- *the last digit of a Mersenne prime greater 3 is 1 or 7.*

Recurrent relations for Mersenne numbers

3.2.3. There exists a simple recurrent relation for the Mersenne numbers:

- *For any positive integer n, it holds $M_{n+1} = 2M_n + 1$, with* $M_1 = 1$.

    □ In fact,

$$2M_n + 1 = 2(2^n - 1) + 1 = (2 \cdot 2^n - 2) + 1 = 2^{n+1} - 1 = M_{n+1}. \ \square$$

Now it is easy to check, that there exists the following linear recurrent relation with constant coefficients of the second order for the Mersenne numbers:

- *for any positive integer n, it holds $M_{n+2} = 3M_{m+1} - 2M_n$, with* $M_1 = 1, M_2 = 3$.

    □ In fact, $M_{n+1} = 2M_n + 1$. So, $M_{n+2} = 2M_{n+1} + 1$. Then

$$M_{n+2} - M_{n+1} = 2M_{m+1} - 2M_n, \quad \text{and } M_{n+2} = 3M_{m+1} - 2M_n. \ \square$$

Divisibility properties of Mersenne numbers

3.2.4. The results, represented above, allow us to obtain some very important properties of divisibility of Mersenne primes.

It is well-known, that *any Mersenne number divides infinitely many other Mersenne numbers:*

- *for positive integers n, m, it holds $M_n | M_{mn}$.*

    □ It is obviously, that

$$2^{mn} - 1 = (2^n)^m - 1 = (2^n - 1)K, \quad K \in \mathbb{N}.$$

So, we get $M_n | M_{mn}$. □

Moreover, it is easy to see, that *a difference of any two Mersenne numbers is divisible by a third Mersenne number:*

- *for positive integers n, m, $m > n$, it holds $M_{m-n} | (M_m - M_n)$.*

    □ In fact, $M_{m-n} | (M_m - M_n)$, as for $m > n$, we have

$$M_m - M_n = 2^m - 2^n = 2^n(2^{m-n} - 1) = 2^n M_{m-n}. \ \square$$

The next divisibility property of Mersenne numbers is connected with the integer division algorithm:

- *if for $m = nq + r$, where $m, n, q, r \in \mathbb{N}$, it holds $d | M_m$ and $d | M_n$, then it holds $d | M_r$.*

□ In fact, if $d|M_m$ and $d|M_n$, then $d|M_m$ and $d|M_{nq}$. As $d|M_m$ and $d|M_{nq}$, then

$$d|(M_m - M_{nq}) = 2^{nq}(2^{m-nq} - 1).$$

Obviously, any divisor of (odd) Mersenne number is odd; so, $gcd(d, 2) = 1$, and, hence, $d|M_{m-nq}$, i.e., $d|M_r$.

The last result allows us to obtain a very important property of greatest common divisor of Mersenne numbers. In fact, *a greatest common divisor of Mersenne numbers is always a Mersenne number.* More exactly,

• *for any positive integers* m, n *it holds* $gcd(M_m, M_n) = M_{gcd(m,n)}$.

□ In order to prove this property, consider the *Euclidean algorithm* for m and n. Without loss of generality, let $m > n$. Then one has

$$m = nq_1 + r_1,$$

$$n = r_1q_2 + r_2,$$

$$r_1 = r_2q_3 + r_3,$$

$$\ldots,$$

$$r_{s-2} = r_{s-1}q_s + r_s,$$

$$r_{s-1} = r_sq_s,$$

where $0 < r_s < \cdots < r_1 < n$. Then $gcd(m, n) = r_s$. As $r_s|m$ and $r_s|n$, then, using the first property of divisibility of Mersenne numbers, we get $M_{r_s}|M_n$, and $M_{r_s}|M_m$. So, M_{r_s} is an common divisor of the numbers M_n and M_m.

On the other hand, let $d|M_n$, and $d|M_m$. Then, using the second proprety of divisibility of Mersenne numbers for the first equality $m = nq_1 + r_1$ of the algorithm, we get that $d|M_{r_1}$. On a similar way, it holds $d|M_{r_2}$, ..., $d|M_{r_s}$. So, any common divisor d of M_m and M_n divides their common divisor M_{r_s}. It gives, that $M_{r_s} = M_{gcd(m,n)}$ is the greatest common divisor of the Mersenne numbers M_m and M_n. □

From the last property we will get two important consequences. First, it is a statement about the relationship between the divisibility of Mersenne numbers and the divisibility of their indexes:

- *for any positive integers* m, n, $m \geq n$, *it holds* $M_n | M_m$ *if and only if* $n | m$.

□ In fact, if $n | m$, then $gcd(m, n) = n$, and $gcd(M_m, M_n) = M_{gcd(m,n)} = M_n$. So, $M_n | M_m$.

On the other hand, if $M_n | M_m$, it holds $gcd(M_n, M_m) = M_n$. So, $M_n = M_{gcd(n,m)}$. We get, that $n = gcd(n, m)$; hence, $n | m$. □

Secondly, it is a very important statement about coprime Mersenne numbers:

- *for any positive integers* m, n, *it holds* $gcd(M_n, M_m) = 1$ *if and only if* $gcd(m, n) = 1$.

□ This fact is now obvious: $gcd(M_n, M_m) = 1$ if and only if $M_{gcd(n,m)} = 1$, if and only if $gcd(m, n) = 1$. □

This property of Mersenne numbers allows us to give one more proof of *infiniteness of the set of prime numbers*:

- *there are infinity many prime numbers.*

□ Indeed, consider the set $\{M_2, M_3, ..., M_{p_n}, ...\}$ of all Mersenne numbers with prime indexes $p_1 = 2$, $p_2 = 3$, ... For each M_{p_i}, denote by q_i the smallest prime divisor of M_{p_i}. On this way, we get a set $\{q_1, q_2, ..., q_i, ...\}$ of prime numbers: $q_1 = 3$, $q_2 = 7$, $q_3 = 31$, etc.

For $i \neq j$, it holds $p_i \neq p_j$, and hence $gcd(p_i, p_j) = 1$; so, $gcd(M_{p_i}, M_{p_j}) = 1$, and $q_i \neq q_j$. Therefore, all elements of the set $\{q_1, q_2, ..., q_i, ...\}$ are different. If the set P of primes is finite, say, $P = \{p_1, p_2, ..., p_k\}$, it should coincides with the set $\{q_1, q_2, ..., q_k\}$. But $p_1 = 2$, while all numbers q_i are odd; a contradiction. It proves, that the set P of primes is infinite. □

Other elemantary properties of Mersenne numbers

3.2.5. Consider now some elementary properties of Mersenne numbers, which allow to find connections between M_n and other classes of special numbers.

For example, it is easy to see, that M_n *cannot be represented as a non-trivial perfect square*:
- *for any $n \geq 2$, it holds $M_n \neq k^2$, where $k \in \mathbb{N}$.*
□ In fact, if $2^n - 1 = k^2$, then k should be odd.

In this case, $k^2 \equiv 1 (mod\, 4)$, while $2^n - 1 \equiv -1 (mod\ 4)$ for any $n \geq 2$; a contradiction. □

So, $M_n \neq k^2$ for $n \geq 2$. It means, that
- *any Mersenne number, greater 1, cannot be a square number.*

Using the same arguments, we can prove that $M_n \neq k^4$ for $n \geq 2$. It means, that
- *any Mersenne number greater 1 cannot be a biquadratic number.*

In fact, it can be proven, that M_n *cannot be any non-trivial power*:
- *for any $n \geq 2$, it holds $M_n \neq k^s$, where $k, s \in \mathbb{N}$, and $s \geq 2$.*
□ As we noted above, if $2^n - 1 = k^s$, then k should be odd.

If s is even, then $k^s \equiv 1 (mod\, 8)$, and $2^n = k^s + 1 = 2(4t + 1)$, a contradiction.

If s is odd, then $2^n = k^s + 1 = (k+1)T$, where T is odd, and, therefore, is equal to 1. Then $2^n = k + 1$, i.e., $s = 1$, a contradiction. □

It is easy to check, that for $n = 1, 2, 4, 12$ the number M_n can be represented as $\frac{k(k+1)}{2}$, where $k \in \mathbb{N}$.

The numbers of the form $\frac{k(k+1)}{2}$, $k \in \mathbb{N}$, are called *triangular*. So, we got the connection between Mersenne numbers and triangular numbers: $M_1 = S_3(1)$, $M_2 = S_3(2)$, $M_4 = S_3(4)$, $M_{12} = S_3(90)$, where $S_3(k) = \frac{k(k+1)}{2}$ is the k-th triangular number.

In fact, it is possible to prove that M_n *is a triangle number only for these four cases*:
- $M_n = \frac{k(k+1)}{2}$ *only for $n = 1, 2, 4, 12$.*
□ The problem of finding all triangular Mersenne numbers is equivalent to the problem of finding all positive integer solutions of the *Ramanujan-Nagell equation* $2^n - 7 = x^2$. The solutions of this equation in positive integers n and x exist just when $n = 3$, 4, 5, 7 and 15 (sequence A060728 in the OEIS) with corresponding $x = 1$, 3, 5, 11, and 181 (sequence A038198 in the OEIS), as was conjectured by an Indian mathematician S. Ramanujan and proved by a Norwegian mathematician T. Nagell.

In this case, if the number $M_v = 2^v - 1$ is triangular, then the values of v are just those of $n - 3$, so that the triangular Mersenne numbers are 1, 3, 15, 4095 (see sequence A076046 in the OEIS) and no more.

The numbers 0, 1, 3, 15, 4095, which are the only non-negative integers which have simultaneously the form $\frac{u(u+1)}{2}$, $u = 0, 1, 2, ...$, and the form $2^v - 1$, $v = 0, 1, 2, ...$, are called *Ramanujan-Nagell numbers*.

3.2.6. It is easy to prove, that *if p is a prime, then Mersenne number M_p is prime or Fermat pseudoprime to the base 2* (see Chapter 2, Section 2.3):

• *if $p \in P$, then $2^{M_p-1} \equiv 1(mod\ M_p)$.*

□ In fact, let p be a prime, and $M_p = 2^p - 1$ be a Mersenne number with prime index p.

If M_p is prime, the congruence $2^{M_p-1} \equiv 1(mod\ M_p)$ holds due to the Fermat's little theorem.

If M_p is composite, then p is an odd prime. By the Fermat's little theorem, $2^{p-1} \equiv 1(mod\ p)$, and, hence,

$$\frac{M_p - 1}{2} \equiv 2^{p-1} - 1 \equiv 0(mod\ p).$$

So, $\frac{M_p-1}{2} = kp$ for some positive integer k. As for any positive integer k it holds

$$2^{kp} - 1 = (2^p - 1)K, \quad \text{with some } K \in \mathbb{N},$$

we get

$$M_p = 2^p - 1 | (2^{kp} - 1) = 2^{\frac{M_p-1}{2}} - 1.$$

It is equivalent to say that $2^{\frac{M_p-1}{2}} \equiv 1(mod\ M_p)$, which implies that $2^{M_p-1} \equiv 1(mod\ M_p)$. □

Similarly, it holds (see Chapter 2, Section 2.3) that

• *all composite divisors of prime-exponent Mersenne numbers M_p are strong pseudoprimes to the base 2.*

On the other hand,

• *a Mersenne prime cannot be a Wieferich prime.*

□ Remember, that a *Wieferich prime* is a prime number p such that p^2 divides $2^{p-1} - 1$.

So, we are going to show that if $p = 2^q - 1$ is a Mersenne prime, then the congruence $2^{p-1} \equiv 1(mod\ p^2)$ does not hold.

By the Fermat's little theorem, $2^{p-1} \equiv 1(mod\ p)$. As by definition, $p = 2^q - 1$, it holds $2^q \equiv 1(mod\ p)$, and hence, $q|(p-1)$. Therefore, one can write $p - 1 = qk$, where $k \in \mathbb{N}$, $k \geq 2$.

If the congruence $2^{p-1} \equiv 1(mod\ p^2)$ is satisfied, then $p^2|(2^{qk} - 1)$, and $p|\frac{2^{qk}-1}{p}$. Therefore, using that $p = 2^q - 1$, we obtain

$$0 \equiv \frac{2^{qk} - 1}{2^q - 1} \equiv 1 + 2^q + 2^{2q} + \cdots + 2^{(k-1)q} \equiv -k(mod\ 2^q - 1).$$

Hence, $2^q - 1|k$, and, therefore, $k \geq 2^q - 1$. This leads to $p - 1 = qk \geq q(2^q - 1) = qp$, which is impossible. □

The additional proofs and some other additional information see, for example, in [Deza17], [Deza18], [DeKo13], [Nage51], [Nage61], [Rama00], [Sier64].

Exercises

1. Check the following congruences:

    (a) $M_n \equiv 1(mod\ 2)$ for $n \geq 1$;   (c) $M_n \equiv 7(mod\ 8)$ for $n \geq 3$;

    (b) $M_n \equiv 3(mod\ 4)$ for $n \geq 2$;   (d) $M_n \equiv 15(mod\ 16)$ for $n \geq 4$.

2. Find the smallest prime divisor for any M_n, $2 \leq n \leq 20$.

3. Check, that $M_n|M_{nm}$, $n = 1, 2, 3, 4$, $m = 1, 2, 3$.

4. Check, that $M_{n-m}|(M_n - M_m)$, $m, n = 1, 2, 3, 4, 5, 6, 7$.

5. Check, that $gcd(M_n, M_m) = M_{gcd(n,m)}$, $m, n = 1, 2, 3, 4, 5, 6, 7$.

6. Prove the following properties:

    (a) $3|M_n \Leftrightarrow 2|n$;   (c) $7|M_n \Leftrightarrow 3|n$;

    (b) $5|M_n \Leftrightarrow 4|n$;   (d) $11|M_n \Leftrightarrow 10|n$;

(e) $13|M_n \Leftrightarrow 12|n$;   (g) $19|M_n \Leftrightarrow 18|n$;

(f) $17|M_n \Leftrightarrow 8|n$;   (h) $23|M_n \Leftrightarrow 11|n$.

7. Prove, that in the binary numeral system any Mersenne number is represented by a *repunit*, i.e., contains only the digit 1.

8. Prove, that any binary *polindromic prime* (see Chapter 2, Section 2.5) is a Mersenne prime.

9. Prove, that any odd positive integer n divides infinitely many Mersenne numbers.

3.3   Mersenne primes: Prime divisors of Mersenne numbers

The principal problem studied in connection with Mersenne numbers is that of their primality (factorization).

The most elementary and obvious method for finding the factors of a given number n is to test it for divisibility by primes less than $\sqrt{n}$; but if the given number is a large prime or has only large prime factors, the corresponding work is almost prohibitive.

One of the earliest used improvements on this method consists of determining certain properties from the prime factors of the numbers in consideration and then testing with only those primes which have this property.

Properties of prime divisors of Mersenne numbers

3.3.1. If we consider the set of Mersenne numbers in context of their primality, we are going to study only Mersenne numbers M_p with odd prime indexes p: otherwise the problem becomes trivial. So, in this section we are going to prove, that all prime divisors of Mersenne numbers with prime indexes are of the special form.

In the previous section we have proven, that any odd prime p divides Mersenne number M_{p-1}. In context of primality problem the result is almost trivial: as $p-1$ is even, the number M_{p-1} is composite for any $p \geq 5$. But what we can say about the number $\frac{p-1}{2}$? In fact, there exists the following divisibility property:

• *if $q \in P$, $q \equiv \pm1(mod\,8)$, then $q | M_{\frac{q-1}{2}}$.*

□ It is easy to see, that for any odd prime q it holds $q \equiv \pm1(mod\,8)$, or $q \equiv \pm3(mod\,8)$. If $q \equiv \pm1(mod\,8)$, we get, by the properties of the *Legendre symbol* (see Chapter 1, Section 1.5), that $\left(\frac{2}{q}\right) = 1$.

So, by the *Euler's criterion* it holds

$$2^{\frac{q-1}{2}} \equiv 1(mod\ q),$$

and $q | M_{\frac{q-1}{2}}$. □

For example, for $q = 23$, one has $\frac{q-1}{2} = 11$, and $23 | M_{11}$; for $q = 47$, one has $\frac{q-1}{2} = 23$, and $47 | M_{23}$.

On the other hand, one can start from $n = 11$; for $n = 11$ one has $q = 2n + 1 = 23$, and $23 | M_{11}$. Similarly, for $n = 23$ one has $q = 2p + 1 = 47$, and $47 | M_{23}$.

So, we have proved the following fact:

• *for $p \in P$, $p \equiv -1(mod\ 4)$, if $q = 2p + 1$ is prime, then M_p is composite.*

□ For $p \equiv -1(mod\ 4)$, we get $2p+1 \equiv \pm1(mod\ 8)$, so, in the case of primality of $q = 2p - 1$, $q | M_p$, and so M_p is composite. □

In particular, if $p = 11, 23, 83, 131, 179, 191, 239, 251$, then M_p is divisible by 23, 47, 167, 263, 359, 383, 479, 503, respectively.

3.3.2. One can show, that any possible prime divisor of a given Mersenne number M_p with odd prime index p, has the form $2pk + 1$:

• *if $q, p \in P\backslash\{2\}$, and $q | M_p$, then $q = 2pk + 1$ for some $k \in \mathbb{N}$.*

□ Indeed, if $q \in P$, and $q | M_p$, we have due to the Fermat's little theorem, that

$$2^p \equiv 1(mod\,q),$$

i.e., the multiplicative order $ord_q\,2$ of 2 modulo q divides p: $ord_q\,2 | p$.

As $ord_q\,2 \neq 1$, it holds $ord_q\,2 = p$.

Therefore, by the properties of multiplicative order, we get that $p | (q - 1)$, i.e., $q \equiv 1(mod\,p)$.

Obviously, $q \equiv 1 (mod\,2)$. So, $q \equiv 1 (mod\,p)$, $q \equiv 1 (mod\,2)$, and we obtain that $q \equiv 1 (mod\,2p)$, i.e., $q = 2pk + 1$, $k \in \mathbb{N}$. $\square$

3.3.3. It is easy to show, that *any prime divisor q of a Mersenne number M_p with odd prime index p is equal to ± 1 modulo 8:*
- *if $q, p \in P\backslash\{2\}$, and $q|M_p$, then $q \equiv \pm 1 (mod\,8)$.*

$\square$ Consider $q \in P$, $q|M_p$. As $p \in P\backslash\{2\}$, then p is odd, i.e., $p = 2k + 1$. Hence, $q|(2^{2k+1} - 1)$, or $q|(2(2^k)^2 - 1)$, i.e.,

$$q| \left(2(2^k x)^2 - x^2 \right)$$

for any integer x.

As $gcd(2^k, q) = 1$, there exist integers x_0, y_0, such that $2^k x_0 - qy_0 = 1$.

Therefore, $2^k x_0 = 1 + qy_0$. It means that

$$q| \left(2(1 + qy_0)^2 - x_0^2 \right),$$

or

$$q| \left((2(qy_0)^2 + 4qy_0) + (2 - x_0^2) \right).$$

As $q|(2(qy_0)^2 + 4qy_0)$, we get that $q|(2 - x_0^2)$.

Consider the decomposition of $\frac{x_0}{q}$ into a *continued fraction* (see Chapter 1, Section 1.7):

$$\frac{x_0}{q} = [a_0, a_1, ..., a_s] = a_0 + \cfrac{1}{a_1 + \cfrac{1}{...+\frac{1}{a_s}}}.$$

Let $[a_0, a_1, ..., a_k] = \frac{P_k}{Q_k}$. Then

$$1 = Q_0 \leq Q_1 < Q_2 < \cdots < Q_s = q.$$

As $q > 1$, there exists an index n, such that $Q_n^2 < q < Q_{n+1}^2$. Then

$$\left| \frac{x_0}{q} - \frac{P_n}{Q_n} \right| \leq \frac{q^2}{Q_{n+1}^2} < q.$$

Therefore,

$$0 \leq (x_0 Q_n - qP_n)^2 \leq \frac{q^2}{Q_{n+1}^2} < q,$$

and we get that

$$-2q < -2Q_n^2 \le (x_0 Q_n - q P_n)^2 - 2Q_n^2$$
$$= (x_0^2 - 2)Q_n^2 + qM < q - 2Q_n^2 < q.$$

As $q|(2 - x_0^2)$, we have that $(x_0^2 - 2)Q_n^2 + qM$ is divisible by q, i.e., $(x_0 Q_n - q P_n)^2 - 2Q_n^2$ is divisible by q. But

$$-2q < (x_0 Q_n - q P_n)^2 - 2Q_n^2 < q.$$

So, $(x_0 Q_n - q P_n)^2 - 2Q_n^2$ can only take values 0 or $-q$.

In the first case we obtain a contradiction, as we have $2 = t^2$, where $t = \frac{x_0 Q_n - q P_n}{Q_n}$ is a rational number.

In the second case, we get the representation

$$q = 2x^2 - y^2, \quad \text{where} \quad x = Q_n, \ y = x_0 Q_n = q P_n.$$

In this situation, y should be odd: $y = 2k + 1$. Therefore,

$$y^2 = 4k^2 + 4k + 1 = 4k(k + 1) + 1 \equiv 1 (mod \, 8).$$

If x is odd, then $2x^2 \equiv 2(mod \, 8)$, and $q \equiv 1(mod \, 8)$.

If x is even, then $2x^2 \equiv 0(mod \, 8)$, and $q \equiv -1(mod \, 8)$. $\square$

3.3.4. Using the information above, we can now get a representation of any possible prime divisor of the Mersenne number M_p with odd prime index p:

• *if* $q, p \in P \backslash \{2\}$, $q|M_p$, *and* $p = 4t \mp 1$, *then* $q \in \{8pk + 1, 8pk + 1 \pm 2p\}$.

$\square$ By the two last properties, we have that $q = 2pt + 1$, and $q \equiv \pm 1(mod \, 8)$.

1. If $q = 2pt + 1$, and $q \equiv 1(mod \, 8)$, then

$$2pt \equiv 0(mod \, 8), \ pt \equiv 0(mod \, 4), \ t \equiv 0(mod \, 4).$$

So, we get that $q = 8pk + 1$.

2. If $q = 2pt + 1$, and $q \equiv -1(mod \, 8)$, then

$$2pt \equiv -2(mod \, 8), \ pt \equiv -1(mod \, 4).$$

If $p \equiv 1(mod \, 4)$, then $t \equiv -1(mod \, 4)$, and $q = 8pk + 1 - 2p$.

Otherwise, $p \equiv -1(mod \, 4)$, $t \equiv 1(mod \, 4)$, and $q = 8pk + 1 + 2p$. $\square$

L. Euler (1771) proved that M_{31} is prime by testing prime numbers of the form $248n + 1$, $248n + 63$ below 46339 as possible factors of M_{31} (there are only 84 such primes).

But even reduced set of possible factors grows too large as M_p increases.

See [Deza17], [DeKo13], [Dick05], [Step01].

Mersenne numbers and Sophie Germain primes

3.3.5. The first property of prime divisors of a Mersenne number with prime index allow as to say, that

   • *if there exist infinitely many primes p for which $q = 2p + 1$ is a prime, then there exist infinite many primes p such that $M_p \in S$.*

A prime number p is called a *Sophie Germain prime* if $2p + 1$ is also prime. These numbers are named after a French mathematician Marie-Sophie Germain.

The number $2p + 1$ associated with a Sophie Germain prime p is called a *safe prime*.

For example, 11 is a Sophie Germain prime because it is a prime and $2 \cdot 11 + 1 = 23$ is also a prime; in turn, 23 is a Sophie Germain prime because it is a prime and $2 \cdot 23 + 1 = 47$ is also a prime.

The first Sophie Germain primes are 2, 3, 5, 11, 23, 29, 41, 53, 83, 89, ... (sequence A005384 in the OEIS).

Hence, the first safe primes are 5, 7, 11, 23, 47, 59, 83, 107, 167, 179, ... (sequence A005385 in the OEIS).

So, we get a *property of compositeness of Mersenne numbers with prime index $p \equiv -1 \pmod 4$*:

   • *let $p \in P$, $p \equiv -1 \pmod 4$, $p > 3$. Then $M_p \in S$ if p is a Sophie Germain prime.*

Sophie Germain primes and safe primes have applications in Public-key Cryptography and primality testing. It has been conjectured that *there are infinitely many Sophie Germain primes*, but this conjecture remains unproven.

Two distributed computing projects, *PrimeGrid* and *Twin Prime Search*, include searches for large Sophie Germain primes.

The largest known Sophie Germain prime (as of 2020) is 137211 941292195 · $2^{171960} - 1$ (Járai *et al.*, 2006). It has 51780 digits.

Some additional information see, for example, in [Buch09], [Deza17], [DeKo18], [Dick05], [Ehrm67], [Sier64], [Step01].

Exercises

1. Find the set of all possible prime divisors of M_p, $p = 7, 13, 19$. Using this information, prove, that M_p is prime.

2. Find the set of all possible prime divisors of M_p, $p = 11, 23, 29, 37$. Using this information, prove, that M_p is composite.

3. Find the first ten primes, which are not Sophie Germain primes.

4. Prove, that any safe prime $q > 7$ divides $3^{\frac{q-1}{2}} - 1$.

5. Let p be a Sophie Germain prime, $p > 3$. Prove, that $p \equiv 2 (mod\ 3)$.

6. Let q be a safe prime, $q > 7$. Prove, that $q \equiv 3 (mod\ 4)$, $q \equiv 5 (mod\ 6)$, $q \equiv 11 (mod\ 12)$.

7. Find several sequences $\{p, 2p + 1, 2(2p + 1) + 1, ...\}$ in which all of the numbers are primes (such sequence is called a *Cunningham chain* of the first kind).

3.4   Mersenne primes: Lucas-Lehmer test

The other method for testing primality of Mersenne numbers is the *Lucas-Lehmer test*.

History of the question

3.4.1. The theory for this test was initiated by É. Lucas in the late 1870's. É. Lucas studied *Fibonacci numbers*, i.e., the positive integers u_n, such that $u_1 = 1$, $u_2 = 1$, and

$$u_{n+2} = u_n + u_{n+1}$$

for any $n \in \mathbb{N}$. While studying Fibonacci numbers, É. Lucas discovered the following fact:

• *if* $n \equiv \pm 3 (mod\, 10)$ *and* n *is a proper divisor of* u_{n+1} *(i.e.,* $n|u_{n+1}$, *and* $n \nmid u_i$ *for* $i < n+1$), *then* n *is a prime; if* $n \equiv \pm 1 (mod\, 10)$ *and* n *is a proper divisor of* u_{n-1} *(i.e.,* $n|u_{n-1}$, *and* $n \nmid u_i$ *for* $i < n - 1$), *then* n *is a prime.*

Using this approach, É. Lucas proved primality of M_{127} in 1876. He wrote: "...I have proved that number $A = 2^{127} - 1$ is prime. Indeed, the number A is of the form $10k - 3$, and I have verified that u_k is never divisible by A for $k = 2^n$, except for $n = 127$."

So, the primality of the Mersenne number

$$M_{127} = 2^{127} - 1 = 170141183460...715884105727$$

was proven in 1876. It was the last case of "hand" checking of the primality of a Mersenne number.

After É. Lucas, originated the theory, D.H. Lehmer, 1930, simplified the primality test of Lucas.

Using a new version of the test, D.H. Lehmer managed to prove, that M_{257} is a composite number.

See [Deza17], [Luca75], [Hogg69], [Voro61], [Wiki20].

Proof of the theorem

3.4.2. In this section we consider the proof of the *Lucas-Lehmer primality test.*

Theorem (Lucas-Lehmer test). *A number* M_p, p *being an odd prime, is a prime if and only if it is a divisor of the* $(p-1)$-*th term of the sequence* $S_1, S_2, ..., S_{p-1}$, *where* $S_1 = 4$, *and* $S_{k+1} = S_k^2 - 2$.

□ Let $a = 1 + \sqrt{3}$, and $b = 1 - \sqrt{3}$, i.e., $a + b = 2$, and $ab = -2$. Let $u_r = \frac{a^r - b^r}{a - b}$, and $v_r = a^r + b^r$.

I. It is easy to check, that *there exist the following identities*:

1. $2u_{r+s} = u_r v_s + v_r u_s$;

2. $u_s v_r - u_r v_s = (-2)^{s+1} u_{r-s}$;

3. $v_r v_s + 12 u_r u_s = 2v_{r+s}$;

4. $u_r v_r = u_{2r}$;

5. $v_r^2 + (-2)^{r+1} = n_{2r}$;

6. $v_r^2 - 12 u_r^2 = (-2)^{r+2}$.

For example,

$$u_r v_s + v_r u_s = \frac{a^r - b^r}{a - b}(a^s + b^s) + \frac{a^s - b^s}{a - b}(a^r + b^r)$$

$$= \frac{2a^{r+s} - 2b^{r+s} + a^r b^s - b^r a^s + a^s b^r - a^r b^s}{a - b}$$

$$= 2\frac{a^{r+s} - b^{r+s}}{a - b} = 2u_{r+s}.$$

II. It holds, that *for a given prime $p > 3$, we have the following congruences*:

$$u_p \equiv \left(\frac{3}{p}\right) (mod\, p), \quad \text{and } v_p \equiv 2(mod\, p).$$

In fact,

$$u_p = \frac{(1 + \sqrt{3})^p - (1 - \sqrt{3})^p}{2\sqrt{3}}$$

$$= \frac{1}{2\sqrt{2}}\left(2\binom{p}{1}\sqrt{3} + 2\binom{p}{2}(\sqrt{3})^3 + \cdots + 2\binom{p}{p}(\sqrt{3})^p\right)$$

$$= \binom{p}{1} + 3\binom{p}{3} + \cdots + \binom{p}{p}3^{\frac{p-1}{2}} \equiv 3^{\frac{p-1}{2}} = \left(\frac{3}{p}\right) (mod\ p),$$

as binomial coefficients $\binom{p}{k} \equiv 0(mod\, p)$ for any $k \in [1, p-1]$. On the same way,

$$v_p = \left(1 + \sqrt{3}\right)^p + \left(1 - \sqrt{3}\right)^p$$

$$= 2 + 2\binom{p}{2}\left(\sqrt{3}\right)^2 + \cdots + 2\binom{p}{p-1}\left(\sqrt{3}\right)^{\frac{p-1}{2}} \equiv 2(mod\ p).$$

III. Let p be a prime, and $p > 3$. Let $p|u_w$, but p does not divides u_r with $r < w$. Then the number $w = w(p)$ is called *rang* of p.

We can prove, that $p|u_r$ *if and only if $w|r$, where p is a prime, $p > 3$, and w is the rang of p.*

In fact, let $X_p = \{r : u_r \equiv 0(mod\, p)\}$. Using the first and second identities of **I**, we obtain the following fact:

if $u_r \equiv 0(mod\, p)$, and $u_s \equiv 0(mod\, p)$, then $u_{r\pm s} \equiv 0(mod\, p)$.

If w is the least number from the set X_p, and $r \in X_p$, then consecutive subtractions give, that $r - kw = 0$, i.e., $w|r$.

IV. We are going to prove, that $w \le p+1$, *where p is a prime, $p > 3$, and w is the rang of p.*

For a proof of this fact, it is sufficiently to show, that $p|u_{p-1}u_{p+1}$.

Take in the first and second identities of **II** the values $r = p$, and $s = 1$.

Then, as $u_1 = 1$, and $v_1 = 2$, we obtain, that

$$2u_{p+1} = v_p + 2u_p, \quad \text{and} \quad 4u_{p-1} = v_p - 2u_p.$$

Therefore one has, using **II**, that

$$8u_{p+1}u_{p-1} = v_p^2 - 4u_p^2 = 4 - 4(\pm 1)^2 \equiv 0(mod\ p).$$

V. Let prove now, that $v_{2^k} = 2^{2^{k-1}}S_k$, i.e., $M_p|S_k$ *if and only if* $M_p|v_{2^k}$.

In fact, for $k = 1$, one has that $v_2 = 8 = 2S_1$. Going from k to $k + 1$, we obtain, using the fifth identity of **I** with $r = 2^k$, that

$$v_{2^{k+1}} = v_{2^k}^2 - 2^{2^k+1} = (2^{2^{k-1}}S_k)^2 - 2^{2^k+1} = 2^{2^k}(S_k^2 - 2) = 2^{2^k}S_{k+1}.$$

VI. Consider *the proof of the first proposition of the theorem: if $M_p \in P$, then $M_p|S_{p-1}$.*

Let $M_p \in P$. It is sufficiently to show, that

$$M_p|v_{2^p-1} = v_{\frac{M_p+1}{2}}.$$

Using the fifth identity of **I** with $r = \frac{M_p+1}{2}$, we obtain that

$$v_{M_p+1} = v_{\frac{M_p+1}{2}}^2 - 4 \cdot 2^{\frac{M_p-1}{2}}.$$

As $M_p \in P$, and as for prime $p > 3$ we have $M_p = 2^p - 1 \equiv -1(mod\ 8)$, then $(\frac{2}{M_p}) = 1$, i.e., $2^{\frac{M_p-1}{2}} \equiv 1(mod\ p)$. Therefore,

$$v_{M_p+1} \equiv v_{\frac{M_p+1}{2}}^2 - 4(mod\ p).$$

On the other hand, using the third identity of **I** with $r = M_p$ and $s = 1$, we obtain, that

$$2v_{M_p+1} = 2v_{M_p} + 12u_{M_p}$$

(as $v_1 = 2$, and $u_1 = 1$). Now, noting that $M_p \equiv -1 (mod\ 4)$ and $M_p \equiv 1 (mod\ 3)$, i.e., that

$$\left(\frac{3}{M_p} \right) = -\left(\frac{M_p}{3} \right) = -\left(\frac{1}{3} \right) = -1,$$

we obtain using **II** that

$$v_{M_p+1} \equiv v_{M_p} + 6u_{M_p} \equiv 2 + 6 \left(\frac{3}{M_p} \right) \equiv 2 - 6 \equiv -4(mod\ M_p).$$

Therefore,

$$v^2_{\frac{M_p+1}{2}} \equiv 0(mod\ M_p).$$

VII. Consider *the proof of the second proposition of the theorem: if $M_n | S_{n-1}$, then M_n is a prime number.*

Let $M_n | S_{n-1}$, i.e., $M_n | v_{2^{n-1}}$. Let $p | M_n$. As for $k > 1$ any $S_k \equiv -1(mod\ 3)$, then $p > 3$. Let w be the rang of p. Using the fourth identity of **I** with $r = 2^{n-1}$, we obtain that

$$M_n | u_{2^n} = u_{2^{n-1}} v_{2^{n-1}}.$$

Therefore, one has that $p | u_{2^n}$, i.e., by **III**, that $w | 2^n$.

It is easy to see that $w = 2^n$. In fact, if $w < 2^n$, then $p | u_{2^{n-1}}$, and using the last identity of **I** with $r = 2^{n-1}$, we obtain that

$$p | \left(v^2_{2^{n-1}} - 12u^2_{2^{n-1}} \right) = (-2)^{2^{n-1}+2},$$

a contradiction.

But, by **IV**, we have that $w \leq p + 1$. Therefore, one has that $p \geq w - 1 = 2^n - 1 = M_n$, where p is a prime dividing M_n, i.e., we obtain that $p = M_n$. $\square$

See [Deza17], [Luca75], [Hogg69], [Voro61], [Wiki20].

Practical algorithms of computation

3.4.3. The Lucas–Lehmer test works as follows.

Let $M_p = 2^p - 1$ be a Mersenne number to test with p, an odd prime. (The primality of p can be efficiently checked with a simple algorithm like trial division since p is exponentially smaller than M_p.)

Define a sequence $\{S_i\}$: $S_1 = 4$, and $S_{i+1} = S_i^2 - 2$ for any positive integer i.

The first terms of this sequence are 4, 14, 194, 37634, 1416317954, 2005956546822746114, ... (sequence A003010 in the OEIS).

Then M_p is prime if and only if
$$S_{p-1} \equiv 0 (mod\ M_p).$$

The number S_{p-1} modulo M_p is called the *Lucas–Lehmer residue* of p.

For example, the Mersenne number $M_3 = 2^3 - 1 = 7$ is prime, as $S_1 = 4$, and $S_2 = 4^2 - 2 = 14 \equiv 0 (mod\ 7)$.

The Mersenne number $M_5 = 2^5 - 1 = 31$ is prime, as $S_1 = 4$, $S_2 \equiv 4^2 - 2 \equiv 14 (mod\ 31)$; $S_3 \equiv 14^2 - 2 \equiv 8 (mod\ 31)$; $S_4 \equiv 8^2 - 2 \equiv 0 (mod\ 31)$.

On the other hand, in order to check the primality of $M_{11} = 2^{11} - 1 = 2047$, we obtain the sequence $S_1, ..., S_{10}$ modulo 2047: $S_1 \equiv 4 (mod\ 2047)$; $S_2 \equiv 4^2 - 2 \equiv 14 (mod\ 2047)$; $S_3 \equiv 14^2 - 2 \equiv 194 (mod\ 2047)$; $S_4 \equiv 194^2 - 2 \equiv 788 (mod\ 2047)$; $S_5 \equiv 788^2 - 2 \equiv 701 (mod\ 2047)$; $S_6 \equiv 701^2 - 2 \equiv 119 (mod\ 2047)$; $S_7 \equiv 119^2 - 2 \equiv 1877 (mod\ 2047)$; $S_8 \equiv 1877^2 - 2 \equiv 240 (mod\ 2047)$; $S_9 \equiv 240^2 - 2 \equiv 282 (mod\ 2047)$; $S_{10} \equiv 282^2 - 2 \equiv 1736 (mod\ 2047)$. So, $S_{10} \not\equiv 0 (mod\ 2047)$, and, hence, M_{11} id composite.

In the Lucas-Lehmer test, starting values S_1 other than 4 are possible. In fact, the possible starting values for S_1 are 4, 10, 52, 724, 970, 10084, 95050, 140452, 1956244, 9313930, ... (sequence A018844 in the OEIS).

The Lucas-Lehmer residue calculated with these alternative starting values will still be zero if M_p is a Mersenne prime. However, the terms of the sequence will be different and a non-zero Lucas-Lehmer residue for composite M_p will have a different numerical value from the non-zero value calculated when $S_1 = 4$.

Starting values like 4 and 10 are universal, that is, they are valid for all (or nearly all) p. In fact, there are infinitely many additional universal starting values.

However, some other starting values are only valid for a subset of all possible p. For example, $S_0 = 3$ can be used if $p \equiv 3 (mod\ 4)$.

This starting value was often used in the era of hand computation, including by É. Lucas in proving primality of M_{127}.

3.4.4. The Lucas-Lehmer test happens to be ideally suited for binary computers, as the computation of S_k does not involve division and can be done using only multiplication (rotation) and addition, which binary computers do quickly.

The sequence S_n is computed modulo $M_p = 2^p - 1$ to save time. Taking each S_k modulo M_p is easy in binary, too, because M_p is a string of the unities in binary.

After É. Lucas, the next larger Mersenne prime was discovered in 1952 by Raphael M. Robinson with computer assistance. It was

$$M_{521} = 686479766013...291115057151.$$

Today the Lucas–Lehmer test is the primality test used by the Great Internet Mersenne Prime Search (GIMPS) to locate large primes. This search has been successful in locating many of the largest primes known to date.

3.4.5. Since they are prime numbers, Mersenne primes are divisible only by 1 and by themselves.

However, not all Mersenne numbers with prime indexes are Mersenne primes, and the composite Mersenne numbers may be factored non-trivially.

Mersenne numbers are very good test cases for the special *number field sieve algorithm*, so often the largest number factorized with this algorithm has been a Mersenne number. As of 2020, $M_{1193} = 2^{1193} - 1$ is the record-holder, having been factored with a variant of the special number field sieve that allows the factorization of several numbers at once.

On the other hand, as of 2020, the Mersenne number $M_{1277} = 2^{1277} - 1$ is the smallest composite Mersenne number with no known factors; it has no prime factors below 2^{67}.

The number of factors for the first 500 Mersenne numbers can be found at sequence A046800 in the OEIS. The first few elements of this sequence are 0, 0, 1, 1, 2, 1, 2, 1, 3, 2, 3, 2, 4, 1, 3, 3, 4, 1, 4, 1, ...

The other proofs of the Lucas-Lehmer theorem and some additional information see, for example, in [Buch09], [Cohn65], [DeKo13], [DeKo18], [Prim20], [Wiki20].

Exercises

1. Prove, that $u_0 + u_1 + u_2 + \cdots + u_n = u_{n+2} - 1$.

2. Prove, that $u_1 + u_3 + \cdots + u_{2n-1} = u_{2n}$.

3. Prove, that $u_2 + u_4 + \cdots + u_{2n} = u_{2n+1} - 1$.

4. Prove, that $u_1^2 + u_2^2 + \cdots + u_n^2 = u_n \cdot u_{n+1}$.

5. Let $\alpha = \frac{1+\sqrt{5}}{2}$, $\beta = \frac{1-\sqrt{5}}{2}$; prove the *Binet's formula* $u_n = \frac{\alpha^n - \beta^n}{\sqrt{5}}$, $n \in \mathbb{N}$.

6. Let $\alpha = 2 + \sqrt{3}$, and $\beta = 2 - \sqrt{3}$; prove that $S_{i+1} = \alpha^{2^i} + \beta^{2^i}$ for any $i \geq 0$.

7. Prove the Lucas-Lehmer theorem, using the numbers $\alpha = 2+\sqrt{3}$, and $\beta = 2 - \sqrt{3}$.

8. Prove the primality of $M_7 = 2^7 - 1 = 127$ using the Lucas-Lehmer primality test.

9. Prove the primality of $M_{13} = 2^{13} - 1 = 8191$ using the Lucas-Lehmer primality test.

10. Prove the primality of $M_{17} = 2^{17} - 1 = 131071$ using the Lucas-Lehmer primality test.

11. Prove the primality of $M_3 = 2^3 - 1 = 7$ using the Lucas-Lehmer primality test with $S_1 = 10$; with $S_1 = 52$.

12. Prove the primality of $M_5 = 2^5 - 1 = 31$ using the Lucas-Lehmer primality test with $S_1 = 10$; prove the primality of $M_7 = 2^7 - 1 = 127$ using the Lucas-Lehmer primality test with $S_1 = 10$.

3.5 Mersenne numbers in the family of special numbers

Mersenne primes and perfect numbers

3.5.1. Mersenne numbers are connected to a very old mathematical problem.

Consider so called *perfect numbers*: they are positive integers, which are equal to the sum of their proper divisors (i.e., the positive integer divisors different from the number itself).

The first perfect numbers (sequence A000396 in OEIS) are

6, 28, 496, 8128, 33550336, 8589869056, 137438691328,

2305843008139952128, 2658455991569831744654692615953842176,

191561942608236107294793378084303638130997321548169216, ...

The *Euclid–Euler theorem* states that
- *an even natural number is perfect if and only if* $n = 2^{k-1}(2^k - 1)$, *where* $2^k - 1 \in P$.

In other words, it gives an one-to-one relationship between even perfect numbers and Mersenne primes; each Mersenne prime generates one even perfect number, and vice versa:
- *the n-th even perfect number has the form* $n = 2^{p-1}M_p$, *where* M_p *is the n-th Mersenne prime.*

So, we obtain a new even perfect number if and only if we obtain a new Mersenne prime.

As for odd perfect numbers, mathematicians cannot find one. If odd perfect numbers exist, they should be very large and very special.

For example, if there exists one, then:

- it is a perfect square times an odd power of a single prime;

- it is divisible by at least eight primes and has at least 75 prime factors (not necessarily distinct) with at least 9 distinct;

- it has at least 300 decimal digits;

- it has a prime divisor greater that 1020, etc.

As of 2020, there exist 51 known Mersenne prime. It means, that there exists 51 known (even) perfect numbers. The four smallest perfect numbers are

$$6 = 2^1(2^2 - 1) = 2 \cdot M_2; \quad 28 = 2^2(2^3 - 1) = 2^2 \cdot M_3;$$
$$496 = 2^4(2^5 - 1) = 2^4 \cdot M_5; \quad 8128 = 2^6(2^7 - 1) - 2^6 \cdot M_7.$$

The four biggest known perfect numbers are

$$2^{57885160}(2^{57885161} - 1) = 2^{57885160} \cdot M_{57885161}(34850339 \quad \text{digits});$$

$$2^{74207280}(2^{74207281} - 1) = 2^{74207280} \cdot M_{274207281}(44677235 \quad \text{digits});$$

$$2^{77232916}(2^{77232917} - 1) = 2^{77232916} \cdot M_{277232917}(46498850 \quad \text{digits});$$

$$2^{82589932}(2^{82589933} - 1) = 2^{82589932} \cdot M_{282589933}(49724095 \quad \text{digits}).$$

Some additional information see, for example, in [Buch09], [DeKo13], [Deza17], [Deza18], [Dick05].

Mersenne numbers and figurate numbers

3.5.2. There are some interesting connextions between Mersenne numbers ans figurate numbers.

It was shown before, that $M_n = \frac{k(k+1)}{2}$ only for $n = 1, 2, 4, 12$. As any *triangular number*

$$S_3(n) = 1 + 2 + 3 + \cdots + n, n \in \mathbb{N},$$

has the form $S_3(n) = \frac{n(n+1)}{2}$, we can say, that

• *a Mersenne number M_n is a triangular number $S_3(k)$ if and only if $n = 1, 2, 4, 12$.*

In fact, $M_1 = S_3(1)$, $M_2 = S_3(2)$, $M_4 = S_3(4)$, $M_{12} = S_3(90)$.

We have proven, that $M_n \neq k^2$ for any positive integer $k > 1$. As any *square number*

$$S_4(n) = 1 + 3 + 5 \cdots + (2n - 1), n \in \mathbb{N},$$

has the form $S_4(n) = n^2$, we can say, that

• *there is no Mersenne number M_n, $n > 1$, which is a square number.*

We have proven also, that $M_n \neq k^s$ for any positive integers k and s, such that $k, s > 1$. But for $s = 3$ we obtain from k^s a *cubic number*, an 3-dimensional generalization of the notion of square numbers.

For $s = 4$, the formula k^4 gives us so-called *polytope numbers*, an 4-dimensional generalization of the 2-dimensional notion of square numbers and 3-dimensional notion of cubic numbers.

In general, the formula k^s gives us s-dimensional *s-polytope numbers*, an s-dimensional generalization of the 2-dimensional notion of square numbers and 3-dimensional notion of cubic numbers. So, we can say, that

• *there is no Mersenne number M_n, $n > 1$, which is a cubic number, a polytope number and, in general, an s-polytope number, $s \in \mathbb{N}, s \geq 3$.*

See for some additional information [DeDe12], [Nage61], [Rama00], [Wiki20].

Mersenne numbers and Pascal's triangle

3.5.3. It is obviously, that

• *the sum of all entries of the first n rows of Pascal's triangle is equal to n-th Mersenne number M_n.*

□ In fact, it is well-known that the sum of all elements of the i-th row of Pascal's triangle is 2^i, $i = 0, 1, 2, 3, \ldots$. Then the sum of all elements of the first n rows of Pascal's triangle is equal to $2^0 + 2^1 + \cdots + 2^{n-1} = 2^n - 1$. That concludes the proof. □

3.5.4. The Mersenne numbers arise also in the Sierpiński shive, given by the representation of Pascal's triangle modulo 2. It was proven in Chapter 2, Section 2.5, that, for given prime p, the *p-th row of Pascal's triangle is divisible by p*.

It means, that in the p-th row of Pascal's triangle, if p is a prime number, all the terms except the unities are multiples of p.

So, for any Mersenne prime M_p, we get the following property:

• *given Mersenne prime M_p, the M_p-th row of the Pascal's triangle is divisible by M_p.*

In particular, the third row

$$1, 3, 3, 1$$

is divisible by $M_2 = 2^2 - 1 = 3$, the 7-th row

$$1, 7, 21, 35, 35, 21, 7, 1$$

is divisible by $M_3 = 2^3 - 1 = 7$, etc.

Considering Pascal's triangle modulo 2, we can find in this construction all even perfect numbers. In fact, the n-th row of Pascal's triangle is divisible by 2 if and only if $n = 2^k$. In this case all $2^k - 1$ inner entries of the row are even. So, exactly $2^k - 2$ central entries of the next $(2^k + 1)$-th row will be even; exactly $2^k - 3$ central entries of the $(2^k + 2)$-th row will be even, ... , exactly 1 central element of the $(2^{k+1} - 2)$-th row will be even. On this way, we obtain in the Pascal's triangle modulo 2 a black subtriangle, which contains

$$1 + 2 + \cdots + (2^k - 1) = \frac{(2^k - 1)2^k}{2} = 2^{k-1}(2^k - 1)$$

elements. So, for Mersenne prime M_p, the number of the elements in the corresponding black "even subtriangle" will be a perfect number.

Some additional information see, for example, in [Bond93], [Buch09], [DeKo13], [Deza17], [Deza18], [Uspe76], [Wiki20].

Double Mersenne numbers

3.5.5. A *double Mersenne number* is a Mersenne number of the form

$$M_{M_n} = 2^{2^n - 1} - 1, \ n \in \mathbb{N}.$$

The first few double Mersenne numbers are $M_{M_1} = 1$, $M_{M_2} = M_3 = 7$, $M_{M_3} = M_7 = 127$, $M_{M_4} = M_{15} = 32767$ (sequence A077586 in the OEIS).

A double Mersenne number that is prime is called a *double Mersenne prime*.

Since a Mersenne number M_n can be prime only if n is prime, a double Mersenne number M_{M_p} can be prime only if M_p is prime.

The first values of p for which M_p is a prime, are $p = 2, 3, 5, 7, 13, 17, 19, 31$. Of these, M_{M_p} is known to be prime for $p = 2$,

$3, 5, 7$; for $p = 13$, $17, 19$, and 31, M_{M_p} is composite, and its explicit factors have been found.

If another double Mersenne prime will be ever found, it would almost certainly be the largest known prime number.

3.5.6. There exists another interesting subsequence of Mersenne numbers.

Remind (see Chapter 2, Section 2.5), that a *Woodall number* W_n is a positive integer of the form

$$W_n = n \cdot 2^n - 1, \ n \in \mathbb{N}.$$

The first few Woodall numbers are 1, 7, 23, 63, 159, 383, 895, 2047, 4607, 10239, ... (sequence A003261 in the OEIS).

It is easy to check, that for a positive integer m,

- the Woodall number W_{2^m} may be prime only if $2^m + m$ is prime.

It means, that a Woodall prime W_{2^m} should coincide with a Mersenne prime M_{2^m+m}. As of 2020, the only known primes that are both Woodall primes and Mersenne primes are $W_2 = M_3 = 7$, and $W_{512} = M_{521}$.

Some additional information see, for example, in [Deza18], [Sier64], [Prim20], [Wiki20].

Mersenne numbers, Carol numbers and Kynéa numbers

3.5.7. A *Carol number* is an integer of the form

$$CN(n) = 4^n - 2^{n+1} - 1, \ n \in \mathbb{N}.$$

An equivalent formula is

$$CN(n) = (2^n - 1)^2 - 2, \ n \in \mathbb{N}.$$

The first few Carol numbers are -1, 7, 47, 223, 959, 3967, 16127, 65023, 261119, 1046527, ... (sequence A093112 in OEIS).

These numbers were first encountered by C. Emmanuel in 1994, who consequently named them after a friend, Carol G. Kirnon.

For $n > 2$, the binary representation of the n-th Carol number is $n - 2$ consecutive unities, a single zero in the middle, and $n + 1$ more consecutive unities, or to put it algebraically,

$$CN(n) = \sum_{i=1, i \neq n+2}^{2n} 2^{i-1}.$$

So, for example, 47 is 101111 in binary, while 223 is 11011111 in binary, etc.

The difference between the n-th Mersenne number and the n-th Carol number is 2^{n+1}:

$$M_n - CN(n) = 2^{n+1} = M_{n+1} + 1.$$

Starting with 7, every third Carol number is a multiple of 7. Thus, for a Carol number $CN(n)$ to also be a prime number, its index n cannot be of the form $3t + 2$ for a positive integer t.

The first few Carol numbers that are also primes are 7, 47, 223, 3967, 16127, 1073676287, 68718952447, 274876858367, ... (sequence A091516 in the OEIS).

As of 2020, the largest known Carol number that is also a prime is the Carol number for $n = 695631$. It has 418812 digits.

It was found by M. Rodenkirch in 2016 using the programs *CKSieve* and *PrimeFormGW*. In fact, it is the 44-th Carol prime.

3.5.8. A *Kynéa number* is an positive integer of the form

$$KN(n) = 4^n + 2^{n+1} - 1, \ n \in \mathbb{N}.$$

An equivalent formula is

$$KN(n) = (2^n + 1)^2 - 2, \ n \in \mathbb{N}.$$

So, a Kynèa number is the n-th power of 4 plus the $(n + 1)$-th Mersenne number:

$$KN(n) = 4^n + M_{n+1} = M_{n+1} + M_{2n} + 1.$$

The first few Kynèa numbers are 7, 23, 79, 287, 1087, 4223, 16639, 66047, 263167, 1050623, ... (sequence A093069 in the OEIS).

The binary representation of the n-th Kynèa number is a single leading unity, followed by $n - 1$ consecutive zeros, followed by $n + 1$ consecutive unities, or, to put it algebraically:

$$KN(n) = 2^{2n} + \sum_{i=0}^{n} 2^i.$$

So, for example, 23 is 10111 in binary, 79 is 1001111 in binary, etc.

The difference between the n-th Kynèa number and the n-th Carol number is the $(n + 2)$-th power of two:

$$KN(n) - CN(n) = 2^{n+2} = M_{n+2} + 1.$$

Starting with 7, every third Kynèa number is a multiple of 7. Thus, for a Kynèa number to also be a prime number, its index n cannot be of the form $3t + 1$ for a positive integer t.

The first few Kynèa numbers that are also prime are 7, 23, 79, 1087, 66047, 263167, 6785407, 1073807359, 17180131327, 68720001023, ...(sequence A091514 in the OEIS).

As of 2006, the largest known Kynèa number that is also a prime is the Kynèa number $KN(281621)$. Approximately

$$KN(281621) = 5.455289117190661 \cdot 10^{169552}.$$

It was found by C. Emmanuel in 2005, using *k-Sieve* from *Phil Comody* and *OpenPFGW*. This is the 46-th Kynèa prime.

3.5.9. Like Carol and Kynèa numbers, Mersenne numbers have very special binary representation; in fact,

• *in binary, the n-th Mersenne number consists from $n-1$ unities*:

$$M_n = 2^n - 1 = 2^{n-1} + 2^{n-2} + \cdots + 2^1 + 1 = \sum_{i=0}^{n-1} 2^i.$$

It means, that any Mersenne number is a *repunit* in binary; so, in binary it gives a simplest example of *palindromic number*.

Moreover, any Mersenne prime gives a *binary palindromic prime*. Remind (see Chapter 2, Section 2.5), that the sequence of binary palindromic primes begins (in binary) from 11, 101, 111, 10001, 11111, 1001001, 1101011, 1111111, 100000001, 100111001, ...

(sequence A117697 in the OEIS) and contains all *Mersenne primes* and all *Fermat primes*.

As for *permutable primes* (see Chapter 2, Section 2.5), in base 2 only repunits can be permutable primes, because any 0 permuted to the last place results in an even number. Therefore,

- *the base 2 permutable primes are the only Mersenne primes.*

3.5.10. The simplest version of *generalized Mersenne primes* can be obtained by consideration of prime numbers of the form $f(2^n)$, where $f(x)$ is a low-degree polynomial with small integer coefficients.

An example is $2^{64} - 2^{32} + 1$; in this case, $n = 32$, and $f(x) = x^2 - x + 1$.

Another example is $2^{192} - 2^{64} - 1$; in this case, $n = 64$, and $f(x) = x^3 - x - 1$.

By this approach, the Kynèa primes and the Carol primes can be consider as examples of generated Mersenne primes, with the polynomials $f(x) = x^2 \pm 2x - 1$.

Some additional information see, for example, in [Deza17], [DeKo13], [Wiki20].

Generalized Mersenne numbers

3.5.11. It is also natural to try to generalize primes of the form $2^n - 1$ to primes of the form $b^n - 1$ for $b \neq 2$. However, $b^n - 1$ is always divisible by $b - 1$, so unless the latter is a unit, the former is not a prime.

But we can regard a ring of "integers" on complex numbers instead of real numbers, like *Gaussian integers* and *Eisenstein integers*.

If we regard the ring $(\mathbb{Z}[i], +, \cdot)$ of *Gaussian integers*,

$$\mathbb{Z}[i] = \{a + bi \mid a, b \in \mathbb{Z}\}, \qquad \text{where } i^2 = -1,$$

we get the cases $b = 1 + i$, and $b = 1 - i$, and can ask for which n the number $(1 + i)^n - 1$ is a Gaussian prime which will then be called a *Gaussian Mersenne prime*.

In fact, $(1+i)^n - 1$ is a Gaussian prime for the $n = 2, 3, 5, 7, 11,$ $19, 29, 47, 73, 79, ...$ (sequence A057429 in the OEIS).

Like the sequence of exponents for usual Mersenne primes, this sequence contains only (rational) prime numbers. As for all Gaussian primes, the norms (that is, squares of absolute values) of these numbers are rational primes $5, 13, 41, 113, 2113, 525313, 536903681,$ $140737471578113, ...$ (sequence A182300 in the OEIS).

We can also regard the ring $(\mathbb{Z}[\omega], +, \cdot)$ of *Eisenstein integers*,

$$\mathbb{Z}[\omega] = \{a + b\omega \mid a, b \in \mathbb{Z}\}, \qquad \text{where } \omega = \frac{-1 + i\sqrt{3}}{2};$$

here we get the cases $b = 1 + \omega$, and $b = 1 - \omega$. So, we can ask for what n the number $(1 + \omega)^n - 1$ is an Eisenstein prime which will then be called an *Eisenstein Mersenne prime*.

In fact, $(1 + \omega)^n - 1$ is an Eisenstein prime for the $n = 2, 5, 7, 11,$ $17, 19, 79, 163, 193, 239, ...$ (sequence A066408 in the OEIS). The norms (that is, squares of absolute values) of these Eisenstein primes are (rational) primes: $7, 271, 2269, 176419, 129159847, 1162320517,$ $...$ (sequence A066413 in the OEIS).

3.5.12. The other way to deal with the fact that $b^n - 1$ is always divisible by $b - 1$, it is to simply take out this factor and ask which values of n make

$$M_n(b) = \frac{b^n - 1}{b - 1}$$

be prime.

Least n such that $M_n(b) = \frac{b^n - 1}{b - 1}$ is prime are (starting with $b = 2$, with 0 if no such n exists) are $2, 3, 2, 3, 2, 5, 3, 0, 2, 17, ...$ (sequence A084740 in the OEIS).

If we take $b = 10$, we get $n = 2, 19, 23, 317, 1031, 49081, 86453,$ $109297, 270343, ...$ (sequence A004023 in the OEIS), corresponding to primes

$$11, 1111111111111111111, 11111111111111111111111, ...$$

(sequence A004022 in the OEIS). These primes are called *repunit primes*.

3.5.13. Another sets of *generalized Mersenne numbers* can be constructed using the formula

$$M_n(a, b) = \frac{a^n - b^n}{a - b},$$

where a and b are coprime integers, such that $a > 1$, and $-a < b < a$.

For example, if $(a, b) = (3, 2)$, the least n such that $M_n(3, 2) = \frac{a^n - b^n}{a - b}$ is prime are 2, 3, 5, 17, 29, 31, 53, 59, 101, 277, 647, 1061, 2381, 2833, 3613, 3853, 3929, 5297, 7417, 90217, ... (sequence A057468 in the OEIS); if $(a, b) = (10, 3)$, the least n such that $M_n(10, 3) = \frac{a^n - b^n}{a - b}$ is prime are 2, 3, 5, 37, 599, 38393, 51431, ... (sequence A128026 in the OEIS); for $(a, b) = (8, 1)$, the number $M_n(8, 1) = \frac{a^n - b^n}{a - b}$ is prime (in fact, 73) only for $n = 3$.

Some additional information see, for example, in [Buch09], [CoGu96], [Deza17], [DeKo13], [Wiki20].

Exercises

1. Check the equality $M_n = S_3(k)$ for all $n, k \in \{1, 2, 3, 4, 5, 6, 7, 8, 9, 10\}$.

2. Check the equality $M_n = S_4(k)$ for all $n, k \in \{1, 2, 3, 4, 5, 6, 7, 8, 9, 10\}$.

3. Check the equality $M_n = S_5(k)$ for all $n, k \in \{1, 2, 3, 4, 5, 6, 7, 8, 9, 10\}$.

4. Find all Mersenne numbers less than 1000, which are centered triangular numbers, i.e., represent triangles with a dot in the center and all other dots surrounding the center in successive triangular layers.

5. Find all Mersenne numbers less than 1000, which are pentagonal numbers; centered pentagonal numbers.

6. Find first five Mersenne primes in the Pascal's triangle. Can we find in the Pascal's triangle any Mersenne number?

7. Prove, that starting with 7, every third Carol number is a multiple of 7.

8. Prove, that starting with 7, every third Kynéa number is a multiple of 7.

9. Find first three generalized Mersenne primes of the form $M_n(a,b) = \frac{a^n - b^n}{a - b}$ for $(a,b) = (3,1)$; for $(a,b) = (3,-1)$; for $(a,b) = (4,3)$; for $(a,b) = (4,-3)$.

10. Check, that there is only one generalized Mersenne prime of the form $M_n(a,b) = \frac{a^n - b^n}{a - b}$ for $(a,b) = (4,1)$; for $(a,b) = (8,1)$; for $(a,b) = (8,-1)$; for $(a,b) = (9,4)$.

3.6 Open problems

Many fundamental questions about Mersenne primes remain unresolved. It is not even known whether the set of Mersenne primes is finite or infinite. It is also not known whether infinitely many Mersenne numbers with prime exponents are composite.

In this section we consider some most important unsolved problems and conjectures of the Theory of Mersenne numbers.

Infiniteness of the set of Mersenne primes

3.6.1. The most important open problem of the Theory of Mersenne numbers is represented by the following question:

• *are there infinitely many Mersenne primes?*

Equivalently we could ask:

• *are there infinitely many even perfect numbers?*

The answer is probably "yes". In fact, consider the *Prime number theorem* (1896, independently by Hadamard and de la Vallée Poussin):

$$\pi(x) = \sum_{p \le x} 1 = \frac{x}{\log x} + o\left(\frac{x}{\log x}\right).$$

It means, that

$$\lim_{x \to \infty} \frac{\pi(x)}{x / \log x} = 1.$$

So, by the Prime number theorem, the n-th prime number is approximatively equal to $n \log n$ (more exactly, $cn \ln n < p_n < Cn \ln n$, where $C > c > 0$ are some positive absolute constants). Then we can prove that the probability of a "random" number n being prime is at most $\frac{A}{\log n}$ for some positive absolute constant A.

Easy to see, that

$$\frac{A}{\log M_n} > \frac{A}{\log 2^n} = \frac{A}{n \log 2} = \frac{C}{n},$$

where $C = \frac{A}{\log 2}$ is an absolute constant. Summing over all positive integers we get the following estimation:

$$\sum_{n=1}^{\infty} \frac{A}{\log M_n} > \sum_{n=1}^{\infty} \frac{A}{\log 2^n} = \sum_{n=1}^{\infty} \frac{A}{n \log 2} = \sum_{n=1|}^{\infty} \frac{C}{n} = C \sum_{n=1}^{\infty} \frac{1}{n}.$$

In other words, it is proven, that $\sum_{n=1}^{\infty} \frac{A}{\log M_n}$ is greater than a divergent sum $\sum_{n=1}^{\infty} \frac{1}{n}$ (*harmonic series*); so, it seems likely that there are infinitely many Mersenne primes.

However, our reason cannot be used as a proof of infiniteness of the set of Mersenne primes, since the Mersenne numbers are not "random".

For the proof of the Prime number theorem and for some additional information see, for example, [Deza17], [Ingh32], [Kara83], [Titc87].

Lenstra-Pomerance-Wagstaff and Gillies' conjectures

3.6.2. *Lenstra–Pomerance–Wagstaff conjecture* states that there is an infinite number of Mersenne primes; more precisely, that

• *the number of Mersenne primes less than x is asymptotically approximated by*

$$e^{\gamma} \cdot \log_2 \log_2 x,$$

where γ is the Euler–Mascheroni constant.

In other words, the number of Mersenne primes with a given exponent p less than x is asymptotically $e^{\gamma} \cdot \log_2 x$.

This means that there should on average be about $e^\gamma \cdot \log_2 10 \approx$ 5.92 primes p of a given number of decimal digits such that M_p is prime. The conjecture is fairly accurate for the first 40 Mersenne primes, but between 220000000 and 285000000 there are at least 12, rather than the expected number which is around 3.7.

More generally, the Lenstra–Pomerance–Wagstaff conjecture states that

- *the number of primes $p \le y$ such that the generalized Mersenne number $M_n(a,b) = \frac{a^p - b^p}{a-b}$ is prime is asymptotically*

$$(e^\gamma + m \cdot \log 2) \cdot \log_a y;$$

here a, b are coprime integers, $a > 1$, $-a < b < a$, a and b are not both perfect r-th powers for any natural number $r > 1$, $-4ab$ is not a perfect fourth power; the number m is the largest nonnegative integer such that a and $-b$ are both perfect 2^m-th powers.

We get the case of Mersenne primes for $(a, b) = (2, 1)$: $M_n = M_n(2, 1)$.

3.6.3. *Gillies' conjecture* is a conjecture about the distribution of prime divisors of Mersenne numbers; it was made by D.B. Gillies in 1964:

- *if $x < y < \sqrt{M_p}$, as $\frac{x}{y} \to \infty$, and $M_p \to \infty$, the number of prime divisors of M_p in the interval $[x, y]$ is Poisson-distributed with*

$$mean \ \sim \ \begin{cases} \log\left(\frac{\log y}{\log x}\right), & if \ x \ge 2p, \\ \log\left(\frac{\log y}{\log 2p}\right), & if \ x < 2p. \end{cases}$$

He noted that his conjecture would imply that

- the number of Mersenne primes less than x is $\sim \frac{2}{\log 2} \log\log x$;
- the expected number of Mersenne primes M_p with $x \le p \le 2x$ is ~ 2;
- the probability that M_p is prime is $\sim \frac{2\log 2p}{p\log 2}$.

The Lenstra–Pomerance–Wagstaff conjecture gives different values:

- the number of Mersenne primes less than x is $\frac{e^\gamma}{\log 2} \log\log x$;

- the expected number of Mersenne primes M_p with $x \leq p \leq 2x$ is $\sim e^\gamma$;

- the probability that M_p is prime is $\frac{e^\gamma \log 2p}{p \log 2}$ if $p \equiv 3 (mod\ 4)$, and is $\frac{e^\gamma \log 6p}{p \log 2}$, otherwise.

Asymptotically these values are about 11% smaller.

While Gillies' conjecture remains open, several papers have added empirical support to its validity, including Ehrman's 1964 paper.

See [BSW89], [Dick05], [Ehrm67], [Wiki20].

New Mersenne conjecture

3.6.4. In the connection with the problem of infiniteness of the set of Mersenne primes, mathematicians are very interested in the following question:

- *is the New Mersenne Conjecture true?*

We remind, that the *New Mersenne conjecture* of P.T. Bateman, J.L. Selfridge and Jr.S.S. Wagstaff (1989) states that

- *for any odd positive integer p, if any two of the following conditions hold, then so does the third:*

- $p = 2^k \pm 1$ or $p = 4k \pm 3$ for some natural number k (sequence A122834 in the OEIS);

- $2^p - 1$ is a prime (a *Mersenne prime* M_p; sequence A000043 in the OEIS);

- $\frac{2^p + 1}{3}$ is a prime (a *Wagstaff prime*; sequence A000978 in the OEIS).

The first odd primes of the form $2^k \pm 1$ or $4^k \pm 3$ are 3, 5, 7, 13, 17, 19, 31, 61, 67, 127, ... (sequence A122834 in the OEIS).

The exponents p which give Mersenne primes are 2, 3, 5, 7, 13, 17, 19, 31, ... (sequence A000043 in the OEIS).

The first (odd) primes p such that the numbers $\frac{2^p + 1}{3}$ are (Wagstaff) primes, are 3, 5, 7, 11, 13, 17, 19, 23, 31, 43, (sequence A000978 in the OEIS).

The known numbers p for which all three conditions hold are 3, 5, 7, 13, 17, 19, 31, 61, 127 (sequence A107360 in the OEIS).

It is also a conjecture that
- *no number which is greater than* 127 *satisfies all three conditions.*

As of 2020, all the Mersenne primes up to $2^{82589933} - 1$ are known, and for none of these does the third condition hold except for the ones just mentioned.

Note that the two primes for which the original Mersenne conjecture is false (67 and 257) satisfy the first condition of the new conjecture ($67 = 2^6 + 3$, $257 = 2^8 + 1$), but not the other two. 89 and 107, which were missed by Mersenne, satisfy the second condition, but not the other two.

Mersenne may have thought that $2^p - 1$ is prime only if $p = 2^k \pm 1$ or $p = 4^k \pm 3$ for some natural number k, but if he thought it was "if and only if" he would have included 61.

R. Lifchitz has shown that the New Mersenne Conjecture is true for all positive integers less than or equal to 30402456 by systematically testing all primes for which it is already known that one of the conditions holds. His website documents the verification of results up to this number.

For some additional information see, for example, [BSW89], [Deza17], [Dick05], [Wiki20].

Other open questions

3.6.5. Besides the problem of the infiniteness of the set of Mersenne primes, we have an important open question about the infiniteness of Mersenne composites:
- *are there infinitely many composite Mersenne numbers M_p with prime indexes?*

L. Euler proved the following fact:
- *if $k > 1$ and $p = 4k + 3$ is prime, then $2p + 1$ is prime if and only if $2^p \equiv 1 (mod\ 2p + 1)$.*

So if $p = 4k + 3$ and $2p + 1$ are primes (in fact, a *Sophie Germain prime* and a *safe prime*, correspondingly), then the Mersenne number $2^p - 1$ is composite, because it has a prime divisor $2p + 1$.

As it seems reasonable to conjecture that *there are infinitely many prime pairs $(p, 2p + 1)$*, we can expect, that

• *there are infinitely many composite Mersenne numbers M_p with prime index p.*

3.6.6. The following question is still open

• *if every Mersenne number M_p is square free?*

This falls more in the category of an open question (to which we do not know the answer), rather than a conjecture (which we guess is true).

It is easy to show (see Chapter 3, Section 3.2) that

• *if the square of a prime p divides a Mersenne number, then p is a Wieferich prime.*

But the Wieferich primes, i.e., by definition, primes p such that p^2 divides $2^{p-1} - 1$, are very rare.

The only known Wieferich primes are 1093 (Meissner, 1913) and 3511 (Beeger, 1922); see sequence A001220 in OEIS.

If any others exist, they must be greater than $1.25 \cdot 10^{15}$.

Neither of these two primes squared divide a Mersenne number.

3.6.7. Consider a sequence

$$CC_0 = 2, \ CC_1 = 2^{CC_0} - 1, \ CC_2 = 2^{CC_1} - 1,$$
$$CC_3 = 2^{CC_2} - 1, ..., CC_{n+1} = 2^{CC_n} - 1, ...$$

Starting with CC_1, this sequence form a subset of the set of Mersenne numbers. An open question is:

• *are all the numbers CC_n primes?*

These numbers grow very quickly. In fact, the numbers

$$CC_0 = 2; \ CC_1 = 2^2 - 1 = 3; \ CC_2 = 2^3 - 1 = 7;$$
$$CC_3 = 2^7 - 1 = 127;$$
$$CC_4 = 2^{127} - 1 = 170141183460469231731687303715884105727$$

are primes. It is unknown, if the number

$$CC_5 > 10^{51217599719369681875006054625051616349}$$

is prime.

It seems very unlikely that CC_5 (or many of the larger terms) would be prime, so this is no doubt another example of Guy's strong law of small numbers.

Notice that if there is even one composite term in this sequences, then all of the following terms are composite, too.

3.6.8. One more open question about the behaviour (the infiniteness) of some subset of the set of Mersenne numbers is the following:
- *are there more double Mersenne primes?*

The double Mersenne number MM_n is a Mersenne number of the form

$$MM_n = M_{M_n} = 2^{2^n - 1} - 1, \ n \in \mathbb{N}.$$

The first few double Mersenne numbers are $MM_1 = 1$, $MM_2 = M_3 = 7$, $MM_3 = M_7 = 127$, $MM_4 = M_{15} = 32767$ (sequence A077586 in the OEIS).

An early misconception was that
- *if $n = M_p$ is prime, then so is M_n.*

Indeed each of the first four such numbers is prime:

$$MM_2 = 2^3 - 1 = 7;$$

$$MM_3 = 2^7 - 1 = 127;$$

$$MM_5 = 2^{31} - 1 = 2147483647;$$

$$MM_7 = 2^{127} - 1 = 170141183460469231731687303715884105727.$$

However, the next four ($MM_{13}, MM_{17}, MM_{19}$, and MM_{31}) all have known factors, so they are composite.

One more question:
- *are there any more primes in this sequence?*

Probably not, but it remains an open question.

Note that the sequence of numbers $CC_n, n = 2, 3, 4, ...$, discussed above, is a subsequence of the sequence of the double Mersenne numbers, and a positive answer to the question of primality of these numbers will give a positive answer to the questions regarding the numbers MM_p.

Some additional information see, for example, in [Deza17], [Dick05], [Ehrm67], [Guy94], [BSW89].

Exercises

1. Find the smallest square-full Mersenne number.

2. Find the smallest square-full Mersenne number M_n with odd n.

3. Prove, that if p is an odd composite number, then $2^p - 1$ and $\frac{2^p+1}{3}$ are both composite.

4. Check, that for the numbers 3, 5, 7, 13, 17, 19, 31, 61, 127 all three conditions of the New Mersenne conjecture hold.

5. Check, that the numbers 2, 3, 5, 7, 11, 13, 17, 19, 23, 31, 43, 61, 67, 79, 89 satisfy at least one condition of the New Mersenne conjecture.

Chapter 3: References

[Anke57], [Anto85], [Apos86], [Avan67], [BaCo87], [Bond93], [CoGu96], [DeDe12], [Deza17], [Deza18], [Dick27], [Dick05], [Diop74], [Ehrm67], [Gard61], [Gard88], [Gill64], [Wief09], [Guy94] [HaWr79], [Lagr70], [Lege30], [LiNi96], [Luca75], [Nico26], [Hogg69], [Pasc54], [Ries94], [SlPl95], [Sloa20], [Shan93], [Sier64], [Weis99], [Wiki20].

Chapter 4

Fermat numbers

4.1 History of the question

4.1.1. A *Fermat number* F_n is a positive integer of the form

$$F_n = 2^{2^n} + 1, \ n \in \mathbb{Z}, \ n \geq 0.$$

The first Fermat numbers are $3, 5, 17, 257, \ 65537, 4294967297,$ $18446744073709551617, \ldots$ (sequence A000215 in the OEIS).

Fermat numbers were studied by the XVII-th century French lawyer and mathematician Pierre de Fermat (1607–1665).

P. Fermat conjectured (1650) that

- *these numbers would always give a prime for $n = 0, 1, 2, \ldots$* .

In fact, the first five Fermat numbers are primes: $F_0 = 3$, $F_1 = 5$, $F_2 = 17$, $F_3 = 257$, $F_4 = 65537$.

But in 1732 L. Euler found that F_5 is a composite number:

$$F_5 = 2^{2^5} + 1 = 4294967297 = 641 \cdot 6700417,$$

where $641, 6700417 \in P$.

To show it, consider $2^{32}+1 = 2^4 \cdot 2^{28} + 1$, and write $641 = 5^4 + 2^4 = 5 \cdot 2^7 + 1$. Therefore, $5^4 \equiv -2^4 (mod\ 641)$, and $5 \cdot 2^7 \equiv -1 (mod\ 641)$. Then we obtain, that $5^4 \cdot 2^{28} \equiv 1 (mod\ 641)$, or $-2^4 \cdot 2^{28} \equiv 1 (mod\ 641)$, or $2^{32}+1 \equiv 0 (mod\ 641)$. It means, that $641 | F_5$, i.e., F_5 is composite.

Other proof is even more elementary. Denote $a = 2^7$, and $b = 5$. Then, noting that $a - b^3 = 3$, we obtain that

$$2^{2^5} + 1 = 2^4(2^7)^4 + 1 = 2^4 a^4 + 1 = (1 + 3b)a^4 + 1$$
$$= (1 + (a - b^3)b)a^4 + 1$$
$$= (1 + ab - b^4)a^4 + 1 = (1 + ab)a^4 + (1 - a^4 b^4)$$
$$= (1 + ab)a^4 + (1 - a^2 b^2)(1 + a^2 b^2)$$
$$= (1 + ab)a^4 + (1 + ab)(1 - ab)(1 + a^2 b^2)$$
$$= (1 + ab)(a^4 + (1 + ab)(1 + a^2 b^2)).$$

Therefore, $1 + ab = 1 + 5 \cdot 2^7 = 641$ divides $2^{2^5} + 1$, i.e., $641 | F_5$.

So, *Fermat's conjecture* that F_n is always prime is clearly false. Moreover, beyond F_4, no further Fermat primes have been found.

4.1.2. The discovery that established further interest in these numbers came in 1796 when Carl Friedrich Gauss proved the constructibility of the regular 17-gon and announced a relationship between Fermat numbers and constructible regular polygons. Five years later, he developed the theory of Gaussian periods, which allowed him to formulate a sufficient condition for the constructibility of regular polygons. Gauss stated without proof that this condition was also necessary, but never published his proof. A full proof of necessity was given by Pierre Wantzel in 1837. The result is known as the *Gauss–Wantzel theorem*:

  • *a regular n-gon can be constructed with compass and straightedge if and only if n is the product of a power of 2 and any number of distinct Fermat primes (including none):*

$$n = 2^r p_1 p_2 ... p_k, \quad where \ r, k \geq 0,$$

$$and \ \ p_1 < ... < p_k \ \ are \ Fermat \ primes.$$

Since there are 31 combinations of five known Fermat primes, there are 31 known constructible polygons with an odd number of sides.

In fact, a regular n-gon is constructible if $n = 3, 4, 5, 6, 8, 10, 12, 15, 16, 17, ...$ (sequence A003401 in the OEIS), while a regular

n-gon is not constructible with compass and straightedge if $n = 7, 9, 11, 13, 14, 18, 19, 21, 22, 23, \ldots$ (sequence A004169 in the OEIS).

4.1.3. In XIX-th and XX-th centuries many mathematicians considered different aspects of the theory of the Fermat numbers. Among them: F.É.A. Lucas, J.F.T. Pépin, T. Clausen (proved the compositeness of F_6, 1855), I.M. Pervushin (proved the compositeness of F_{12}, 1877, and F_{23}, 1878), Seelhoff (proved the compositeness of F_{36}, 1886), A.J. Cunningham (proved the compositeness of F_{11}, 1899), Western (proved the compositeness of F_9, F_{12} and F_{18}, 1903), J. Cullen (proved the compositeness of F_{38}, 1903), Morehead (proved the compositeness of F_{73}, 1906), M. Kratchik (proved the compositeness of F_{15}, 1925), etc.

Summary of factoring status for Fermat numbers F_n (as of 2020) is given in the following table.

Summary of factoring status for Fermat numbers

Prime	$n = 0, 1, 2, 3, 4$
Completely factored	$n = 5, 6, 7, 8$(two factors each), 9(3 factors), 10(4 factors), 11(5 factors)
Six prime factors known	$n = 12$
Four prime factors known	$n = 13$
Three prime factors known	$n = 15, 19, 25, 52, 287$
Two prime factors known	$n = 16, 17, 18, 27, 30, 36, 38, 39, 42, 77, 147, 150, 284, 416, 417$
Only one prime factor known	$n = 14, 21, 22, 23, 26, 28, 29, 31, 32, 37, 43$ and 268 values of n with $43 < n \leq 18233956$
Composite but no factor known	$n = 20, 24$
Character unknown	$n = 33, 34, 35, 40, 41, 44, 45, 46, 47, 49, 50, \ldots$

So, as of 2020, the only known Fermat primes are F_0, F_1, F_2, F_3, and F_4, i.e., the numbers $3, 5, 17, 257, 65537$ (sequence A019434 in the OEIS).

It means, that $F_4 = 65537$ is the largest known Fermat prime.

Moreover, it is known that F_n is composite for $5 \leq n \leq 32$, although of these, complete factorizations of F_n are known only for $0 \leq n \leq 11$, and there are no known prime factors for $n = 20$ and $n = 24$.

It means, that $F_{11} = 2^{2^{11}} + 1 = 2^{2048} + 1$ is the largest known Fermat number with all known factors, and F_{33} is the smallest Fermat number with unknown character.

In fact, for $n = 5, 6, 7, 8, 9$ we have the following decompositions.

$$
\begin{aligned}
F_5 &= 2^{2^5} + 1 = 2^{32} + 1 \\
&= 4294967297 \\
&= (5 \cdot 2^{5+2} + 1) \cdot (52347 \cdot 2^{5+2} + 1) \\
&= 641 \cdot 6700417.
\end{aligned}
$$

It has 10 digits. It is a product of two prime numbers with 3 and 7 decimal digits, correspondingly. It was fully factored in 1732 (Euler).

$$
\begin{aligned}
F_6 &= 2^{2^6} + 1 = 2^{64} + 1 \\
&= 18446744073709551617 \\
&= (1071 \cdot 2^{6+2} + 1) \cdot (262814145745 \cdot 2^{6+2} + 1) \\
&= 274177 \cdot 67280421310721.
\end{aligned}
$$

It has 20 digits. It is a product of two prime numbers with 6 and 14 decimal digits, correspondingly. It was fully factored in 1855 (Clausen).

$$
\begin{aligned}
F_7 &= 2^{2^7} + 1 = 2^{128} + 1 \\
&= 340282366920938463463374607431768211457 \\
&= (116\,503\,103\,764\,643 \cdot 2^{7+2} + 1) \\
&\quad \cdot (11\,141\,971\,095\,088\,142\,685 \cdot 2^{7+2} + 1) \\
&= 59\,649\,589\,127\,497\,217 \cdot 5\,704\,689\,200\,685\,129\,054\,721.
\end{aligned}
$$

It has 39 digits. It is a product of two prime numbers with 17 and 22 decimal digits, correspondingly. It was fully factored in 1970 (Morrison and Brillhart).

$$F_8 = 2^{2^8} + 1 = 2^{256} + 1$$

$$= 1157920892373161954235709850086879078532699846656405640$$
$$3945758400791312963993 7$$

$$= (3853149761 \cdot 157 \cdot 2^{8+3} + 1)$$
$$\cdot (1057372046781162536274034354686893329625329$$
$$\cdot 31618624099079 \cdot 13 \cdot 7 \cdot 5 \cdot 3 \cdot 2^{8+3} + 1)$$

$$= 1238926361552897 \cdot 93461639715357977769163558199606896584$$
$$0512375416381885802803 21.$$

It has 62 digits. It is a product of two prime numbers with 16 and 62 decimal digits, correspondingly. It was fully factored in 1980 (Brent and Pollard).

$$F_9 = 2^{2^9} + 1 = 2^{512} + 1$$

$$= 13407807929942597099574024998205846127479365...66903427$$
$$690031858186486050853753882811946569946433649006084097$$

$$= (37 \cdot 2^{9+7} + 1) \cdot (43226490359557706629 \cdot 1143290228161321$$
$$\cdot 82488781 \cdot 47 \cdot 19 \cdot 2^{9+2} + 1)$$
$$\cdot (16975143302271505426897585653131126520182328037821729$$
$$720833840187223 \cdot 17338437577121 \cdot 40644377 \cdot 26813 \cdot 1129$$
$$\cdot 2^{9+2} + 1)$$

$$= 2424833 \cdot 7455602825647884208337395736200454918783366344$$
$$2657$$
$$\cdot 741640062627530801524787141901937474059940781097519023$$
$$90582131614441575950470500809281871169394073 7.$$

It has 155 digits. It is a product of three prime numbers with 7, 49 and 99 decimal digits, correspondingly. It was fully factored in 1990

(Lenstra, Manasse at all).

$$F_{10} = 2^{2^{10}} + 1 = 2^{1024} + 1$$
$$= 17976931348623159077290...304835356329624224137217$$
$$= 45592577 \cdot 6487031809 \cdot 4659775785220018543264560743076\!7$$
$$78192897$$
$$\cdot 13043987440548818972748...8062178207531270144244577.$$

It has 309 digits. It is a product of four prime numbers with 8, 10, 40 and 252 decimal digits, correspondingly. It was fully factored in 1995 (Brent).

$$F_{11} = 2^{2^{11}} + 1 = 2^{2048} + 1$$
$$= 3231700607131100730071148...19355585361105959623065\!7$$
$$= 319489 \cdot 974849 \cdot 167988556341760475137 \cdot 35608419064458\!3$$
$$3920513$$
$$\cdot 173462447179147555430258...49138244172330659883417\!7.$$

It has 617 digits. It is a product of five prime numbers with 6, 6, 21, 22 and 564 decimal digits, correspondingly. It was fully factored in 1988 (Brent and Morain).

In general, 310 Fermat numbers known to be composite. The largest known composite Fermat number (Propper, Reynolds, Penné and Fougeron, 2020) is $F_{18233954}$; it has the prime factor $p = 7 \cdot 2^{18233956} + 1$.

4.1.4. The Fermat numbers grow very quickly: the 9-th number F_9 has 155 digits and hence is more than *googol* 100^{100}; the 334-th number F_{334} is more than the *gugolplex* $10^{\text{googol}} = 10^{10^{100}}$.

It is of interest to observe the enormous bigness of the numbers F_n even with relatively small index n. For example, in his *Mathematical Recreations and Essays*, W.W.R. Ball remarks that if the number F_{73} "... were printed in full with the type and number of pages used in this book, many more volumes would be required than are contained in all the public libraries of the world."

To put it differently and much more strongly we may say that if it were printed in full with the type and format of the Encyclopedia Britannica, eleventh edition, it would require more volumes than would be contained in 10000000000000 full sets of twenty-nine volumes each.

If printed on ordinary 400-page volumes it would make a library of more than two million volumes for each man, woman and child in the world.

4.1.5. As the consequence, many of prime factors of Fermat numbers are huge, too. For example, we have shown already, that the largest Fermat number known to be composite is $F_{18233954}$. Its prime factor, $7 \cdot 2^{18233956} + 1$, is a *megaprime*, i.e., has more than one million decimal digits.

Last years, the distributed computing project *Fermat Search* is searching for new factors of Fermat numbers.

The set of Fermat factors starts with the numbers 3, 5, 17, 257, 641, 65537, 114689, 274177, 319489, 974849, ... (sequence A023394 in OEIS; see also sequence A050922 in OEIS).

The following factors of composite Fermat numbers (represented in the table below) were known before 1950 (since the 1950's, digital computers have helped find more factors).

Small prime factors of composite Fermat numbers

Year	Finder	Fermat number	Factor
1732	Euler	F_5	$5 \cdot 2^7 + 1$
1732	Euler	F_5	$52347 \cdot 2^7 + 1$
1855	Clausen	F_6	$1071 \cdot 2^8 + 1$
1855	Clausen	F_6	$262814145745 \cdot 2^8 + 1$
1877	Pervushin	F_{12}	$7 \cdot 2^{14} + 1$
1878	Pervushin	F_{23}	$5 \cdot 2^{25} + 1$
1886	Seelhoff	F_{36}	$5 \cdot 2^{39} + 1$
1899	Cunningham	F_{11}	$39 \cdot 2^{13} + 1$
1899	Cunningham	F_{11}	$119 \cdot 2^{13} + 1$
1903	Western	F_9	$37 \cdot 2^{16} + 1$
1903	Western	F_{12}	$397 \cdot 2^{16} + 1$
1903	Western	F_{12}	$973 \cdot 2^{16} + 1$
1903	Western	F_{18}	$13 \cdot 2^{20} + 1$
1903	Cullen	F_{38}	$3 \cdot 2^{41} + 1$
1906	Morehead	F_{73}	$5 \cdot 2^{75} + 1$
1925	Kraitchik	F_{15}	$579 \cdot 2^{21} + 1$

As of 2020, 354 prime factors of Fermat numbers are known (and 310 Fermat numbers are known to be composite). Several new Fermat factors are found each year.

Some additional information see, for example, in [BaCo87], [BrMo75], [DeKo13], [Prim20], [Wiki20].

4.2   Elementary properties of Fermat numbers

In this section, we give the large list of simplest properties of Fermat numbers. All of them can be proven using elementary properties of divisibility of integers.

Elementary conditions of primality of Fermat numbers

4.2.1. First of all, consider elementary properties of Fermat numbers, connected with their prime divisors.

It is easy to see, that a number of the form $2^k + 1$ can be a prime only if k is a power of 2:

- *for a positive integer k, if $2^k + 1 \in P$, then $k = 2^n$, $n \in \mathbb{Z}$, $n \geq 0$.*

□ In fact, consider $k \in \mathbb{N}$. Let $k = 2^n \cdot d$, where $n \in \mathbb{Z}$, $n \geq 0$, and d is an odd positive integer. Then

$$2^k + 1 = 2^{2^n \cdot d} + 1 = (2^{2^n})^d + 1^d = (2^{2^n} + 1) \cdot K, \ K \in \mathbb{N}.$$

So, the number $2^k + 1$ has a positive integer divisor $2^{2^n} + 1$. It means, that $2^k + 1$ can be prime only if it coincides with $2^n + 1$, i.e., in the case $d = 1$, and $k = 2^n$. □

By the similar way, one can prove, that

- *if $n^n + 1$ is prime, there exists an integer k such that $n = 2^{2^k}$.*

It means, that

- *all primes of the form $n^n + 1$ are Fermat primes: in this case, $n^n + 1 = F_{2^k + k}$.*

4.2.2. The main question about a given Fermat number is his possible primarily. So, we are going to find primes which cannot to be divisors of F_n. In fact, as far back as 1730 it was observed that

- *no F_n has a proper divisor less than* 100;

a fact easily verified directly by the aid of congruences.

Obviously, the first prime 2 cannot divide any Fermat number F_n:

- *for $n \geq 1$, it holds $F_n \equiv 1(mod\ 2)$, i.e., 2 not divides F_n.*

It is easy to check, that 3 cannot divide F_n, $n \geq 1$:

- *for $n \geq 1$, it holds $F_n \equiv 2(mod\ 3)$, i.e., 3 not divides F_n.*

□ Indeed, for $n \geq 1$, one has $2^n = 2l$, $l \in \mathbb{N}$, and we get

$$2^{2^n} + 1 \equiv (-1)^{2l} + 1 \equiv 1 + 1 \equiv 2(mod\ 3).\ □$$

Similarly, the second odd prime number 5 cannot divide any Fermat number except F_1:

- *for $n \geq 2$, it holds $F_n \equiv 2(mod\ 5)$, i.e., 5 not divides F_n.*

□ In fact, for $n \geq 2$, one has $2^n = 4l$, $l \in \mathbb{N}$; as $2^4 \equiv 1(mod\ 5)$, we get

$$2^{2^n} + 1 \equiv 2^{4k} + 1 \equiv (2^4)^k + 1 \equiv 1^k + 1 \equiv 1 + 1 \equiv 2(mod\ 5).\ □$$

We can obtain a similar result for the third odd prime 7: it cannot divide any Fermat number F_n:

- *for $n \geq 0$, it holds $F_n \equiv 3(mod\ 7)$ if n is even, and $F_n \equiv 5(mod\ 7)$ if n is odd, i.e., 7 not divides F_n.*

□ For $n = 0, 1$ it is elementary: $F_0 \equiv 3(mod\ 7)$, and $F_1 \equiv 5(mod\ 7)$.

Consider now an even $n \geq 2$. In this case, $n = 2l$, $l \in \mathbb{N}$. For such n one has $2^n = 2^{2l} = 4^l$, i.e., $2^n \equiv 1(mod\ 3)$, or $2^n = 3k + 1$, $k \in \mathbb{N}$. Then

$$2^{2^n} + 1 \equiv 2^{3k+1} + 1 \equiv 2 \cdot (2^3)^k + 1 \equiv 2 \cdot 8^k + 1$$

$$\equiv 2 \cdot 1 + 1 \equiv 3(mod\ 7).$$

At last, consider an odd $n \geq 3$. In this case, $n = 2l + 1$, $l \in \mathbb{N}$. For such n one has $2^n = 2^{2l+1} = 2 \cdot 4^l$, i.e., $2^n \equiv 2 \cdot 1 \equiv 2(mod\ 3)$, or $2^n = 3k + 2$, $k \in \mathbb{N}$. Then

$$2^{2^n} + 1 \equiv 2^{3k+2} + 1 \equiv 2^2 \cdot (2^3)^k + 1 \equiv 4 \cdot 8^k + 1 \equiv 4 \cdot 1 + 1 \equiv 5(mod\ 7).\ □$$

It is checked, that

• *no Fermat number F_n has a factor less than 10^6 other than factors known at present.*

Moreover, it is possible to prove (Křížek, Luca and Somer, 2002), that

• *the series of reciprocals of all prime divisors of Fermat numbers is convergent.*

For some additional information see, for example, [Carm19], [Deza17], [DeKo13], [Dick05], [KLS01], [Madd05], [Tsan10], [Sier64].

Last decimal digits of Fermat numbers

4.2.3. Consider now possible reminders of Fermat numbers modulo 10 (modulo 100).

It will give us the information about last digit (last two digits) of Fermat number F_n.

In fact, with the (one-digit) exceptions $F_0 = 3$ and $F_1 = 5$, the last digit of a Fermat number is 7:

• *for $n \geq 2$, it holds $F_n \equiv 7(mod\ 10)$, i.e., the last digit of F_n is 7.*

□ Really, for $n \geq 2$ it holds $F_n \equiv 2(mod\ 5)$, i.e., $F_n = 5t + 2$, $t \in \mathbb{N}$.

Obviously, $F_n \equiv 1(mod\ 2)$.

So, we get the congruence $5t + 2 \equiv 1(mod\ 2)$. It is equivalent to the congruence $t \equiv 1(mod\ 2)$, which gives $t = 2t_1 + 1$, $t_1 \in \mathbb{Z}$. So,

$$F_n = 5t + 2 = 5(2t_1 + 1) + 2 = 10t + 7, \quad \text{or } F_n \equiv 7(mod\ 10). \ \square$$

One can easy obtain the similar result modulo 100: with the (one-digit) exceptions $F_0 = 3$ and $F_1 = 5$, the last two digit of a Fermat number are 17, 37, 57 or 97:

• *for $n \geq 2$, $n = 4l$, it holds $F_n \equiv 37(mod100)$, i.e., the last two digits of F_n are 37;*

• *for $n \geq 2$, $n = 4l + 1$, it holds $F_n \equiv 97(mod100)$, i.e., the last two digits of F_n are 97;*

- *for $n \geq 2$, $n = 4l + 2$, it holds $F_n \equiv 17 \pmod{100}$, i.e., the last two digits of F_n are 17;*

- *for $n \geq 2$, $n = 4l + 3$, it holds $F_n \equiv 57 \pmod{100}$, i.e., the last two digits of F_n are 57.*

□ The reader can easily check these facts, solving the congruence $x \equiv 2^{2^n} + 1 \pmod{2^2 \cdot 5^2}$ (see Chapter 1). □

For some additional information see, for example, [Carm19], [Deza17], [DeKo13], [Dick05], [KLS01], [Madd05], [Tsan10], [Sier64].

Recurrent relations for Fermat numbers

4.2.4. Consider now some important recurrent relations, connected with Fermat numbers.

First of all, it is easy to obtain the Fermat number with index n, using the Fermat numbers with all previous indexes $0, 1, 2, ..., n - 1$:

- *for $n \geq 1$, it holds $F_n = F_0 \cdot F_1 \cdot ... \cdot F_{n-1} + 2$.*

□ In fact, for $n \geq 1$, one has

$$F_n - 2 = 2^{2^n} - 1 = (2^{2^{n-1}} + 1)(2^{2^{n-1}} - 1)$$
$$= (2^{2^{n-1}} + 1)(2^{2^{n-2}} + 1)(2^{2^{n-2}} - 1) = ...$$
$$= (2^{2^{n-1}} + 1)(2^{2^{n-2}} + 1)...(2^{2^1} + 1)(2^{2^0} + 1)(2^{2^0} - 1)$$
$$= F_{n-1} \cdot F_{n-2} \cdot ... \cdot F_0. \quad \square$$

There are several other recurrences between Fermat numbers. For example,

- *for $n \geq 1$, it holds $F_n = (F_{n-1} - 1)^2 + 1$.*

□ We can prove it using elementary reasons:

$$F_n - 1 = 2^{2^n} = (2^{2^{n-1}})^2 = (F_{n-1} - 1)^2. \quad \square$$

Moreover, the Fermat numbers satisfy the following recurrence relation:

- *for $n \geq 2$, it holds $F_n = F_{n-1} + 2^{2^{n-1}} F_0 \cdot F_1 \cdot ... \cdot F_{n-2}$.*

☐ As before, it is easy to see, that

$$F_n - F_{n-1} = 2^{2^n} - 2^{2^{n-1}} = 2^{2^{n-1}}(2^{2^{n-1}} - 1)$$

$$= 2^{2^{n-1}}(2^{2^{n-2}} + 1)(2^{2^{n-2}} - 1) = \ldots$$

$$= 2^{2^{n-1}}(2^{2^{n-2}} + 1)\ldots(2^{2^1} + 1)(2^{2^0} + 1)(2^{2^0} - 1)$$

$$= 2^{2^{n-1}} F_{n-2} \cdot \ldots \cdot F_0.$$

Note, that the last formula can be rewrite as

$$F_n = F_0 \cdot F_1 \cdot \ldots \cdot F_{n-2}(F_{n-1} - 1) + F_{n-1}. \ \square$$

At last, we can prove the following fact:

• *for $n \geq 2$, it holds $F_n = F_{n-1}^2 - 2(F_{n-2} - 1)^2$.*

☐ Really, elementary (but accurate) work with the powers of 2 gives us, that

$$F_{n-1}^2 - 2(F_{n-2} - 1)^2 = (2^{2^{n-1}} + 1)^2 - 2 \cdot (2^{2^{n-2}})^2$$

$$= (2^{2^{n-1}})^2 + 1 + 2 \cdot 2^{2^{n-1}} - 2 \cdot 2^{2 \cdot 2^{n-2}}$$

$$= 2^{2^n} + 1 + 2^{2^{n-1}+1} - 2^{2^{n-1}+1}$$

$$= 2^{2^n} + 1 = F_n. \ \square$$

Note, that the pair $(F_{n-1}, F_{n-2}-1)$ gives a representation of the n-th Fermat number F_n in the form $x^2 - 2y^2$, where x and y are positive integers. One can check (see Chapter 8), that

• *for $n \geq 2$, every Fermat number has infinitely many representations in the form $x^2 - 2y^2$, where x and y are positive integers.*

For some additional information see, for example, [Carm19], [Deza17], [DeKo13], [DeMo10], [Dick05], [KLS01], [Madd05], [Tsan10], [Sier64].

Divisibility properties of Fermat numbers

4.2.5. The results, represented before, allow us to obtain a very important property of divisibility of Fermat primes.

It is well-known, that *any two different Fermat numbers form a pair of coprime positive integers:*

• *for $n \neq m$, it holds $\gcd(F_n, F_m) = 1$, i.e., the numbers F_n and F_m have no common prime divisors.*

□ In order to prove this formula, let assume that $gcd(F_n, F_m) = d$, $d > 1$. Then there exists a prime number p, such that $p|d$. In this case, it holds that $p|F_n$, and $p|F_m$.

Without loss of generality, let $n > m$. As $p|F_m$, then $p|F_0 \cdot F_1 \cdot ... \cdot F_m \cdot ... \cdot F_{n-1}$. From the recurrence $F_n = F_0 \cdot F_1 \cdot ... \cdot F_{n-1} + 2$, using relations $p|F_n$, $p|F_0 \cdot F_1 \cdot ... \cdot F_m \cdot ... \cdot F_{n-1}$, we get $p|2$, i.e., $p = 2$.

As any Fermat number is an odd positive integer, it is not divisible by 2; it gives a contradiction. So, it is proven, that $gcd(F_n, F_m) = 1$ for $n \neq m$. □

This fact allow as to give one more proof of infiniteness of the set of prime numbers:

• *there are infinity many prime numbers.*

□ Indeed, consider the set $\{F_0, F_1, F_2, ..., F_n, ...\}$ of all Fermat numbers. It is obviously infinite. For each F_n, denote by q_n the smallest prime divisor of F_n. On this way, we get a set $\{q_0, q_1, q_2, ..., q_n, ...\}$ of prime numbers: $q_0 = 3$, $q_1 = 5$, $q_2 = 17$, etc. As for $n \neq m$ it holds $q_n \neq q_m$, all elements of the set $\{q_0, q_1, q_2, ..., q_n, ...\}$ are different odd prime numbers, and we obtain an infinite subset of the set P of primes. It proves, that the set P itself is infinite. □

For some additional information see, for example, [Carm19], [Deza17], [DeKo13], [Dick05], [KLS01], [Madd05], [Tsan10], [Sier64].

Composite relatives of Fermat numbers

4.2.6. As we could see before, the question about primarily of Fermat numbers, i.e., the numbers of the form $2^{2^n} + 1$, is really difficult.

On the other hand, we can construct several classes of numbers, which are relative to the set of Fermat numbers, but all are composite.

For example, what can we obtain, if change the unity in the formula $2^{2^n} + 1$ to the other odd positive integer (if instead of 1 we put an even number, the situation became trivial, all such numbers will be even)?

For the smallest such shift the situation is not obvious; we immediately get prime numbers of the form $2^{2^n}+3$ (i.e., of the form F_n+2): $2^{2^0}+3 = 5$, $2^{2^1}+3 = 7$, and $2^{2^2}+3 = 19$.

However, one can prove, that

• *among the numbers $2^{2^n}+3$ there are infinitely many composite ones.*

□ For example, the number $F_3 + 2 = 259 = 7 \cdot 37$.

Moreover, for any Fermat number with odd index it holds $7|(F_n + 2)$: it was proven, that for odd n we have the congruence $F_n \equiv 5(mod\ 7)$; hence, $F_n + 2 \equiv 0(mod\ 7)$.

More exactly, let $n = 2k + 1$, $k \in \mathbb{N}$. As $2k$ is even, we get

$$2^n \equiv 2^{2k+1} \equiv 2 \cdot 2^{2k} \equiv 2(mod\ 3),$$

i.e., $2^n = 3l + 2$, and

$$2^{2^n} + 3 \equiv 2^{3l+2} + 3 \equiv 2^2 \cdot (2^3)^2 + 3 \equiv 4 \cdot 1 + 3 \equiv 0(mod\ 7).\ \square$$

On the other hand, the next possible shift gives us only composite numbers:

• *for any $n \geq 1$, the number $F_n + 4 = 2^{2^n} + 5$ is composite.*

□ It is easy to check, that all numbers $9, 21, 261, 65541, \dots$ of the form $2^{2^n}+5$, $n \geq 1$, are divisible by 3. In fact, for such n the number 2^n is even: $2^n = 2l$, $l \in \mathbb{N}$. So, it holds

$$F_n + 4 \equiv 2^{2^n} + 5 \equiv (-1)^{2l} + 5 \equiv 1 + 5 \equiv 6 \equiv 0(mod\ 3).\ \square$$

Some shifts of such kind we can use also for squares of Fermat numbers. For example, we can prove the following fact:

• *for any $n \geq 1$, the number $(F_n)^2+2 = (2^{2^n}+1)^2+2$ is composite.*

□ In fact, the sequence of the numbers $(2^{2^n}+1)^2+2$, $n \geq 1$, starts from the numbers $27, 291, 66051, \dots$. It is easy to see, that all these numbers are divisible by 3. Really, we proved, that $F_n \equiv 2(mod\ 3)$ for any $n \geq 1$. So, for such n it holds

$$(F_n)^2 + 2 \equiv 2^2 + 2 \equiv 6 \equiv 0(mod\ 3).\ \square$$

The same arguments allow us to prove the similar result for the numbers of the form $(F_n)^2 + 8$:

• *for any $n \geq 1$, the number $(F_n)^2+8 = (2^{2^n}+1)^2+8$ is composite.*

Using similar property of Fermat numbers modulo 7, we obtain, that all numbers of the form $(F_n)^3 + 8$ are divisible by 7:

• *for any* $n \geq 1$, *the number* $(F_n)^3 + 8 = (2^{2^n} + 1)^3 + 8$ *is composite.*

□ In fact, the sequence of the numbers $(2^{2^n} + 1)^3 + 8$, $n \geq 1$, starts from the numbers $133 = 7 \cdot 19$, $4921 = 7 \cdot 703$, and $16974601 = 7 \cdot 2424943$.

In general, one has proven, that $F_n \equiv 3 (mod\ 7)$ or $F_n \equiv 5 (mod\ 7)$ for any $n \geq 1$. As $3^3 \equiv 27 \equiv -1 (mod\ 7)$, and $5^3 \equiv 125 \equiv -1 (mod\ 7)$, we have, that $(F_n)^3 \equiv -1 (mod\ 7)$ for any $n \geq 1$. So, it holds

$$(F_n)^3 + 8 \equiv (-1) + 8 \equiv 7 \equiv 0 (mod\ 7).\ \square$$

For some additional information see, for example, [Carm19], [Deza17], [DeKo13], [Dick05], [KLS01], [Madd05], [Tsan10], [Sier64].

Additive properties of Fermat numbers

4.2.7. Consider some additive properties of Fermat numbers.

For example, it is easy to show, that any Fermat number except $F_1 = 5$ cannot be represented as a sum of two primes:

• *for any* $n \neq 1$ *and any primes* p *and* q, *it holds* $F_n \neq p + q$.

□ As F_n is odd for any $n \geq 0$, it cannot be represented as a sum of two odd primes p, q. So, in each representation of $F_n = p + q$ as one of the primes, say p, is equal to 2. It is impossible for $F_0 = 3$; it is possible for F_1: in fact, $5 = 2 + 3$. In general, if $F_n = 2 + q$ for $n \geq 2$, it holds $q = F_n - 2$. As $F_n \equiv 2 (mod\ 3)$, then $F_n - 2 \equiv 0 (mod\ 3)$, i.e., $q \equiv 0 (mod\ 3)$. It is possible only for $q = 3$. In this case we have that $p + q = 5$, a contradiction to the case $n \geq 2$. □

It can be proven also a similar result about squares of prime numbers:

• *for any* $n \geq 0$ *and any primes* p *and* q, *it holds* $F_n \neq p^2 + q^2$.

□ Without loss of generality, let $n \geq 2$: it is easy to see, that $p^2 + q^2 \geq 8 > F_1 = 5$.

Note, that for $F_n = p^2 + q^2$ one of primes p, q (say, q) should be even: for two odd numbers p, q (as well as for two even numbers), we get even sum $p^2 + q^2$.

In this case, it holds $F_n = p^2 + 2^2$, or $F_n - 4 = p^2$.

For $n \geq 2$, we have the congruence $F_n \equiv 2(mod\ 5)$; so, $F_n - 4 \equiv 2 - 4 \equiv 3(mod\ 5)$.

On the other hand, for any prime number p it holds $p \equiv 0, \pm 1, \pm 2(mod\ 5)$, and, hence, $p^2 \equiv 0, \pm 1(mod\ 5)$; a contradiction. □

Moreover, it is easy to see, that

• *for any $n \geq 0$ and any primes p and q, it holds $F_n \neq p^2 + q^2 + 1$.*

□ Without loss of generality, let $n \geq 2$: it is easy to see, that $p^2 + q^2 + 1 \geq 9 > F_1 = 5$.

In order to prove the last formula, note, that equality $F_n = p^2 + q^2 + 1$ is possible only for two odd primes p and q: if exactly one of these primes is even, we will get an even number $p^2 + q^2 + 1$, a contradiction; if both such numbers are even, then $p^2 + q^2 + 1 = 9$; it is not a Fermat number.

Moreover, if both such numbers are equal to 3, then $p^2 + q^2 + 1 = 19$; it is not a Fermat number, too.

If only one of the primes (say, q) is equal to 3, we get the equality $2^{2^n} + 1 = p^2 + 3^2 + 1$, or the equality $(2^{2^{n-1}})^2 - 3^2 = p^2$. In this case it holds $(2^{2^{n-1}} - 3)(2^{2^{n-1}} + 3) = p^2$; it is possible only for $2^{2^{n-1}} - 3 = 1$, and $2^{2^{n-1}} + 3 = p^2$. From the first condition we get $n = 2$; but in this case $2^{2^{n-1}} + 3 = 7 \neq p^2$.

So, if $F_n = p^2 + q^2 + 1$, then p, q are odd primes greater than 3. It is well-known (see Chapter 1), that for any odd prime number $p > 3$ it holds $p \equiv \pm 1(mod\ 6)$; so, $p^2 \equiv 1(mod\ 6)$. Therefore,

$$p^2 + q^2 + 1 \equiv 1 + 1 + 1 \equiv 3(mod\ 6).$$

On the other hand, for any $n \geq 1$ it holds $F_n \equiv -1(mod\ 6)$. In fact, for such n the number 2^n is even, and $2^{2^n} \equiv 4(mod\ 6)$; hence,

$$2^{2^n} + 1 \equiv 4 + 1 \equiv 5 \equiv -1(mod\ 6).$$

This gives a contradiction. □

It can be proven also, that any Fermat number, except F_2, cannot be decomposed into a sum of squares of three primes numbers:

• *for any $n \neq 2$ and any prime numbers p, q, r, it holds $F_n \neq p^2 + q^2 + r^2$.*

□ In fact, $F_0 = 3$ and $F_1 = 5$ cannot be represent as sums of squares of three primes numbers; $F_2 = 17$ is represented as $3^2 + 2^2 + 2^2$.

For $n \geq 3$, there are two cases: all three primes are odd, or exactly two of them (say, q and r), are even. In the last case, we have, for $p = 3$, the sum $3^2 + 2^2 + 2^2$; if $p > 3$, we get the congruence

$$p^2 + q^2 + r^2 \equiv 1 + 4 + 4 \equiv 9 \equiv 3 (mod\ 6).$$

It gives a contradiction with the property $F_n \equiv -1 (mod\ 6)$.

Now, let all prime numbers be odd. If all three of them are equal to 3, we get the number $3^2 + 3^2 + 3^2 = 27$; it is not a Fermat number. If exactly two of them equal to 3, we get the congruence

$$p^2 + 3^2 + 3^2 \equiv 1 + 9 + 9 \equiv 19 \equiv 1 (mod\ 6).$$

It gives again a contradiction with the property $F_n \equiv -1 (mod\ 6)$.

If exactly one of them is equal to 3, we get

$$2^{2^n} + 1 = p^2 + q^2 + 3^2, \quad \text{or} \quad p^2 + q^2 = 2^{2^n} - 8.$$

For $n \geq 1$, $2^{2^n} - 8 \equiv 0 (mod\ 4)$; but for odd prime p it holds $p^2 \equiv 1 (mod\ 4)$ (see Chapter 1), so, $p^2 + q^2 \equiv 2 (mod\ 4)$; a contradiction.

At last, if all three primes are greater than 3, we have the congruence

$$p^2 + q^2 + r^2 \equiv 1 + 1 + 1 \equiv 3 (mod\ 6);$$

It gives again a contradiction with the property $F_n \equiv -1 (mod\ 6)$. □

One more interesting additive property is connected with Fermat primes. In fact, *no Fermat prime can be expressed as the difference of two p-th powers, where p is an odd prime*:

- if $F_n \in P$, then $F_n \neq a^p - b^p$, where $a, b \in \mathbb{N}$, $p \in P$, $p > 2$.

□ Assume for contradiction that there is such a Fermat prime. Then

$$F_n = a^p - b^p = (a - b)(a^{p-1} + a^{p-2}b + ... + ab^{p-1} + b^{p-1}),$$

where $a > b$, and p is an odd prime. Since F_n is a prime, it must be the case $a - b = 1$.

Moreover, by the *Fermat's little theorem*, $a^p \equiv a(mod\ p)$, and $b^p \equiv b(mod\ p)$. Thus,

$$F_n \equiv a^p - b^p \equiv a - b \equiv 1(mod\ p).$$

This implies $p|(F_n - 1)$, i.e., $p|2^{2^n}$, a contradiction. $\square$

For some additional information see, for example, [Carm19], [Deza17], [DeKo13], [Dick05], [KLS01], [Madd05], [Tsan10], [Sier64].

Other elementary properties of Fermat numbers

4.2.8. Consider now some elementary properties of Fermat numbers, which allows us to find connections between F_n and other classes of special numbers.

For example, it is easy to see, that F_n *cannot be represented as a perfect square*:

- *for any $n \geq 0$, it holds $F_n \neq k^2$, where $k \in \mathbb{N}$.*

$\square$ This fact is almost obvious. It was proven before, that $F_n \equiv 7(mod\ 10)$.

But only a number that is congruent to $0, 1, 4, 5, 6$, or 9 modulo 10 can be a perfect square: so, $F_n \neq k^2$.

The other proof uses elementary divisibility properties. If $2^m + 1 = k^2$, then k is odd, and $2^m = (k-1)(k+1)$. Therefore, $2^t = k - 1$, $2^{m-t} = k + 1$, where $t \in \mathbb{N}$, $t < m - t$. In this case one has

$$2^{m-t} - 2^t = 2^t(2^{m-2t} - 1) = (k + 1) - (k - 1) = 2,$$

i.e., $t = 1$, $k = 3$, and $m = 3$.

Therefore, the only perfect square of the form $2^m + 1$ is 9, which is not a Fermat number. $\square$

It means, that

- *any Fermat number cannot to be a square number.*

Similarly, it is easy to see, that F_n *cannot to be a perfect cube*:

- *for any $n \geq 0$, it holds $F_n \neq k^3$, where $k \in \mathbb{N}$.*

$\square$ Thus, any integer k is congruent to $0, \pm 1, \pm 2$, or ± 3 modulo 7, so, $k^3 \equiv 0, 1, 6(mod\ 7)$. But for any $n \geq 1$, it holds $F_n \equiv 3, 5(mod\ 7)$.

For $n = 0$, $F_0 = 3$ also is not a cube of an integer. Therefore, $F_n \neq k^3$. $\square$

It means, that

• *any Fermat number cannot be a cubic number.*

If we consider modulo 3, we can prove for the same way, that $F_n \neq k^4$. (Of course, as $k^4 = (k^2)^2$, we can use in this case the previous result $F_n \neq k^2$.)

It means, that

• *any Fermat number cannot to be a biquadratic number.*

In general, it is proven, that F_n cannot to be any non-trivial power:

• *for any $n \geq 0$, it holds $F_n \neq k^s$, where $k, s \in \mathbb{N}$, and $s \geq 2$.*

$\square$ In fact, as $2^m + 1, m \neq 3$, is not a perfect square, then in the equality $2^m + 1 = k^s$ the number s should be odd. Then

$$2^m = k^s - 1 = (k - 1)(k^{s-1} + k^{s-2} + \cdots + k + 1).$$

But it is impossible, as the number $(k^{s-1} + k^{s-2} + \cdots + k + 1)$ is odd. $\square$

Similar arguments allow to get the following result:

• *for any $n \geq 0$, it holds $F_n \neq \frac{k(k+1)}{2}$, where $k \in \mathbb{N}$.*

$\square$ In fact, for $n \geq 2$ it holds

$$F_n = 2^{2^n} + 1 \equiv 4^{2^{n-1}} + 1 \equiv (-1)^{2^{n-1}} + 1 \equiv 1 + 1 \equiv 2 (mod\ 5),$$

while the number $\frac{k(k+1)}{2} \not\equiv 2 (mod\ 5)$: for $k \equiv 0, 4 (mod\ 5)$ we get $\frac{k(k+1)}{2} \equiv 0 (mod\ 5)$; for $k \equiv 1, 3 (mod\ 5)$ we get $\frac{k(k+1)}{2} \equiv 1 (mod\ 5)$; for $k \equiv 2 (mod\ 5)$ we get $\frac{k(k+1)}{2} \equiv 3 (mod\ 5)$. For $n = 1$ the number $F_1 = 5$ is not a triangular number too. $\square$

The numbers of the form $\frac{k(k+1)}{2}$, $k \in \mathbb{N}$, are called *triangular*. So, we got the connection between Fermat numbers and triangular numbers:

• *any F_n cannot be a triangular number.*

It is proven (Luca, 2000), that

• *any Fermat number cannot be a perfect number or part of a pair of amicable numbers.*

4.2.9. There exist simple connections between Fermat numbers and Mersenne numbers. Obviously, it holds

$$F_n - 2 = M_{2^n}.$$

From this formula one can obtain the following elementary fact:

• *the pairs* $(3,5)$ *and* $(5,7)$ *are the only pair of twin primes, contain one Fermat number and one Mersenne number.*

□ In fact, if $(p, p+2)$ is a pair of twin primes with $p = M_k$, $p + 2 = F_n$, then $M_k + 2 = F_n$, i.e., $2^k = 2^{2^n}$, and $k = 2^n$. The index of prime number M_k should be prime; so, the only possible case is $k = 2$, and $n = 1$. So, the pair $(3, 5)$ is the only pair (F_n, M_k) of twin primes.

On the other hand, if $(p, p+2)$ is a pair of twin primes with $p = F_n$, $p + 2 = M_k$, then $M_k - 2 = F_n$. It holds for $k = 2$, and $n = 1$. For odd $k \geq 3$, we get $M_k \equiv 1, 2 \pmod{5}$, and $M_k - 2 \equiv 4, 0 \pmod{5}$. But for $n \geq 2$, $F_n \equiv 2 \pmod{5}$. So, the pair $(5, 7)$ is the only pair (M_k, F_n) of twin primes. □

Using the Fermat's little theorem, we can prove, that *any prime Fermat number* F_n *divides infinitely many Mersenne numbers*:

• *if* p *is a Fermat prime, then* $p | M_{k(p-1)}$, *where* $k = 1, 2, 3, \ldots$.

□ In fact, for a given odd prime p, we have $2^{p-1} \equiv 1 \pmod{p}$; therefore, $2^{m(p-1)} \equiv 1 \pmod{p}$ for any positive integer m, i.e., $p | (2^{m(p-1)} - 1)$. □

In particular, *any Fermat prime* p *divides the Mersenne number* M_{p-1}. On the other hand,

• *if* p *is a Fermat prime, then* $p^2 \nmid M_{p-1}$.

□ Let p be a Fermat prime, i.e., $p = 2^{2^n} + 1$. So, it holds $p - 1 = 2^{2^n}$. If the congruence $2^{p-1} \equiv 1 \pmod{p^2}$ holds, then $2^{2(p-1)} \equiv 1 \pmod{p^2}$, and $p^2 | (2^{2(p-1)} - 1)$. Therefore, $p | \frac{2^{2(p-1)} - 1}{p}$. It means, that

$$\frac{2^{2(p-1)} - 1}{p} \equiv 0 \pmod{p}, \quad \text{or} \quad \frac{2^{2 \cdot 2^n} - 1}{2^{2^n} + 1} \equiv 0 \pmod{p}.$$

It is ease to see, that

$$\frac{2^{2 \cdot 2^n} - 1}{2^{2^n} + 1} = \frac{(2^{2^n})^2 - 1}{2^{2^n} + 1} = \frac{(2^{2^n} - 1)(2^{2^n} + 1)}{2^{2^n} + 1} = 2^{2^n} - 1 = p - 2.$$

Hence,

$$\frac{2^{2 \cdot 2^n} - 1}{2^{2^n} + 1} \equiv p - 2 \equiv -2 \pmod{p}, \quad \text{and} \quad -2 \equiv 0 \pmod{p}.$$

It means, that $p | 2$, i.e., $p = 2$; a contradiction. □

4.2.10. Remind, that a *Wieferich prime* is defined as a prime number p such that p^2 divides $2^{p-1} - 1$. In other words, we have proven that

- *Fermat prime cannot be a Wieferich prime.*

On the other hand, we have proven, that for prime F_n it holds $F_n | (2^{F_n - 1} - 1)$, or, which is the same, $F_n | (2^{F_n} - 2)$. In fact, this property is true for all Fermat numbers:

- *for $n \geq 0$, it holds $F_n | 2^{F_n} - 2$.*

□ It is easy to see, that $F_0 | 2^{F_0} - 2$. For $n \geq 1$ we get

$$2^{F_n} - 2 = 2(2^{F_n - 1} = 1) = 2(2^{2^{2^n}} - 1) = 2(F_{2^n} - 2).$$

Since for any $k \geq 1$ it holds $F_k - 2 = F_0 \cdot F_1 \cdot \ldots F_{k-1}$, we have, that

$$2^{F_n} - 2 = 2F_0 \cdot F_1 \cdot \ldots F_{2^n - 1}, \quad \text{and} \quad F_k | (2^{F_n} - 2) \quad \text{for any} \quad k \leq 2^n - 1.$$

Since for any positive integer n it holds $n \leq 2^n - 1$, we have proven, that in this case $F_n | 2^{F_n} - 2$. □

This property means, that

- *composite Fermat numbers are Fermat pseudoprimes with base 2.*

As by definition, a *Poulet number* is a Fermat pseudoprime with base 2, we have proven, that

- *any composite Fermat number is a Poulet number.*

Moreover, it can be proven that *if for a natural number k number $m = 2^k + 1$ satisfies the relation $m | 2^m - 2$, that m is a Fermat number:*

- *if $2^k + 1 | (2^{2^k + 1} - 2)$, $k \in \mathbb{N}$, then $k = 2^n$ for some $n \geq 0$.*

In other words, *if $2^k + 1$ is a pseudoprime to the base 2, then k is a power of* 2.

□ In fact, let $2^{2^k} \equiv 1 \pmod{2^k + 1}$. Obviously,

$$2^k \equiv -1 \pmod{2^k + 1}, \quad \text{so,} \quad 2^{2k} \equiv 1 \pmod{2^k + 1}.$$

Let $e = ord_{2^k + 1} 2$. Then $e \geq k + 1$; otherwise we have $2^e \leq 2^k < 2^k + 1$. Moreover, $e | 2k$, so it follows that $e = 2k$. But $e | 2^k$, which is only possible if k is a power of 2. □

However, in general it is not true for other bases. For examples, $F_5 = 4294967297$ is not a pseudoprime to the bases 5 or 6, since we have

$$5^{4294967296} \equiv 2179108346 (mod\ 4294967297),\quad \text{and}$$

$$6^{4294967296} \equiv 3029026160 (mod\ 4294967297).$$

Hence, it is possible to use the Fermat's little theorem to test the primality of a Fermat number as long as we do not choose 2 to be our base.

In 1904, M. Cipolla showed that

• *the product $F_{n_1} \cdot F_{n_2} \cdot \ldots \cdot F_{n_k}$ of at least two distinct Fermat numbers is a Fermat pseudoprime to base 2 if and only if $2^k > n_1$.*

For some additional information see, for example, [Carm19], [Deza17], [Deza18], [DeKo13], [Dick05], [KLS01], [Luca00], [Luca01], [Madd05], [Tsan10], [Sier64], [Prim20], [Wiki20].

Exercises

1. Prove, that $F_n + 10$ is composite for any $n \geq 1$.

2. Prove, that $F_n \equiv 2 (mod\ 15)$, $F_n \equiv 1 (mod\ 16)$, $F_n \equiv 17, 41 (mod\ 72)$ for any $n \geq 2$.

3. Check, that $F_8 < 10^{100}$, but $F_9 > 10^{100}$.

4. Prove, that F_{73} has more than $24 \cdot 10^{20}$ digits.

5. Prove, that F_{334} has more than 210^{100} digits.

6. Prove, that F_{1945} has more than $24 \cdot 10^{582}$ digits.

7. Prove, that last two digits of any Fermat number, except F_0 and F_1, are 17, 37, 57 or 97.

8. Check, that last three digits of the numbers F_{73} and F_{1945} are 897 and 297, correspondingly.

4.3   Fermat primes: Prime divisors of Fermat numbers

4.3.1. The principal problem studied in connection with Fermat numbers is that of their primality (factorization). The most elementary and obvious method for finding the factors of a given number n is to test it for divisibility by primes less than $\sqrt{n}$; but if the given number is a large prime or has only large prime factors it is obviously that the corresponding work is almost prohibitive.

One of the earliest used improvements on this method consists in determining certain properties of the prime factors of the numbers in consideration and then testing with only those primes which have this property.

Thus, L. Euler showed that all factors of F_n, $n \geq 1$, are of the form $2^{n+1} \cdot k + 1$. Later É. Lucas proved that this k should be even; in fact, he proved, that all factors of F_n, $n \geq 2$, are of the form $2^{n+2} \cdot k + 1$.

At a time when it was still unknown whether F_6 is prime or composite, É. Lucas remarked that if it is prime the demonstration of this fact' by aid of Euler's theorem would require a calculator the enormous period of three thousand years.

Properties of prime divisors of Fermat numbers

4.3.2. So, we are going to prove, that all prime divisors of Fermat numbers are of the special form:

- *any prime divisor p of $F_n, n \geq 1$, has the form $p = 2^{n+1}k + 1$.*
- □ In fact, if $p|F_n$, $n \geq 1$, then

$$2^{2^n} \equiv -1 (mod\ p), \quad \text{and} \quad (2^{2^n})^2 \equiv 1 (mod\ p),$$

i.e., $2^{2^{n+1}} \equiv 1 (mod\ p)$. Therefore, $ord_p 2 = 2^{n+1}$. As $ord_p a | (p-1)$, we obtain, that

$$2^{n+1} | (p-1), \quad \text{and} \quad p = 2^{n+1}k + 1. \ \square$$

For $n \geq 2$ this result cam be improved:

- *any prime divisor p of $F_n, n \geq 2$, has the form $p = 2^{n+2}k + 1$.*
- □ In fact, for $n \geq 2$, we have that

$$p = 2^{n+1}k + 1 \equiv 1 (mod\ 8),$$

hence, the Legendre symbol $(\frac{2}{p}) = 1$, i.e., by the Euler's criterion, it holds

$$2^{\frac{p-1}{2}} \equiv 1 (mod\ p), \quad \text{and} \quad 2^{n+1} = ord_p\, 2 \Big| \frac{p-1}{2}, \quad \text{i.e.,} \quad p = 2^{n+2}k+1. \quad \square$$

P. Fermat was probably aware of the form of the factors later proved by L. Euler, so it seems curious that he failed to follow through on the straightforward calculation to find the factor. One common explanation is that P. Fermat made a computational mistake.

For some additional information see, for example, [Carm19], [Deza17], [DeKo13], [DeKo18], [Dick05], [KLS01], [Madd05], [Tsan10], [Sier64], [Step01].

History of the search of prime divisors of Fermat numbers

4.3.3. This properties of prime divisors of F_n are used in investigations, whether a given Fermat number is prime or not.

Thus, any prime divisor of F_5 must be of the form $2^7 \cdot k + 1 = 128k + 1$. We obtain the primes for $k = 2$ and $k = 5$ only: they are the numbers 257 and 641, respectively. The prime 641 divides F_5, and F_5 id the product of two different primes:

$$F_5 = 2^{2^5} + 1 = 4294967297 = 641 \cdot 6700417,$$
$$\text{where} \quad 641, 6700417 \in P.$$

This fact was discovered by L. Euler in 1732.

F. Landry, 1880, proved, that $2741|F_6$. He checked primes of the form $2^8 \cdot k + 1 = 256k + 1$. The prime number $2741 = 256k + 1$ for $k = 1071$. In fact, F_6 has 20 digits and is a product of two primes of the form $256k + 1$ with $k = 1071$ and $k = 262814145745$:

$$F_6 = 274177 \cdot 67280421310721,$$
$$\text{where} \quad 274177, 67280421310721 \in P.$$

This method was not so easy for 39-digital F_7 and 78-digital F_8. Only in 1975 it was obtained by M.A. Morrison and J. Brillhart, that F_7 has a prime divisor $2^9 \cdot 116503103764643 + 1$. More exactly, F_7 has

39 digits and is a product of two primes of the form $512k + 1$ with $k = 11141971095088142685$ and $k = 116503103764643$:

$$F_6 = 9649589127497217 \cdot 5704689200685129054721, \quad \text{where}$$

$$9649589127497217, 5704689200685129054721 \in P.$$

The decomposition of 62-digital number F_8 into two primes, one of which has 16 digits and is of the form $2^{11} \cdot k + 1$ with

$$k = 3853149761 \cdot 157 = 604944512477,$$

and the other has 62 digits and is of the form $2^{11} \cdot k + 1$ with

$$k = 10573720467811625362740343546868933329625329$$
$$\cdot\, 31618624099079 \cdot 13 \cdot 7 \cdot 5 \cdot 3$$
$$= 93461639715357977769163558199606896584051237541638188858$$
$$0280321,$$

was obtained in 1980 by R.P. Brent and J.M. Pollard.

A.E. Western (1913) found prime divisor 2424833 of the form $2^{11} \cdot k + 1$ with $k = 2^5 \cdot 37 = 1184$ for the 155-digital F_9. Two others prime divisors of it, 49-digital prime $2^{11} \cdot k + 1$ with

$$k = 43226490359557706629 \cdot 1143290228161321 \cdot 82488781 \cdot 47 \cdot 19,$$

and 99-digital prime $2^{11} \cdot k + 1$ with

$$k = 169751433022715054268975856531311265201823280378217297 2$$
$$0833840187223 \cdot 17338437577121 \cdot 40644377 \cdot 26813 \cdot 1129$$

were obtained in 1990 (Lenstra, Manasse at all).

J.L. Selfridge (1953) proved that $p = 2^{12} \cdot 11131 + 1$ divides 309-digital F_{10}. Other three prime divisors of it were obtained in 1962 (Brillhart, $q = 2^{12} \cdot 395937 + 1$), and in 1995 (Brent).

Two prime divisors, $2^{13} \cdot 39 + 1$ and $2^{13} \cdot 119 + 1$, of 617-digital F_{11} were obtained in 1899 by Cunningham; other three prime divisors were ontained in 1988 (Brent and Morain).

The prime divisor $2^{14} \cdot 7 + 1$ of F_{12} was found in 1877 by I.M. Pervishin and, independently, by É. Lucas. In 1903, A.E. Western found two more its prime divisors, $2^{16} \cdot 397 + 1$, and $2^{16} \cdot 973 + 1$.

The first prime divisor, $2^{16} \cdot 41365885 + 1$, of F_{13} was found only in 1974 (Hallyburton and Brillhart).

Some additional informstion about small prime divisors of Fermat numbers, which were found before 1950, is given in the Chapter 4, section 4.1.

See also [Kell20], as well as [Carm19], [Deza17], [DeKo13], [Dick05], [KLS01], [Madd05], [Tsan10], [Sier64].

Primality of numbers $k \cdot 2^m + 1$

4.3.4. As any prime divisor of F_n, $n \geq 1$, has the form $k \cdot 2^{n+1} + 1$, the natural question is: *which number of the form $k \cdot 2^m + 1$, $m \in \mathbb{N}$, $k = 1, 2, \ldots$, is a prime?*

If $k = 1$, we have only five known primes of the form $2^m + 1$ (for $m = 1, 2, 4, 8, 16$). They are all known Fermat primes $F_0 - F_4$.

If $k = 2$, we have only four known primes of the form $2 \cdot 2^m + 1$ (for $m = 1, 3, 7, 15$). They also are Fermat primes $F_1 - F_4$.

If $k = 3$, we have 19 known primes of the form $3 \cdot 2^m + 1$ (for $m = 1, 2, 5, 6, 8, 12, 18, 35, \ldots$).

If $k = 4$, we have only three known primes of the form $4 \cdot 2^m + 1$ (for $m = 2, 6, 14$). They are Fermat primes $F_2 - F_4$.

If $k = 5$, there are 17 known primes of the form $5 \cdot 2^m + 1$ (for $m = 1, 3, 7, 13, 15, 25, 39, 55, \ldots$).

Moreover, for any $k < 3061$ we know at least one m such that $k \cdot 2^m + 1$ is a prime number.

For $k = 3061$, the number $k \cdot 2^m + 1$ is prime for any $m < 17008$.

On the other hand, W. Sierpiński had shown in 1960, that

• *there exist infinitely many $k \in \mathbb{N}$, such that $k \cdot 2^m + 1$ is composite for any $m \in \mathbb{N}$.*

□ In fact, we know, that F_n is prime for $n \in \{0, 1, 2, 3, 4\}$, while F_5 is a product of two primes, 641 and p, where $p > F_4$.

The *Chinese Remainder theorem* allows to state, that there are infinitely many positive integers k, such that

$$k \equiv 1 (mod \, (2^{32} - 1) \cdot 641), \quad \text{and} \quad k \equiv -1 (mod \, p).$$

It is easy to show, that for all such numbers, greater than p, all numbers $k \cdot 2^n + 1$, $n \in \mathbb{N}$, are composite.

Consider $n = 2^s(2t + 1)$, where $s \in \{0, 1, 2, 3, 4\}$, and $t \geq 0$ is some integer. In this case, it holds

$$k \cdot 2^n + 1 \equiv 2^{2^s(2t+1)} + 1 (mod\ (2^{32} - 1));$$

as $F_s | (2^{2^s(2t+1)} + 1)$, and $F_s | (2^{32} - 1)$, then $F_s | k \cdot 2^n + 1$. Moreover, $F_s < k$. So, the number $k \cdot 2^n + 1$ is composite.

Consider now $n = 2^5(2t + 1)$, where $t \geq 0$ is some integer. Then

$$k \cdot 2^n + 1 \equiv 2^{2^5(2t+1)} + 1 (mod\ 641).$$

As $641 | (2^{2^5} + 1)$, and $2^{2^5} + 1 | (2^{2^5(2t+1)} + 1)$, then $641 | (k \cdot 2^n + 1)$, and $641 < k \cdot 2^n + 1$. So, $k \cdot 2^n + 1$ is composite.

At last, consider $n = 2^6 \cdot t$, where t is some positive integer. In this case

$$k \cdot 2^n + 1 \equiv -2^{2^6 t} + 1 (mod\ p).$$

As $p | (2^{2^5} + 1)$, $2^{2^5} + 1 | (2^{2^6} - 1)$, $2^{2^6} - 1 | (2^{2^6 t} - 1)$, it holds $p | (k \cdot 2^n + 1)$, and $p < k \cdot 2^n + 1$. So, the number $k \cdot 2^n + 1$ is composite. $\square$

A *Sierpiński number* is defined as an odd natural number k such that $k \cdot 2^m + 1$ is composite for all natural numbers m. So, it is proven, that

• *there are infinity many Sierpiński numbers.*

The sequence of currently known Sierpiński numbers begins with 78557, 271129, 271577, 322523, 327739, ... (sequence A076336 in the OEIS).

The first number of this sequence, 78557, was proved to be a Sierpiński number by J. Selfridge in 1962. No smaller such numbers have been discovered.

It is now believed that 78557 is the smallest Sierpiński number.

4.3.5. If for number $k \cdot 2^m + 1$ we put $k = m = n$, we obtain a *Cullen number*

$$C_n = n \cdot 2^n + 1, \ n \in \mathbb{N}.$$

They were first studied by J. Cullen in 1905.

The only known Cullen primes are those for $n = 1, 141, 4713,$ $5795, 6611, 18496, 32292, 32469, 59656, 90825, 262419, 361275, 481899,$ $1354828, 6328548, 6679881$ (sequence A005849 in the OEIS).

It has been shown (Hooley, 1976) that
- *almost all Cullen numbers are composite.*

Still, it is conjectured that
- *there are infinitely many Cullen primes.*

For some additional information see [Sier64], [Hool76], as well as [Carm19], [Deza17], [DeKo13], [Dick05], [Kell20], [KLS01], [Madd05], [Tsan10], [Wiki20].

Exercises

1. Prove, that F_3 is prime, by checking of all primes $p \leq \sqrt{F_3}$ of the form $32k + 1$.

2. Prove, that F_4 is prime, by checking of all primes $p \leq \sqrt{F_4}$ of the form $64k + 1$.

3. Prove, that F_5 is composite, by checking of all primes $p \leq \sqrt{F_5}$ of the form $128k + 1$.

4. Prove, that F_6 is composite, by checking of all primes $p \leq \sqrt{F_6}$ of the form $512k + 1$.

5. For $k = 4$, find the first ten elements of the sequence $k \cdot 2^n + 1$, $n \in \mathbb{N}$. How many primes are found?

6. For $k = 4$, find the first ten elements of the sequence $k \cdot 2^n + 1$, $n \in \mathbb{N}$. How many primes are found?

7. For $k = 5$, find the first ten elements of the sequence $k \cdot 2^n + 1$, $n \in \mathbb{N}$. How many primes are found?

8. Prove, that $k = 4, 5, 6, 7, 8, 9, 10$ are not Sierpiński nubers.

9. Find the first ten Cullen numbers $n \cdot 2^n + 1$, $n \in \mathbb{N}$. How many primes are found?

4.4 Fermat primes: Pépin's test

4.4.1. An other method to check primality of Fermat numbers is the *Pépin's test: Fermat number F_n, $n \geq 1$, is a prime if and only if it holds $F_n | 3^{\frac{F_n - 1}{2}} + 1$.*

It is a variant of the *Proth's test*. The test is named for a French mathematician T. Pépin.

This method consists not just in seeking the factors of F_n but also in the inverse process of ascertaining whether it is itself a factor of a number in a certain sequence.

Proof of the theorem

4.4.2. Consider the proof of the *Pépin's theorem*.

Theorem (Pépin's test). *For $n \geq 1$, $F_n \in P$ if and only if $3^{\frac{F_n-1}{2}} \equiv -1(mod\ F_n)$.*

□ To prove this theorem, note, that for $n \geq 1$ one has $F_n = 12k + 5$, as $4^k \equiv 4(mod\ 12)$ for any positive integer k.

I. Let $F_n \in P$. Then, using properties of *Legendre symbol* (see Chapter 1, Section 1.5) and the congruence $F_n \equiv 1(mod\ 4)$, we get

$$\left(\frac{3}{F_n}\right) = \left(\frac{F_n}{3}\right) = \left(\frac{-1}{3}\right) = -1.$$

Therefore, by the *Euler's criterion*,

$$3^{\frac{F_n-1}{2}} \equiv -1(mod\ F_n).$$

II. If $3^{\frac{F_n-1}{2}} \equiv -1(mod\ F_n)$, then $3^{F_n-1} \equiv 1(mod\ F_n)$.

Let p be a prime divisor of F_n. As $p|F_n$, then

$$3^{\frac{F_n-1}{2}} \equiv -1(mod\ p), \quad \text{and} \quad 3^{F_n-1} \equiv 1(mod\ p).$$

If $ord_p\,3 = \gamma$, then $\gamma|(F_n - 1)$, and $\gamma \nmid \frac{F_n-1}{2}$. In other words, it holds

$$\gamma|2^{2^n}, \quad \text{and} \quad \gamma \nmid 2^{2^n-1}.$$

Therefore, $\gamma = 2^{2^n}$. It is well-known, that $ord_p\,a|(p-1)$. So, $\gamma|(p-1)$, i.e., $2^{2^n}|(p-1)$. Therefore, it holds

$$F_n - 1|(p-1), \quad \text{and} \quad p|F_n.$$

It is possible only for $p = F_n$. In fact, if $p \neq F_n$, one has $p = 2^{2^n} \cdot X + 1$ for some positive integer $X > 1$, and $p|(2^{2^n} + 1)$. In this case, it holds that $p|2^{2^n}(X-1)$, i.e., for an odd prime p, that $p|(X-1)$. In this case, $X - 1 \geq p$, or $X - 1 \geq 2^{2^n}X + 1$, a contradiction. □

See, for example, [Buch09], [DeKo13], [DeKo18], [Step01].

Practical algorithms of calculation

4.4.3. The expression $3^{\frac{F_n-1}{2}}$ can be evaluated modulo F_n by repeated squaring. If we denote by $rest(t)$ the remainder $rest(t, F_n)$ obtained as F_n divides an integer t, and set

$$r_1 = 3, \ r_{k+1} = rest(r_k^2),$$

we get, that

$$F_n | (3^{2^{k-1}} - r_k) \text{ for any } k = 1, 2, \dots .$$

In fact, $3^{2^0} - r_1 = 0$, and $F_n | (3^{2^0} - r_1)$. Let $F_n | (3^{2^{k-1}} - r_k)$. Then

$$(3^{2^k} - r_{k+1}) \equiv (3^{2^{k-1}})^2 - rest(r_k^2) \equiv (3^{2^{k-1}})^2 - r_k^2$$

$$\equiv (3^{2^{k-1}} - r_k)(3^{2^{k-1}} + r_k) \equiv 0 (mod \ F_n),$$

and it holds $F_n | (3^{2^k} - r_{k+1})$.

For $k = 2^n$, we have $3^{2^{k-1}} = 3^{\frac{F_n-1}{2}}$, and it holds

$$F_n | (3^{\frac{F_n-1}{2}} - r_{2^n}).$$

Therefore, $3^{\frac{F_n-1}{2}} \equiv r_{2^n} (mod \ F_n)$, and we can use for the Pépin's test the number r_{2^n}, obtained as the last term of the sequence

$$r_1 = 3, \ r_2 \equiv r_1^2 (mod \ F_n), \ r_3 \equiv r_2^2 (mod \ F_n), \ \dots ,$$

$$r_{2^n} \equiv r_{2^n-1}^2 (mod \ F_n).$$

This procedure makes the Pépin's test a fast polynomial-time algorithm, if F_n is known.

4.4.4. É. Lucas (1879) used this method for study of the number F_6.

This is the very method by which the numbers F_7 (Morehead and Western 1905) and F_8 (Morehead and Western, 1908), have been proved to be composite.

However, Fermat numbers grow so rapidly that only a handful of Fermat numbers can be tested in a reasonable amount of time and space.

Because of the sparsity of the Fermat numbers, the Pépin's test has only been run eight times for Fermat numbers whose primality statuses were not already known.

Fermat numbers checked by Pépin's test

Year	Provers	Fermat number	Pépin test result
1905	Morehead and Western	F_7	composite
1909	Morehead and Western	F_8	composite
1952	Robinson	F_{10}	composite
1960	Paxson	F_{13}	composite
1961	Selfridge and Hurwitz	F_{14}	composite
1987	Buell and Young	F_{20}	composite
1993	Crandall, Doenias, Norrie and Young	F_{22}	composite
1999	Mayer, Papadopoulos and Crandall	F_{24}	composite

Note, that the method gives no information of the number into a product of two factors greater than 1. This is why we do not know any such decomposition of the number F_{20} and F_{24}.

As of 2020, the smallest untested Fermat number with no known prime factor is F_{33}, which has 2585827973 digits.

4.4.5. Instead of the number 3 we can take any odd integer number k, such that the Legendre symbol $(\frac{k}{F_n}) = -1$. in particular, k can be equal to 5 or 10.

□ In fact, as for any $n \geq 2$ it holds $F_n \equiv 1 (mod\ 4)$, and $F_n \equiv 2 (mod\ 5)$, we get for a prime number F_n, that

$$\left(\frac{5}{F_n}\right) = \left(\frac{F_n}{5}\right) = \left(\frac{2}{3}\right) = -1.$$

On the same way, using in additional the congruence $F_n \equiv 1 (mod\ 8)$, $n \geq 2$, we have

$$\left(\frac{10}{F_n}\right) = \left(\frac{2}{F_n}\right)\left(\frac{5}{F_n}\right) = 1 \cdot \left(\frac{5}{F_n}\right) = \left(\frac{F_n}{5}\right) = \left(\frac{2}{3}\right) = -1. \ □$$

It is easy to check, that bases which can be used for the Pépin's test, are 3, 5, 6, 7, 10, 12, 14, 20, 24, 27, 28, ... (sequence A129802 in the OEIS).

The primes in the above sequence are called *Elite primes*, they are 3, 5, 7, 41, 15361, 23041, 26881, 61441, 87041, 163841, ... (sequence A102742 in the OEIS).

4.4.6. In fact, for Fermat numbers, Pépin's test is the same as the *Solovay–Strassen primality test*. It is deterministic, since the Jacobi symbol $\left(\frac{b}{F_n}\right)$ is equal to -1, i.e., there are no Fermat numbers which are *Euler-Jacobi pseudoprimes* to any base b listed above.

On the other hand, the Pépin's test is a particular case of the (deterministic) *Lucas test*.

In 1878, a generalization of Pépin's test, now called *Proth's theorem*, was obtained. (It is named after a French mathematician F. Proth.)

Theorem (Proth's theorem). *Given $n = h \cdot 2^k + 1$, with h odd and $h < 2^k$, if there exists an integer a, such that $a^{\frac{n-1}{2}} \equiv -1 \pmod{n}$, then n is a prime.*

We get one of propositions of Pépin's test for $h = 1$, and $a = 3$.

Proth theorem is a primality test for *Proth numbers*, i.e., the numbers of the form $h \cdot 2^k + 1$ with h odd and $h < 2^k$.

In the case of primality one obtains a *Proth prime*. This is a practical test because if p is prime, any chosen a has about a 50 % chance of working.

The first Proth primes are $3, 5, 13, 17, 41, 97, 113, 193, 241, 257, \dots$ (sequence A080076 in the OEIS).

As of 2020, the largest known Proth prime (P. Szabolcs, 2016) is $110223 \cdot 2^{31172165} + 1$. It has 9383761 digits. It is also the largest known non-Mersenne prime.

In turn, the *Proth's theorem* is a particular case of the *Pocklington criterion* (see Chapter 5, Section 5.5).

Theorem (Pocklington criterion). *Given $n = h \cdot q^k + 1$ with $q \in P$, $\gcd(h, q) = 1$, and $h < q^k$, if there exists an integer a, such that $a^{n-1} \equiv 1 \pmod{n}$, and $\gcd(a^{\frac{n-1}{q}} - 1, n) = 1$, then n is a prime.*

We get the *Proth theorem* in the simplest case $q = 2$.

For additional proofs see Chapter 5, Section 5.5. See also [Buch09], [Carm19], [Deza17], [DeKo13], [DeKo18], [Dick05],

[Kell20], [KLS01], [Knut68], [Madd05], [Tsan10], [Sier64], [Step01], [Prim20], [Wiki20].

Exercises

1. Prove, that the numbers $6, 7, 12, 14, 20, 24, 27, 28$ can be used as the base of the Pépin's test.

2. Prove, that the numbers $8, 9, 10, 11, 13, 15, 17, 18, 19, 21, 22, 23$ cannot be used as the base of the Pépin's test.

3. Check, that numbers F_3 and F_4 are primes, using the algorithm of the Pépin's test.

4. Prove, that numbers $3, 5, 7, 41$ are elite primes. Prove, that there are no other elite primes less then 50.

5. Check, that the Pépin's test is a particular case of the Lucas primality test.

6. Prove the Proth' theorem.

7. Prove the Pocklington criterion.

8. Prove that the first Proth prime, which is not a Fermat number, is 13.

9. Check the primality of 229 using the Pocklington criterion.

4.5 Fermat numbers in the family of special numbers

Fermat numbers and perfect numbers

4.5.1. A positive integer n is called a *perfect number*, it is equal to the sum of its proper divisors, i.e., its positive divisors, excluding the number itself. In terms of the sigma function, it holds $\sigma(n) = 2n$.

The first perfect number is 6: it has divisors 1, 2 and 3 (excluding 6 itself), and it holds

$$1 + 2 + 3 = 6.$$

The next three are 28, 496, and 8128.

Amicable numbers are two different positive integers n, m related in such a way that the sum of the proper divisors of each is equal to the other number. In terms of the sigma function, it holds $\sigma(n) - n = m$, and $\sigma(m) - m = n$, or $\sigma(n) = \sigma(m) = n + m$.

The smallest pair of amicable numbers is $(220, 284)$.

It can be proven (Luca, 2000), that

- *Fermat number cannot be a perfect number or part of a pair of amicable numbers.*

4.5.2. There exists a connection between even perfect numbers, Fermat numbers and so-called *trapezoidal numbers*.

An positive integer is called a *trapezoidal number*, if it can be represented as some isosceles trapezoid.

For example, a trapezoidal representation of 18 is given on the picture below.

The first trapezoidal numbers are 5, 7, 9, 11, 12, 13, 14, 15, 17, 18, ... (sequence A165513 in the OEIS).

By definition, any trapezoidal number is a sum of two or more consecutive positive integers, greater than 1:

$$Tr(n, k) = n + (n + 1) + (n + 2) + \cdots + (n + (k - 1)), \ n \neq 1, \ k \neq 1.$$

So, one obtains, that *any trapezoidal number is a difference of two non-consecutive triangular numbers*:

$$Tr(n, k) = n + (n + 1) + (n + 2) + \cdots + (n + (k - 1))$$
$$= (1 + 2 + \cdots + (n + (k - 1))) - (1 + 2 + \cdots + (n - 1))$$
$$= S_3(n + k - 1) - S_3(n - 1).$$

Moreover, the above definition implies that *any trapezoidal number is a particular case of polite number*.

In fact, a *polite number* is a positive integer that can be represented as the sum of two or more consecutive positive integers.

So, if such *polite representation* starts with 1, we obtain a triangular number, otherwise one gets a trapezoidal number.

The first polite numbers are 1, 3, 5, 6, 7, 9, 10, 11, 12, 13, ... (sequence A138591 in the OEIS).

Obviously, a given positive integer can have several polite representations. In the case of 18, one more such representation is given below.

The *politeness* of a positive integer number is the number of ways it can be expressed as a sum of consecutive positive integers.

It is easy to show that

• *for any positive integer x, the politeness of x is equal to the number of odd divisors of x that are greater than one.*

□ In fact, suppose a number x has an odd divisor $y > 1$. Then y consecutive integers, centered on $\frac{x}{y}$ (so that their average value is $\frac{x}{y}$), have x as their sum:

$$x = \left(\frac{x}{y} - \frac{y-1}{2}\right) + \cdots + \frac{x}{y} + \cdots + \left(\frac{x}{y} + \frac{y-1}{2}\right).$$

Some of the terms in this sum may be zero or negative. However, if a term is zero, it can be omitted, and any negative terms may be used to cancel positive ones, leading to a polite representation for x. The requirement that $y > 1$ corresponds to the requirement that a polite representation have more than one term. For instance, the polite number $x = 18$ has two non-trivial odd divisors, 3 and 9. It is therefore the sum of 3 consecutive numbers centered at $\frac{18}{3} = 6 : 18 = 5 + 6 + 7$. On the other hand, it is the sum of 9 consecutive integers centered at $\frac{18}{9} = 2 : 18 = (-3) + (-1) + 0 + 1 + 2 + 3 + 4 + 5 + 6$, or $18 = 3 + 4 + 5 + 6$.

Conversely, every polite representation of x can be formed by this construction. If a representation has an odd number y of terms, then the middle term can be written as $\frac{x}{y}$, and $y > 1$ is a non-trivial odd divisor of x. If a representation has an even number $2l$ of terms and its minimum value is m, it may be extended, in an unique way, to a longer sequence with the same sum and an odd number $y = 2(m+l)-1$ of terms, by including the $(2m-1)$ numbers $-(m-1)$, $-(m-2)$, ..., -1, 0, 1, ..., $m-2$, $m-1$. After this extension the middle term of the new sequence can be written as $\frac{x}{y}$, and $y > 1$ is a non-trivial odd divisor of x. By this construction, the polite representations of a number and its odd divisors greater than one may be placed into an one-to-one correspondence, that completes the proof. □

From this consideration it follows, that the *inpolite numbers*, i.e., positive integers which are not polite, are exactly the powers of two.

So, the first few inpolite numbers are 1, 2, 4, 8, 16, 32, 64, 128, 256, 512, ... (sequence A000079 in the OES), and the politeness of the numbers 1, 2, 3, 4, 5, 6, 7, 8, 9, 10, ... is 0, 0, 1, 0, 1, 1, 1, 0, 2, 1, ... (sequence A069283 in the OEIS).

Therefore, the only polite numbers that may be non-trapezoidal are the triangular numbers with only one non-trivial odd divisor, because for those numbers, according to the bijection described above, the odd divisor corresponds to the triangular representation and there can be no other polite representation.

Thus, polite non-trapezoidal numbers must have the form of a power of two multiplied by a prime number.

It is easy to show that *there are exactly two types of triangular numbers with this form:*

- *the even perfect numbers $2^{k-1}(2^k - 1)$ formed by the product of a Mersenne prime $2^k - 1$ with half the nearest power of two;*

- *the products $2^{k-1}(2^k + 1)$ of a Fermat prime $2^k + 1 = 2^{2^n} + 1$ with half the nearest power of two.*

□ Indeed, if we have equality $\frac{n(n+1)}{2} = 2^{k-1} \cdot p$ with odd prime number p, then $n(n+1) = 2^k \cdot p$, and, hence, $p|n(n+1)$. So, we get that either $p|n$, or $p|(n+1)$.

If $p|n$, then, by elementary arguments, $p = n$ and $2^k = n+1$. Therefore, it holds $p = 2^k - 1$, i.e., p is a Mersenne prime, and the corresponding triangular number $2^{k-1}(2^k - 1)$ is an even perfect number.

If $p|(n+1)$, then, by elementary arguments, $p = n+1$ and $2^{t+1} = n$. Hence, it holds $p = 2^{t+1} + 1$, i.e., $p = 2^{2^n} + 1$ is a Fermat prime, and the corresponding triangular number $2^{k-1}(2^k + 1)$ has the form $2^{2^n-1}(2^{2^n} + 1)$. □

So, we have proven, that *there are only three types of non-trapezoidal numbers:*

- *the even perfect numbers $2^{k-1}(2^k - 1)$ formed by the product of a Mersenne prime $2^k - 1$ with half the nearest power of two;*

- *the products $2^{2^n-1}(2^{2^n} + 1)$ of a Fermat prime $2^{2^n} + 1$ with half the nearest power of two;*

- *powers of two, in particular, all numbers of the form $M_n + 1$, where M_n is the n-th Mersenne number, and all numbers of the form $F_n - 1$, where F_n is the n-th Fermat number.*

Some additional information see, for example, in [Deza18], [JoLo99], [Luca00], [Smit97], [Hons91].

Fermat numbers and Pascal's triangle

4.5.3. The *Gauss–Wantzel theorem* states, that *an n-sided regular polygon can be constructed with compass and straightedge if and only if n is the product of a power of 2 and distinct Fermat primes.*

Since there are exactly 5 known Fermat primes, we know exactly 31 numbers that are products of distinct Fermat primes, and hence 31 constructible odd-sided regular polygons.

They are 3, 5, 15, 17, 51, 85, 255 , 257, 771, 1285, 3855, 4369, 13107, 21845, 65535, 65537, 196611, 327685, 983055, 1114129, 3342387, 5570645, 16711935, 16843009, 50529027, 84215045, 252645135, 286331153, 858993459, 1431655765, 4294967295 (sequence A045544 in the OEIS).

These numbers, when written in binary, are equal to the first 32 rows of the modulo 2 Pascal's triangle, minus the top row, which corresponds to a monogon. Because of this, the unities in such a list form an approximation to the Sierpiński triangle.

This pattern breaks down after this, as the next Fermat number is composite. On the first picture one can see the algorithm of this construction.

			modulo 2					base 2		

$$
\begin{array}{ccccccccc}
 & & 1 & & & \to & & 1 & & \to & 1_2 = 1 \\
 & 1 & & 1 & & \to & & 1 \quad 1 & & \to & 11_2 = 1 \cdot 2 + 1 = 3 \\
 1 & & 2 & & 1 & \to & 1 & 0 & 1 & \to & 101_2 = 1 \cdot 2^2 + 0 \cdot 2 + 1 = 5 \\
1 & 3 & & 3 & 1 & \to & 1 \; 1 \; 1 \; 1 & & \to & 1111_2 = 1 \cdot 2^3 + 1 \cdot 2^2 + 1 \cdot 2 + 1 = 15 \\
1 & 4 & 6 & 4 & 1 & \to & 1 \; 0 \; 0 \; 0 \; 1 & & \to & 10001_2 = 1 \cdot 2^4 + 0 \cdot 2^3 + 0 \cdot 2^2 + 0 \cdot 2 + 1 = 17
\end{array}
$$

On the second picture several additional rows of the Pascal's triangle modulo 2 are represented.

$$
\begin{array}{ccccccccc}
 & & & & 1 & & & & & = 1 \\
 & & & 1 & & 1 & & & & = 3 \\
 & & 1 & & 0 & & 1 & & & = 5 \\
 & 1 & & 1 & & 1 & & 1 & & = 15 \\
1 & & 0 & & 0 & & 0 & & 1 & = 17 \\
1 & 1 & & 0 & & 0 & & 1 & 1 & = 51 \\
1 & 0 & 1 & & 0 & & 1 & 0 & 1 & = 85 \\
1 & 1 & 1 & 1 & & 1 & 1 & 1 & 1 & = 255
\end{array}
$$

For other connections between Fermat numbers (as well as generalized Fermat numbers) and Pascal's triangle see Chapter 8.

For some additional information see, for example, [Bond93], [Deza17], [Deza18], [DeMo10], [KLS01], [Madd05], [Tsan10], [Sier64], [Uspe76].

Generalized Fermat numbers

4.5.4. There are several possibilities to construct some generalizations of Fermat numbers.

So, numbers of the form

$$ F_n(a, b) = a^{2^n} + b^{2^n} $$

with positive coprime integers a and b, such that $a > b$, are called *generalized Fermat numbers*.

Unfortunately, this construction is too large. In fact, it is easy to show, that

• *an odd prime p is a generalized Fermat number $F_n(a, b)$, $n \geq 1$, if and only if p is congruent to 1 modulo 4.*

□ Indeed, it is well-known, that an odd prime p is represented as a sum of two perfect squares if and only if $p \equiv 1 (mod\ 4)$. If $p = x^2 + y^2$, it holds that x and y are positive coprime integers, such that $x > y$, and one can write that $p = F_1(x, y)$. □

Here we consider only the case $n \geq 1$, so $3 = 2^{2^0} + 1$ is not a counterexample.

4.5.5. It is more convenient to consider numbers $F_n(a, b)$ only with $b = 1$. By analogy with the ordinary Fermat numbers, it is common to write generalized Fermat numbers $F_n(a, 1)$, i.e., numbers of the form $a^{2^n} + 1$, as $F_n(a)$. For example,

$$F_4(10) = 10^{2^4} + 1 = 10000000000000001.$$

If $F_n(a) = a^{2^n} + 1$ is a prime, it is called a *Fermat prime to base a*.

Generalized Fermat numbers $F_n(a)$ can be prime only for even a, because if a is odd then every generalized Fermat number will be divisible by 2.

If we require $n \geq 1$, then Landau's fourth problem *"are there infinitely many primes p such that $p - 1$ is a perfect square?"* asks if there are infinitely many generalized Fermat primes $F_n(a)$.

For the base 4, the generalized Fermat number $F_n(4) = F_{n+1}$; so, there exist exactly four known generalized Fermat primes to base 4; they are $F_1 = 5$, $F_2 = 17$, $F_3 = 257$, and $F_4 = 65537$.

Similarly, there exist exactly three known generalized Fermat primes to base 16.

For the base 6, there exist exactly three known generalized Fermat primes $F_n(6)$; they are $F_0(6) = 7$, $F_1(6) = 37$, and $F_4(6) = 1297$.

Obviously, there are no generalized Fermat primes $F_n(8)$ to base 8: any number of the form $8^k + 1$, $k \geq 1$, is divisible by 3.

In general, if a is a perfect power with an odd exponent, then all generalized Fermat numbers $F_n(a)$, $n \geq 1$, can be algebraic factored, so they cannot be primes.

For the base 10, there exist exactly two known generalized Fermat primes $F_n(10)$; they are $F_0(10) = 11$, and $F_2(10) = 101$.

The smallest prime number $F_n(a)$ with $n > 4$ is $F_5(30)$:

$$F_5(30) = 30^{2^5} + 1$$

$$= 1853020188851841000000000000000000000000000000001.$$

For $n = 0, 1, 2, 3, ...$, the smallest bases a such that $F_n(a) = a^{2^n} + 1$ is prime are 2, 2, 2, 2, 2, 30, 102, 120, 278, 46, ... (sequence A056993 in the OEIS).

In fact, the numbers $2^{2^0} + 1$, $2^{2^1} + 1$, $2^{2^2} + 1$, $2^{2^3} + 1$, $2^{2^4} + 1$, $30^{2^5} + 1$, $102^{2^6} + 1$, $120^{2^7} + 1$, $278^{2^8} + 1$, $46^{2^9} + 1$ are primes.

For $n = 1, 2, 3, ...$, the smallest k such that $(2n)^k + 1$ is prime are 1, 1, 1, 0, 1, 1, 2, 1, 1, 2, ... (sequence A079706 in the OEIS; see also A228101, and A084712).

The following table gives a list of the 5 largest known generalized Fermat primes $F_n(a)$. They are all megaprimes. Each of them was discovered by participants in the *PrimeGrid* project.

The largest known generalized Fermat primes

	Prime number	Notation	Number of digits	Year
1	$1059094^{1048576} + 1$	$F_{20}(1059094)$	6317602	2018
2	$919444^{1048576} + 1$	$F_{20}(919444)$	6253210	2017
3	$3214654^{524288} + 1$	$F_{19}(3214654)$	3411613	2019
4	$2985036^{524288} + 1$	$F_{19}(2985036)$	3394739	2019
5	$2877652^{524288} + 1$	$F_{19}(2877652)$	3386397	2019

On the *Prime Pages* one can find the current top 100 generalized Fermat primes.

4.5.6. There are many properties of generated Fermat numbers $F_n(a)$, which are similar to the corresponding propreties of F_n. So,

• *any prime divisor of $F_n(a) = a^{2^n} + 1$ has the form $k \cdot 2^m + 1$ with odd k and $m > n$.*

Moreover, the different numbers $F_n(a)$ and $F_m(a)$ are coprime:

• *$gcd(F_n(a), F_m(a)) = 1$ for $n \neq m$.*

However, for different bases it is not true. For example, 641 divides $10^{16} + 1$, $20^{32} + 1$, and $40^8 + 1$.

4.5.7. Besides, a *half generalized Fermat number* to an odd base a is defined as $\frac{a^{2^n}+1}{2}$.

It is expected that there will be only finitely many half generalized Fermat primes for each odd base.

4.5.8. A *Mersenne–Fermat number* $MF(p,r)$ is defined, for a positive integer r and a prime p, as

$$MF(p,r) = \frac{2^{p^r} - 1}{2^{p^{r-1}} - 1}.$$

When $r = 1$, it is a Mersenne number. When $p = 2$, it is a Fermat number.

The only known *Mersenne–Fermat primes* with $r > 1$ are $MF(2,2)$, $MF(2,3)$, $MF(2,4)$, $MF(2,5)$, $MF(3,2)$, $MF(3,3)$, $MF(7,2)$, and $MF(59,2)$.

For some additional information see, for example, [Bond93], [Buch09], [Carm19], [Deza17], [Deza18], [DeKo18], [KLS01], [Land09], [Madd05], [Tsan10], [Sier64], [SlP195], [Sloa20], [Prim20], [Wiki20].

Exercises

1. Prove, that any Fermat number is a palindromic number in binary.

2. Prove, that the sequence of binary palindromic primes begins (in binary) from 11, 101, 111, 10001, 11111, 1001001, 1101011, 1111111, 100000001, 100111001, ... (sequence A117697 in the OEIS) and contains all Mersenne primes and Fermat primes.

3. Find the first five trapezoidal numbers; the first five non-trapezoidal numbers; first five numbers of the form $2^{n-1}(2^n + 1)$, where $2^n - 1$ is a prime. Find all possible intersections of the obtained sets.

4. Construct the first 32 rows of Pascal's triangle. Construct the first 32 rows of Pascal's triangle modulo 2. Find in the obtained construction Fermat numbers; perfect numbers: Mersenne numbers.

5. Find all generalized Fermat primes $F_n(a)$ with $1 \leq n \leq 4$, and $a \leq 30$.

6. Prove, that any generalized Fermat number $F_n(a)$ with $a = 8$ is composite. Check the similar statement for $a = 64$; for $a = 216$.

7. Prove, that any generalized Fermat number $F_n(a)$ with $a = 4$ is a Fermat number. Check the similar statement for $a = 16$; for $a = 256$.

8. Prove, that Mersenne-Fermat number $MF(p, r)$ is a Mersenne number for $r = 1$.

9. Prove, that Mersenne-Fermat number $MF(p, r)$ is a Fermat number for $p = 2$.

10. Check, that numbers $MF(2, 2)$, $MF(2, 3)$, $MF(2, 4)$, $MF(2, 5)$, $MF(3, 2)$, $MF(3, 3)$ and $MF(7, 2)$ are primes.

4.6   Open problems

There exist many open problems relating to Fermat numbers. In this section we consider the most important of them.

Finiteness of the set of Fermat primes

4.6.1. First of all, mathematicians are interested in the main question, concerning to the Theory of Fermat numbers:

- *if there exist infinitely many Fermat primes?*

On the othe hand, we need to get an answer to the following question:

- *if there exist infinitely many Fermat composite numbers?*

However, there exists a really simple question that has not yet been answered:

- *is there at least one Fermat prime greater than F_4?*

Heuristic reasons suggest that *there is only a finite number of Fermat primes.*

Indeed, by the *Prime number theorem*, it holds that the number $\pi(x)$ of primes, less than or equal to the given positive real number x, is approximatively equal to $\frac{x}{\log x}$; so, the n-th prime number is approximatively equal to $n \log n$ (more exactly, $cn \ln n < p_n < Cn \ln n$, where $C > c > 0$ are some positive absolute constants). Then we can guess the probability of a "random" number n to be prime is at most $\frac{A}{\log n}$ for some positive absolute constant A.

So, summing

$$\frac{A}{\log F_n} < \frac{A}{2^n \log 2}.$$

over the non-negative integers n, we should expect at most $\frac{2A}{\log 2}$ Fermat primes, as

$$\sum_{n=0}^{\infty} 2^{-n} = 1 + \frac{1}{2} + \frac{1}{2^2} + \cdots + \frac{1}{2^n} + \cdots = 2.$$

So, the number of Fermat primes is bounded by the constant $\frac{2A}{\log 2}$.

This argument is not a rigorous proof. Here we assume that Fermat numbers behave enough like random numbers to make the above argument. But they are not random numbers; for example, they have special divisibility properties.

4.6.2. Many years there existed conjecture about primality of all numbers of the form

$$2 + 1, \ 2^2 + 1, \ 2^{2^2} + 1, \ 2^{2^{\cdot^{\cdot^{\cdot^2}}}} + 1, \ \dots.$$

It was disproved in 1953, when it was shown by J. Selfridge, that the number F_{16} (having 19729 decimal digits) is composite; it has a prime divisor $1575 \cdot 2^{18} + 1 = 825753601$.

4.6.3. W. Sierpiński (1958) introduced the numbers

$$S_n = n^n + 1, \ n \in \mathbb{N} \backslash \{1\}.$$

He proved that if the number S_n is a prime, then $n = 2^{2^m}$, i.e., S_n is a Fermat number:

$$S_n = F_{m+2^m}.$$

Primes of this form are very rare. In fact, there are only two such primes that do not exceed $3 \cdot 10^{12}$: for $m = 0, 1$ we get primes $F_1 = 5$ and $F_3 = 257$, for $m = 2, 3, 4, 5$ we get the composite numbers F_6, F_{11}, F_{20}, F_{37}; for $m = 6$ we get the number F_{70}, which already has more than $3 \cdot 10^{20}$ decimal digits.

However, the open question is

- *if there are infinitely many prime numbers of the form $n^n + 1$?*

If the set of primes of the form $n^n + 1$ is finite, *there exist infinity many composite Fermat numbers.*

4.6.4. Mathematicians study also the sequence

$$G_m = F_m \cdot 2^{F_m} - 1, \; m \geq 0.$$

It is easy to see, that the numbers $G_0 = 11$, $G_1 = 79$, and $G_2 = 1114111$ are primes. The number G_3 is also prime.

So, it was conjectured (1968) that *all numbers of this form are primes.*

However, W. Keller (1992) showed that G_4, having 19734 decimal digits, is composite; it is divided by 16267.

The open question is

• *what if there are infinitely many of such prime numbers (at least, one more prime of this form)?*

Other open questions

4.6.5. Still open is also the question

• *if any Fermat number is squarefree?*

It is easy to show that if the square of a prime number p divides Fermat number, then

$$2^{p-1} \equiv 1 (mod \; p^2).$$

That is, the number p is a *Wieferich prime.*

Although the congruence $2^{p-1} \equiv 1 (mod \; p)$ takes place for all odd primes p (due to the Fermat's little theorem), Wieferich congruence is very rare.

Only two such primes, 1093 and 3511, are known up to $4 \cdot 10^{12}$. The first of them was discovered by W. Meissner (1913), and the second by N.G.W.H. Biger (1922).

For the primes 1093 and 3511, it was shown that neither of them is a divisor of any Fermat number.

4.6.6. A *Sierpiński number* is defined as an odd natural number k such that $k \cdot 2^m + 1$ is composite for all natural numbers m.

It is proven (Sierpiński, 1960), that there are infinity many of Sierpiński numbers.

The sequence of currently known Sierpiński numbers begins with 78557.

The number 78557 was proved to be a Sierpiński number by J. Selfridge in 1962.

No smaller Sierpiński numbers have been discovered.

The *Sierpiński problem* asks for the value of the smallest Sierpiński number.

In private correspondence with P. Erdős, J. Selfridge conjectured that 78557 was the smallest Sierpiński number.

It is now believed that 78557 is indeed the smallest Sierpiński number.

Some additional information see, for example, in [Buch09], [DeKo13], [Guy94]. For some additional information see, for example, [Carm19], [Deza17], [Deza18], [DeKo13], [Dick05], [Kell20], [KLS01], [Madd05], [Shan93], [Tsan10], [Sier64], [Prim20].

Exercises

1. Prove the infiniteness of primes, using Fermat numbers.

2. Prove the infiniteness of primes in the arithmetic progression $4k + 1$, using Fermat numbers.

3. Using Fermat numbers, prove that $p_n \leq F_{n-5}$ for $n \geq 7$, where p_n is the n-th prime number.

4. Prove that $\pi(x) \geq \log_2 \log_2 x + 4$ for $x \geq 17$.

5. Prove the infiniteness of primes in the arithmetic progressions $2k + 1$; $4k + 1$; $8k + 1$; $16k + 1$.

6. Prove the infiniteness of primes in the arithmetic progressions $6k + 1$; $18k + 1$; $54k + 1$.

7. Check, that Fermat numbers F_0, ..., F_4 are not Wieferich primes.

8. Prove, that any prime number of the form $n^n + 1$ is a Fermat prime.

Chapter 4: References

[Anke57], [Anto85], [Apos86], [Avan67], [BaCo87], [Bond93], [Carm19], [CoGu96], [DeDe12], [Deza17], [Deza18], [Dick27], [Dick05], [Diop74], [Edva77], [Gard61], [Gard88], [Gaus01], [Wief09], [Guy94], [HaWr79], [Kost82] [KLS01], [Lagr70], [Lege30], [LiNi96], [Madd05], [BrMo75], [Pasc54], [SlP195], [Sloa20], [Shan93], [Sier64], [Tsan10], [Weis99], [Weis99], [Wiki20].

Chapter 5

Modern Applications

5.1 On place of prime numbers in Mathematics

5.1.1. The central importance of prime numbers to Number Theory and Mathematics in general stems from the *fundamental theorem of Arithmetics.*

This theorem states that *every integer larger than 1 can be written as a product of one or more primes.* More strongly, this product is unique in the sense that any two prime factorizations of the same number will have the same numbers of copies of the same primes, although their ordering may differ. Primes can thus be considered the "basic building blocks" of the natural numbers.

Many conjectures revolving about primes have been posed. Often having an elementary formulation, many of these conjectures have withstood proof for decades.

One of them is the *Goldbach's conjecture*, which asserts that *every even integer n greater than 2 can be written as a sum of two primes.*

The branch of number theory studying such questions is called the *Additive Number Theory.*

Another type of problem concerns *prime gaps*, the differences between consecutive primes.

The distribution of primes in the large, such as the question *how many primes are smaller than a given x*, is described by the *Prime number theorem*, but no efficient formula for the n-th prime number is known.

The *Dirichlet's theorem on arithmetic progressions* asserts that *linear polynomials $a + bn$ with relatively prime integers a and b take infinitely many prime values.*

But prime numbers are more than just mathematical oddities. They have become extremely important in Cryptography because it is difficult to factor a large composite number into primes.

In fact, for general numbers that are few hundred decimal digits long, factorization is nearly impossible; this gave rise to the *RSA cryptosystem.*

In this section we consider the most important facts of the mathematical Theory of prime numbers.

Distribution of prime numbers

5.1.2. Well-known is the irregularity of the distribution of prime numbers on the number line.

The existence of arbitrarily large prime gaps can be seen by elementary arguments. In fact,

• *for a given N, there exists an interval of length N which does not contain prime numbers.*

□ It is easy to see, that all the numbers $(N + 1)! + k$, $2 \leq k \leq (N+1)!+(N+1)$, are composite. The number $(N+1)!+2$ is divisible by 2, $(N + 2)! + 3$ is divisible by 3, ..., $(N + 1)! + (N + 1)$ is divisible by $(N + 1)$. □

5.1.3. On the other hand, in the list of primes you will quite often see two consecutive odd prime numbers: 3 and 5, 5 and 7, 11 and 13, 17 and 19, ... We call these pairs of primes numbers $(p, p + 2)$ *twin primes*. This name was coined by Stäckel in 1916.

The first such pairs are: (3, 5), (5, 7), (11, 13), (17, 19), (29, 31), (41, 43), (59, 61), (71, 73), (101, 103), (107, 109), ... (sequence A077800 in the OEIS).

A question to which the answer is not known is

• *whether there exist infinitely many twin primes.*

R.P. Brent, 1976, has found that there are 152892 pairs of twin primes less than 10^{11}.

The greatest known pair of twin primes is the pair $2996863034895 \cdot 2^{1290000} \pm 1$ (2016). It has 388342 digits.

5.1.4. Another question to which the answer is not known is
- *whether there exist infinitely many primes p for which $p, p+2$, $p+6$ and $p+8$ are all prime numbers.*

The first six consecutive quadruplets are obtained for $p = 5, 11, 101, 191, 821, 1481$.

As of 2020, the largest known prime quadruplet (Kaiser, 2019) has 10132 digits and starts with

$$p = 667674063382677 \cdot 2^{33608} - 1.$$

For more information see Chapter 2, section 2.6. Moreover, see, for example, [Buch09], [Deza17], [DeKo13], [DeKo18], [Dick05], [Ingh32], [Ore48], [Prim20], [Wiki20].

Prime number theorem

5.1.5. For any positive real number x, let the *prime counting function* $\pi(x)$ be the number of primes not greater than x:

$$\pi(x) = \sum_{p \leq x} 1.$$

We have, that $\pi(1) = 0$, $\pi(2) = 1$, $\pi(3) = \pi(4) = 2$, $\pi(5) = \pi(6) = 3$, ..., $\pi(100) = 25$, $\pi(1000) = 168$, $\pi(10000) = 1229$, $\pi(10^5) = 9592$, $\pi(10^6) = 78498$, $\pi(10^7) = 664579$, $\pi(10^8) = 5761455$, $\pi(10^9) = 50847534$.

In 1972, J. Bohmann calculated that $\pi(10^{10}) = 455052511$. In 1985, J.C. Lagarias, V.S. Miller and A.M. Odlyzko have computed the value $\pi(10^{16}) = 27923834033925$. The value

$$\pi(10^{27}) = 16352460426841680446427399$$

was published in 2015 by D. Baugh and K. Walisch.

Obviously, we have that

$$\pi(p_n) = n.$$

From the *Euclid's theorem* it follows, that

$$\pi(x) \to \infty, \quad \text{as} \quad x \to \infty.$$

L. Euler, 1737, considered the series

$$\zeta(s) = 1 + \frac{1}{2^s} + \frac{1}{3^s} + \dots + \frac{1}{n^s} + \dots = \sum_{n=1}^{\infty} \frac{1}{n^s} \quad \text{for} \quad s \in \mathbb{R}, \ s > 1.$$

He evaluated the sums $\zeta(2k)$ for every $k \geq 1$; in particular, he obtained the values

$$\zeta(2) = \frac{\pi^2}{6}, \quad \text{and} \quad \zeta(4) = \frac{\pi^4}{90}.$$

Euler noticed, that for real $s > 1$ it holds

$$\sum_{n=1}^{\infty} \frac{1}{n^s} = \prod_{p \in P} \left(1 + \frac{1}{p^s} + \frac{1}{p^{2s}} + \dots \right) = \prod_{p \in P} \left(1 - \frac{1}{p^s} \right)^{-1}.$$

Due to this observation he gave an analytic proof of the *infiniteness of prime numbers* (see Chapter 2, section 2.2).

Moreover, he proved that *the sum* $\sum_{p \in P} \frac{1}{p}$ *of the inverses of the prime numbers is divergent*. More exactly, that

$$\sum \frac{1}{p} = \log \log x + O(1).$$

Euler also proved that

$$\pi(x) = o(x), \quad \text{i.e.,} \quad \lim_{x \to \infty} \frac{\pi(x)}{x} = 0.$$

5.1.6. Later C.F. Gauss, 1792, had conjectured that there exists an asymptotic formula for the function $\pi(x)$:

$$\pi(x) \sim \frac{x}{\log x}, \quad \text{as} \ x \to \infty.$$

Later he refined this estimate to

$$\pi(x) \sim \int_2^x \frac{du}{\log u}.$$

C.F. Gauss did not publish this result, which he first mentioned in 1849 (in a letter to J.F. Encke). It was subsequently posthumously published in 1863.

The logarithmic integral $Li(n) = \int_2^n \frac{dx}{\ln x}$ has the asymptotic expansion about of

$$Li(n) = \sum_{k=0}^{\infty} \frac{k!n}{(\ln n)^{k+1}} = \frac{n}{\ln n} + \frac{n}{(\ln n)^2} + \frac{2n}{(\ln n)^3} + \cdots$$

and taking the first three terms has been shown to be a better estimate than alone.

A.-M. Legendre, 1808, conjectured a similar result:

$$\pi(x) \sim \frac{x}{\log x - 1.08366}.$$

5.1.7. First essential success in this problem was obtained by P.L. Chebyshev in 1850. He found the following boundaries for the function $\pi(x)$:

$$a \cdot \frac{x}{\log x} < \pi(x) < b \cdot \frac{x}{\log x},$$

where $x \geq 2$, $a = \frac{2^{1/2}3^{1/3}5^{1/5}}{30^{1/30}}$, and $b = \frac{6}{5}a$; hence, the new boundaries for n-th prime number p_n were found:

$$0.91 \cdot n \log n < p_n < 1.7 \cdot n \log n.$$

P.L. Chebyshev proved also well-known *Bertrand's Postulate: if n is a natural number > 1 then between n and 2n there is at least one prime number.*

5.1.8. The fundamental contribution to further investigations on the problem of distribution of primes was made by B. Riemann in 1859. He returned to the Euler's proof of infiniteness of primes and introduced the function

$$\zeta(s) = \sum_{n=1}^{\infty} \frac{1}{n^s} = 1 + \frac{1}{2^s} + \frac{1}{3^s} + \frac{1}{4^s} + \frac{1}{5^s} + \cdots + \frac{1}{n^s} + \cdots$$

as a function of a complex variable $s = \sigma + it$, $\sigma > 1$. This sum absolutely convergent for $Res = \sigma > 1$ and represents an analytic function in this region. B. Riemann showed that $\zeta(s)$ is analytically continuable on the whole complex plane, and has a simple pole with the residue 1 at the point $s = 1$.

Moreover, B. Riemann showed that the zeta-function satisfies a functional equation from which it follows that it has zeros at

$$s = -2, -4, \ldots, -2n, \ldots$$

(so called *trivial zeros*), and that all other zeros lie in the strip $0 \leq Res \leq 1$, so called *critical strip*.

The *Riemann Hypothesis* asserts that *all the non-trivial zeros of the zeta-function actually lie on the line $Res = \frac{1}{2}$*, which is called *critical line*. The conjecture currently remains unproven.

Numerical computations have been made to check the Riemann Hypothesis for the first $1.5 \cdot 10^9$ zeros.

5.1.9. B. Riemann first realized the connection between the distribution of primes and the distribution of the non-trivial zeros of the zeta-function. Further progress led to proving of the Prime number theorem in 1896, independently by J. Hadamard and Ch.J. de la Vallée Poussin:

$$\pi(x) = \frac{x}{\log x} + o\left(\frac{x}{\log x}\right), \quad \text{i.e.,} \quad \lim_{x \to \infty} \frac{\pi(x)}{x/\log x} = 1.$$

The main part of the proof was establishing that $\zeta(1 + it) \neq 1$ for any $t \in \mathbb{R}$.

More exactly, Vallée Poussin (1899) showed that, for some constant a, it holds

$$\pi(x) = Li(x) + O\left(\frac{x}{\ln x} e^{-a\sqrt{\ln x}}\right).$$

In 1901, H. von Koch showed that if the Riemann Hypothesis is true, then

$$\pi(x) = Li(x) + O\left(\sqrt{x} \ln x\right),$$

which can be rewritten in the slightly weaker form as

$$\pi(x) = Li(x) + O\left(x^{1/2+\epsilon}\right).$$

So, from the Riemann Hypothesis we can obtain good estimation for $\pi(x)$ with an error term of the order $\sqrt{x}$. The best bounds known today are far from this result.

The logarithmic integral $Li(x)$ is larger than $\pi(x)$ for "small" values of x. However, in 1914, J.E. Littlewood proved that this is not always the case. The first value of x where $\pi(x)$ exceeds $Li(x)$ is around $x = 10^{316}$.

As a consequence of the Prime number theorem, one gets an asymptotic expression for the n-th prime number:

$$p_n \sim n \ln n.$$

A better approximation is

$$p_n = n \ln n + n \ln \ln n + O\left(\frac{n}{\ln n} \ln \ln n\right).$$

5.1.10. In the first half of the XX-th century, some mathematicians felt that there exists a hierarchy of techniques in Mathematics, and that the prime number theorem is a "deep" theorem, whose proof requires the Complex Analysis. Methods with only real variables were supposed to be inadequate. G.H. Hardy was one notable member of this group.

The formulation of this belief was somewhat shaken by a proof of the Prime number theorem based on the *Wiener's tauberian theorem*, though this could be circumvented by awarding Wiener's theorem "depth" itself equivalent to the complex methods.

However, P. Erdős and A. Selberg found a so-called "elementary" proof of the Prime number theorem in 1949. This proof uses only number-theoretical means. The Selberg-Erdős work effectively laid rest to the whole concept of "depth", showing that technically "elementary" methods were sharper than previously expected.

Subsequent development of sieve methods showed they had a definite role in the Theory of prime numbers.

For references and some additional information see, for example, [Chan70], [Ingh32], [Ivic85], [Kara83], [KaVo92], [MaSt15], [Titc87].

Exercises

1. Find $\pi(100)$, $\pi(\sqrt{50})$, $\pi(\log_8 121)$, $\pi(p_{13})$ $\pi(p_{17})$; $\pi(p_{17} - 1)$; $\pi(p_{17} + 1)$.

2. Prove, that $\pi(x) \geq \log_2 \log_2 x$.

3. Prove, that $\pi(x) \geq \log_2 \log_2 x + 4$ for any $x \geq 17$.

4. Prove, that $\pi(x) = o(x)$ as $x \to \infty$.

5. Prove, that there are no triples $(p, p+2, p+4)$ of prime numbers.

6. Find first five pairs of cousin primes $(p, p+4)$.

7. Find all twin prime pairs, in which one number is a Mersenne prime, and another is a Fermat prime.

5.2   Problems in Number Theory, connected with Mersenne numbers

Mersenne numbers and perfect numbers: history of the question

The *Euclid–Euler theorem* is a theorem in Mathematics that relates perfect numbers to Mersenne primes. It states that *an even number is perfect if and only if it has the form* $2^{p-1}(2^p - 1)$, *where* $2^p - 1$ *is a prime number*. The theorem is named after Euclid and L. Euler.

5.2.1. In Number Theory, a *perfect number* is a positive integer that is equal to the sum of its positive divisors, excluding the number itself. For instance, 6 has divisors 1, 2 and 3 (excluding itself), and $1 + 2 + 3 = 6$, so 6 is a perfect number.

This definition is ancient, appearing as early as Euclid's *Elements* (VII.22), where it is called *perfect, ideal,* or *complete number*.

The first few perfect numbers are 6, 28, 496 and 8128 (sequence A000396 in the OEIS).

In about 300 BC Euclid proved that $2^{p-1}(2^p - 1)$ *is an even perfect number whenever* $2^p - 1$ *is prime* (*Elements*, Proposition IX.36).

Two millennia later, L. Euler proved that *all even perfect numbers are of this form*. This is known as the *Euclid–Euler theorem*.

The first four perfect numbers were the only ones known to early Greek Mathematics.

Nicomachus noted 8128 as early as around AD 100. He stated also (without proof) that every perfect number is of the form $2^{n-1}(2^n-1)$, where $2^n - 1$ is a prime.

Philo of Alexandria in his first-century book *On the creation* mentioned perfect numbers, claiming that the world was created in 6 days and the moon orbits in 28 days because 6 and 28 are perfect.

Origen added the observation that there are only four perfect numbers that are less than 10000.

St. Augustine considered perfect numbers in *City of God* (Book XI, Chapter 30) in the early V-th century AD, repeating the claim that God created the world in 6 days because 6 is the smallest perfect number.

In XIII-th century, the Egyptian mathematician Ismail ibn Fallus mentioned the next three perfect numbers (33550336, 8589869056, and 137438691328) and listed a few more which are now known to be incorrect.

The first known European mention of the fifth perfect number is a manuscript written between 1456 and 1461 by an unknown mathematician.

In 1588, the Italian mathematician P. Cataldi identified the sixth (8589869056) and the seventh (137438691328) perfect numbers, and also proved that every perfect number obtained from Euclid's rule ends with a 6 or an 8.

Although Nicomachus had stated that all perfect numbers were of the form $2^{n-1}(2^n - 1)$, where $2^n - 1$ is prime, it was not until the XVIII-th century that L. Euler proved that the formula $2^{p-1}(2^p - 1)$ will yield all the even perfect numbers.

An exhaustive search by the GIMPS distributed computing project has shown that the first 47 even perfect numbers are $2^{p-1}(2^p - 1)$ for $p = 2, 3, 5, 7, 13, 17, 19, 31, 61, 89, 107, 127, 521, 607, 1279, 2203, 2281, 3217, 4253, 4423, 9689, 9941, 11213, 19937, 21701, 23209, 44497, 86243, 110503, 132049, 216091, 756839, 859433, 1257787, 1398269, 2976221, 3021377, 6972593, 13466917, 20996011, 24036583, 25964951, 30402457, 32582657, 37156667, 42643801$ and 43112609 (sequence A000043 in the OEIS).

Four higher perfect numbers have also been discovered, namely those for which $p = 57885161$, 74207281, 77232917, and 82589933, though there may be others within this range.

As of December 2020, 51 Mersenne primes, and therefore, 51 even perfect numbers are known. The largest of them is $2^{82589932}(2^{82589933} - 1)$ with 49724095 digits.

It has been conjectured that *there are infinitely many Mersenne primes.*

Although the truth of this conjecture remains unknown, it is equivalent, by the Euclid–Euler theorem, to the conjecture that *there are infinitely many even perfect numbers.*

However, it is also unknown whether there exists even a single odd perfect number.

In fact, odd perfect numbers either do not exist or are rare. There are a number of results on perfect numbers that are actually quite easy to prove but nevertheless superficially impressive; some of them obviously come under Richard Guy's strong law of small numbers.

Some additional information see, for example, in [Deza17], [Deza18], [Dick05], [Prim20].

Proof of the Euclid–Euler theorem

5.2.2. The sum $s(n)$ of divisors of a number n, excluding the number itself, is called its *aliquot sum*, so a perfect number is one that is equal to its aliquot sum: $s(n) = n$.

Equivalently, a perfect number is a number that is half the sum of all of its positive divisors including itself; in symbols,

$$\sigma(n) = 2n, \quad \text{where} \quad \sigma(n) = \sum_{d \mid n} d.$$

For instance, 28 is perfect as $1 + 2 + 4 + 7 + 14 + 28 = 56 = 2 \cdot 28$.

The *Euler's proof* depends on the fact that the *sum of divisors function* $\sigma(n)$ is multiplicative; that is, if a and b are any two relatively prime integers, then

$$\sigma(a \cdot b) = \sigma(a) \cdot \sigma(b).$$

For this formula to be valid, the sum of divisors of a number must include the number itself, not just the proper divisors.

5.2.3. Consider the Euclid's part of the theorem (sufficiency):
- *any number $2^{k-1}(2^k - 1)$ with a prime number $2^k - 1$ is a perfect number.*

□ The part already proved by Euclid immediately follows from the multiplicative property: every Mersenne prime gives rise to an even perfect number. When $2^p - 1$ is prime,

$$\sigma(2^{p-1}(2^p - 1)) = \sigma(2^{p-1})\sigma(2^p - 1).$$

The divisors of 2^{p-1} are $1, 2, 4, 8, ..., 2^{p-1}$. The sum of these divisors is a geometric series whose sum is $2^p - 1$.

Next, since $2^p - 1$ is prime, its only divisors are 1 and itself, so the sum of its divisors is 2^p.

Combining these,

$$\sigma(2^{p-1}(2^p - 1)) = \sigma(2^{p-1})\sigma(2^p - 1)$$
$$= (2^p - 1)(2^p)$$
$$= 2(2^{p-1})(2^p - 1).$$

Therefore, $2^{p-1}(2p - 1)$ is perfect. □

5.2.4. Consider the Euler's part of the theorem (necessity):
- *if n is an even perfect number, then it has the form $n = 2^{k-1}(2^k - 1)$, where $2^k - 1 \in P$.*

□ In this direction, suppose that an even perfect number has been given, and partially factor it as $2^k x$, where x is odd.

For $2^k x$ to be perfect, the sum of its divisors must be twice its value:

$$2^{k+1}x = \sigma(2^k x) = (2^{k+1} - 1)\sigma(x).$$

The odd factor $2^{k+1} - 1$ on the right side is at least 3, and it must divide x, the only odd factor on the left side, so $y = \frac{x}{2^{k+1}-1}$ is a proper divisor of x:

$$x = (2^{k+1} - 1)y.$$

The factor 2^{k+1} on the left side must divide $\sigma(x)$, the only even factor on the right side, so

$$y = \frac{\sigma(x)}{2^{k+1}}$$

is a proper divisor of $\sigma(x)$:

$$\sigma(x) = 2^{k+1}y.$$

Dividing both sides of by the common factor $2^{k+1} - 1$ gives

$$2^{k+1}y = \sigma(x).$$

Taking into account the known divisors $x = (2^{k+1} - 1)y$ and y of x gives

$$\sigma(x) \geq y + x = y + (2^{k+1} - 1)y.$$

For the equality above to be true, there can be no other divisors of x. Therefore, y must be 1, and x must be a prime of the form $2^{k+1} - 1$. $\square$

Some additional information see, for example, in [Buch09], [DeKo13], [Deza17], [Deza18], [Dick05], [Sier64], [Step01].

Properties of perfect numbers

5.2.5. As well as having the form $2^{p-1}(2^p - 1)$, *each even perfect number is the* $(2^p - 1)$*-th triangular number*, and, hence, *is equal to the sum of the integers from 1 to* $2^p - 1$:

$$2^{p-1}(2^p - 1) = S_3(2^p - 1) = 1 + 2 + 3 + \cdots + (2^p - 1).$$

Similarly, $2^{p-1}(2^p - 1)$ *is the* 2^{p-1}*-th hexagonal number*:

$$2^{p-1}(2^p - 1) = S_6(2^p).$$

However, *the even perfect numbers are not trapezoidal numbers* (see Chapter 4, section 4.5); that is, they cannot be represented as the difference of two positive non-consecutive triangular numbers.

Furthermore, *each even perfect number except for* 6 *is the* $(\frac{2^p+1}{3})$-*th centered nonagonal number and is equal to the sum of the first* $2^{\frac{p-1}{2}}$ *odd cubes*:

$$6 = 2^1(2^2 - 1) \qquad = 1 + 2 + 3,$$

$$28 = 2^2(2^3 - 1) \qquad = 1 + 2 + 3 + 4 + 5 + 6 + 7 = 1^3 + 3^3,$$

$$496 = 2^4(2^5 - 1) \qquad = 1 + 2 + 3 + \cdots + 29 + 30 + 31$$
$$= 1^3 + 3^3 + 5^3 + 7^3,$$

$$8128 = 2^6(2^7 - 1) \qquad = 1 + 2 + 3 + \cdots + 125 + 126 + 127$$
$$= 1^3 + 3^3 + 5^3 + 7^3 + 9^3 + 11^3 + 13^3$$
$$+ 15^3,$$

$$33550336 = 2^{12}(2^{13} - 1) \qquad = 1 + 2 + 3 + \cdots + 8189 + 8190 + 8191$$
$$= 1^3 + 3^3 + 5^3 + \cdots + 123^3 + 125^3 + 127^3.$$

5.2.6. A *pernicious number* is a positive integer such that the *Hamming weight* of its binary representation (i.e., the sum of binary digits) is prime.

The first pernicious number is 3, since $3 = 11_2$ and $1 + 1 = 2$, which is a prime. The next pernicious number is 5, since $5 = 101_2$, followed by 6, 7 and 9 (sequence A052294 in the OEIS).

Owing to their form, $2^{p-1}(2^p - 1)$, every even perfect number is represented in binary form as p ones followed by $p - 1$ zeros; for example,

$$6_{10} = 2^2 + 2^1 = 110_2;$$

$$28_{10} = 2^4 + 2^3 + 2^2 = 11100_2;$$

$$496_{10} = 2^8 + 2^7 + 2^6 + 2^5 + 2^4 = 111110000_2;$$

$$8128_{10} = 2^{12} + 2^{11} + 2^{10} + 2^9 + 2^8 + 2^7 + 2^6 = 1111111000000_2.$$

Thus, *every even perfect number is a pernicious number.*

5.2.7. A *practical number* (or *panarithmic number*) is a positive integer n such that all smaller positive integers can be represented as sums of distinct divisors of n.

For example, 12 is a practical number because all the numbers from 1 to 11 can be expressed as sums of its divisors 1, 2, 3, 4, and 6: as well as these divisors themselves, we have

$$5 = 3 + 2, \; 7 = 6 + 1, \; 8 = 6 + 2,$$

$$9 = 6 + 3, \; 10 = 6 + 3 + 1, \; 11 = 6 + 3 + 2.$$

The sequence of practical numbers begins 1, 2, 4, 6, 8, 12, 16, 18, 20, 24, ... (sequence A005153 in the OEIS).

It is easy to show, that *every even perfect number is also a practical number.*

5.2.8. An *Ore number* (or *harmonic divisor number*) is a positive integer whose divisors have a *harmonic mean* that is an integer.

The *harmonic mean* can be expressed as the reciprocal of the arithmetic mean of the reciprocals of the given set of observations.

As a simple example, the harmonic mean of the four divisors 1, 2, 3, 6 of 6 is an integer:

$$\frac{4}{\frac{1}{1} + \frac{1}{2} + \frac{1}{3} + \frac{1}{6}} = 2.$$

The first Ore numbers are 1, 6, 28, 140, 270, 496, 672, 1638, 2970, 6200, 8128, 8190, ... (sequence A001599 in the OEIS).

In fact, the reciprocals of the divisors of any perfect number n must add up to 2. To get this, take the definition of a perfect number, $\sigma(n) = 2n$ and divide both sides by n.

So, for 6 and 28, we have

$$\frac{1}{6} + \frac{1}{3} + \frac{1}{2} + \frac{1}{1} = 2;$$

$$\frac{1}{28} + \frac{1}{14} + \frac{1}{7} + \frac{1}{4} + \frac{1}{2} + \frac{1}{1} = 2.$$

Moreover, the number of divisors of a perfect number n (whether even or odd) must be even, because n cannot be a perfect square.

From these two results it follows that *every perfect number is an Ore number.*

Some additional information see, for example, in [Buch09], [Deza17], [Deza18], [Sier64], [Step01], [Wiki20].

Generalizations of perfect numbers

5.2.9. The sum of proper divisors gives various other kinds of numbers.

Numbers where the sum is less than the number itself are called *deficient*, and where it is greater than the number, *abundant*. These terms, together with *perfect* itself, come from Greek Numerology.

The first few deficient numbers are 1, 2, 3, 4, 5, 7, 8, 9, 10, 11, ... (sequence A005100 in the OEIS), while the first few abundant numbers are 12, 18, 20, 24, 30, 36, 40, 42, 48, 54, ... (sequence A005101 in the OEIS).

A pair of numbers which are the sum of each other's proper divisors are called *amicable*, and larger cycles of numbers are called *sociable*.

The first few amicable pairs are (220, 284), (1184, 1210), (2620, 2924), (5020, 5564), (6232, 6368), (10744, 10856), (12285, 14595), (17296, 18416), (63020, 76084), (66928, 66992), ... (sequence A259180 in the OEIS; see also A002025, and A002046).

A *semiperfect number* is a natural number that is equal to the sum of all or some of its proper divisors.

The first few semiperfect numbers are 6, 12, 18, 20, 24, 28, 30, 36, 40, 42, ... (sequence A005835 in the OEIS).

A semiperfect number that is equal to the sum of all its proper divisors is a perfect number.

So, *any perfect number is semiperfect.*

Most abundant numbers are also semiperfect; abundant numbers which are not semiperfect are called *weird numbers*. They start from 70, 836, 4030, 5830, 7192, 7912, 9272, 10430, 10570, 10792, ... (sequence A006037 in the OEIS).

An *k-hyperperfect number* is a natural number n for which the equality

$$n = 1 + k(\sigma(n) - n - 1)$$

holds. A *hyperperfect number* is an k-hyperperfect number for some integer k.

Hyperperfect numbers generalize perfect numbers, which are *1-hyperperfect.*

The first few numbers in the sequence of k-hyperperfect numbers are 6, 21, 28, 301, 325, 496, 697, 1333, 1909, 2041, ... (sequence A034897 in the OEIS), with the corresponding values of k being 1, 2, 1, 6, 3, 1, 12, 18, 18, 12, ... (sequence A034898 in the OEIS).

The first few k-hyperperfect numbers that are not perfect are 21, 301, 325, 697, 1333, 1909, 2041, 2133, 3901, 10693, ... (sequence A007592 in the OEIS).

A number n is called *k-perfect* (or *k-fold perfect*) if the sum of all positive divisors of n is equal to kn. A number that is k-perfect for a certain k is called a *multiply perfect number*.

So, any perfect number is *2-perfect*, and hence, *multiply perfect*.

As of 2020, k-perfect numbers are known for each value of k up to 11.

The proofs of these properties and some additional information see, for example, in [Buch09], [DeDe12], [Deza17], [Deza18], [DeKo13], [JoLo99], [Wiki20].

Exercises

1. Prove, that even perfect numbers are of the form $S_3(2^p - 1)$, where $S_3(n)$ is the n-th triangle number; find first resulting triangular numbers.

2. Prove, that even perfect numbers greater 6 are of the form $1 + 9 \cdot S_3(\frac{2^p - 2}{3})$, where $S_3(n)$ is the n-th triangle number; find first resulting triangular numbers. Prove, that their last digits are 3 or 5.

3. Prove, that every even perfect number ends in 6 or 28, base 10.

4. Prove, that every even perfect number greater 6 ends in 1, base 9.

5. Check, that adding the digits of any even perfect number greater 6, then adding the digits of the resulting number, and repeating this process until a single digit (called the digital root) is obtained, always produces the number 1.

6. Prove, that the only square-free (even) perfect number is 6.

5.3 Problems in Number Theory, connected with Fermat numbers

The famous mathematical fact, connected with Fermat numbers, is the *Gauss–Wantzel theorem* on construction of a regular polygon with compass and straightedge.

Constructible polygons: history of the question

5.3.1. A *regular polygon* is a polygon that is *equiangular* (all angles are equal in measure) and *equilateral* (all sides have the same length).

A *constructible polygon* is a regular polygon that can be constructed with compass and straightedge.

Some regular polygons are easy to construct with compass and straightedge; others are not.

The ancient Greek mathematicians knew how to construct regular polygons with 3, 4, or 5 sides.

They knew also how to construct a regular polygon with double the number of sides of a given regular polygon. So, they knew the constructions of any regular n-gon with

$$n = 2^{k+2}, \ n = 3 \cdot 2^k, n = 5 \cdot 2^k \ k \geq 0.$$

This led to the question being posed: *is it possible to construct all regular polygons with compass and straightedge? If not, which n-gons (that is polygons with n edges) are constructible and which are not?*

Carl Friedrich Gauss proved the constructibility of the regular 17-gon in 1796. Five years later, he developed the theory of Gaussian periods in his *Disquisitiones Arithmeticae*. This theory allowed him to formulate a sufficient condition for the constructibility of regular polygons: *a regular n-gon can be constructed with compass and straightedge if n is the product of a power of 2 and any number of distinct Fermat primes (including none).*

Gauss stated without proof that this condition was also necessary, but never published his proof. A full proof of necessity was given by

Pierre Wantzel in 1837. The result is known as the *Gauss–Wantzel theorem*.

Theorem (Gauss–Wantzel theorem). *A regular n-gon can be constructed with compass and straightedge if and only if n is the product of a power of 2 and any number of distinct Fermat primes (including none).*

The five known Fermat primes are:

$F_0 = 3$, $F_1 = 5$, $F_2 = 17$, $F_3 = 257$, and $F_4 = 65537$ (sequence A019434 in the OEIS).

The next twenty-eight Fermat numbers, F_5 through F_{32}, are known to be composite.

Thus, a regular n-gon is constructible if $n = 3$, 4, 5, 6, 8, 10, 12, 15, 16, 17, ... (sequence A003401 in the OEIS), while a regular n-gon is not constructible if $n = 7$, 9, 11, 13, 14, 18, 19, 21, 22, 23, ... (sequence A004169 in the OEIS).

Since there are 31 combinations of anywhere from one to five Fermat primes, there are 31 known constructible polygons with an odd number n of sides. These n are 3, 5, 15, 17, 51, 85, 255, 257, 771, 1285, 3855, 4369, 13107, 21845, 65535, 65537, 196611, 327685, 983055, 1114129, 3342387, 5570645, 16711935, 16843009, 50529027, 84215045, 252645135, 286331153, 858993459, 1431655765, 4294967295 (sequence A045544 in the OEIS).

So, there are infinitely many constructible polygons, but only 31 with an odd number of sides are known.

For some additional information see, for example, [Deza17], [Dick05], [Sier64], [Step01], [Wiki20].

Constructible polygons and constructible numbers

5.3.2. In order to reduce a geometric problem to a problem of pure Number Theory, the proof uses the fact that a regular n-gon is constructible if and only if the cosine of its central angle, $\cos \frac{2\pi}{n}$, is a *constructible number*: that is, can be written in terms of the four basic arithmetic operations and the extraction of square roots.

So, the *Gauss–Wantzel theorem* states, that

- $\cos \frac{2\pi}{n}$ *is a constructible number if and only if* n *is the product of a power of* 2 *and any number of distinct Fermat primes (including none).*

Compass and straightedge constructions are known for all known constructible polygons.

It is easy to check, that *if* $n = p \cdot q$ *with* $p = 2$ *and* q *odd, or with* p *and* q *odd coprime, an* n*-gon can be constructed from a* p*-gon and a* q*-gon.*

□ Consider two possibilities.

- If $p = 2$, draw a q-gon and bisect one of its central angles. From this, a $2q$-gon can be constructed.

- If $p > 2$, inscribe a p-gon and a q-gon in the same circle in such a way that they share a vertex. Because p and q are relatively prime, there exists integers a, b such that $ap + bq = 1$. Then $\frac{2a\pi}{q} + \frac{2b\pi}{p} = \frac{2\pi}{pq}$, i.e., the number $\frac{2\pi}{pq}$ is constructible. From this, an pq-gon can be constructed.

It concludes the proof. □

Thus, one only has to find a compass and straightedge construction for n-gons where n is a Fermat prime.

- The construction for an equilateral triangle is simple and has been known since Antiquity. In fact,

$$\cos \frac{2\pi}{3} = \frac{1}{2}.$$

- Constructions for the regular pentagon were described both by Euclid (*Elements*, circa 300 BC), and by Ptolemy (*Almagest*, circa AD 150). In fact,

$$\cos \frac{2\pi}{5} = \frac{1 + \sqrt{5}}{4}.$$

- Although C.F. Gauss proved that the regular 17-gon is constructible, he did not actually show how to do it. The first construction is due to Erchinger, a few years after Gauss' work.

The explicit construction of a heptadecagon was given by Herbert William Richmond in 1893. In fact,

$$16 \cos \frac{2\pi}{17} = -1 + \sqrt{17} + \sqrt{34 - 2\sqrt{17}}$$

$$+ 2\sqrt{17 + 3\sqrt{17} - \sqrt{34 - 2\sqrt{17}} - 2\sqrt{34 + 2\sqrt{17}}}$$

$$= -1 + \sqrt{17} + \sqrt{34 - 2\sqrt{17}}$$

$$+ 2\sqrt{17 + 3\sqrt{17} - \sqrt{170 + 38\sqrt{17}}}.$$

- The first explicit constructions of a regular 257-gon were given by Magnus Georg Paucker (1822) and Friedrich Julius Richelot (1832).

- A construction for a regular 65537-gon was first given by Johann Gustav Hermes (1894). The construction is very complex; J.H. Hermes spent 10 years completing the 200-page manuscript.

For some additional information see, for example, [Deza17], [Dick05], [Gaus01], [IrRo90], [KLS01], [Kost82], [Madd05], [Tsan10], [Sier64], [Shko61], [Step01], [Wiki20].

Other concepts of constructibility

5.3.3. The concept of constructibility as discussed before applies specifically to compass and straightedge construction.

More constructions become possible if other tools are allowed.

The so-called *neusis constructions*, for example, make use of a marked ruler. Neuseis construction have been important because it sometimes provides a means to solve geometric problems that are not solvable by means of compass and straightedge alone. Examples are the *trisection of any angle in three equal parts*, the *doubling of the cube*, and the *construction of a regular heptagon* (7-gon), *nonagon* (9-gon), or *tridecagon* (13-gon).

In fact, *a regular polygon with n sides can be constructed with ruler, compass, and angle trisector if and only if*

$$n = 2^r 3^s p_1 p_2 ... p_k,$$

where $r, s, k \geq 0$, *and* $3 < p_1 < ... < p_k$ *are distinct Pierpont primes.*

Here, a *Pierpont prime* is a prime number of the form $2^u 3^v + 1$ with some nonnegative integers u and v.

The first few Pierpont primes are 2, 3, 5, 7, 13, 17, 19, 37, 73, 97,... (sequence A005109 in the OEIS).

An Pierpont prime $2^u 3^v + 1$ with $v = 0$ is a Fermat prime. So, all constructible polygons can be constructed with ruler, compass, and angle trisector, too.

As 7 is a Pierpont prime, the regular *heptagon*, which is not constructible with compass and straightedge, is constructible with ruler, compass, and angle trisector. It is the smallest regular polygon with this property.

The smallest prime that is not a Pierpont (or Fermat) prime is 11; therefore, the regular *hendecagon* is the smallest regular polygon that cannot be constructed with compass, straightedge and angle trisector (or *origami*, or *conic sections*).

For some additional information see, for example, [Dick05], [KLS01], [Madd05], [Ore48], [Tsan10], [Sier64].

Sketch of a proof of the Gauss–Wantzel theorem

Consider the sketch of the proof of the Gauss–Wantzel theorem.

5.3.4. First notice that constructing a regular n-gon is equivalent to constructing the n-th roots of unity in the complex plane, starting with the point $(1, 0)$.

It is because the n-th roots of unity form a regular n-gon on the complex plane.

Now consider a *primitive n-th root of unity*

$$\zeta_n = e^{\frac{2\pi i}{n}} = \cos \frac{2\pi}{n} + i \sin \frac{2\pi}{n}.$$

So, now all the n-th roots of unity are

$$1, \zeta_n, \zeta_n^2, \ldots, \zeta_n^{n-1}.$$

As the set of constructible numbers is a field, these numbers will be constructible if and only if ζ_n is constructible. Hence a regular n-gon is constructible if and only if ζ_n is a constructible number.

5.3.5. Now by the characterization of constructible numbers, ζ_n will be constructible if and only if there is a chain of field extensions

$$\mathbb{Q} = K_0 \subseteq K_1 \subseteq \ldots \subseteq K_m = \mathbb{Q}(\zeta_n)$$

such that each extension $K_i \subseteq K_{i+1}$ is quadratic, i.e., $[K_{i+1} : K_i] = 2$.

We claim that this happens if and only if $\phi(n)$ is a power of 2, where $\phi(n)$ is the *Euler's totient function*:

$$\phi(n) = 2^m.$$

First we note that $\mathbb{Q}(\zeta_n)$ is the *n-th cyclotomic field,* and hence

$$[\mathbb{Q}(\zeta_n) : \mathbb{Q}] = \phi(n).$$

Assume that such a chain of field extensions exists. Then by the *tower law:*

$$\phi(n) = [\mathbb{Q}(\zeta_n) : \mathbb{Q}] = [K_m : K_{m-1}] \cdot [K_{m-1} : K_{m-2}] \cdot \ldots \cdot [K_1 : K_0]$$
$$= 2 \cdot 2 \cdot \ldots \cdot 2 = 2^m,$$

as desired.

Now assume that $\phi(n) = 2^m$, for some m. The field $\mathbb{Q}(\zeta_n)$ is the splitting field of *n-th cyclotomic polynomial,* $\Phi_n(x)$, and hence $\mathbb{Q}(\zeta_n)/\mathbb{Q}$ is a *Galios extension.*

Now the order of the *Galios group* $G = Gal(\mathbb{Q}(\zeta_n)/\mathbb{Q})$ is just

$$[\mathbb{Q}(\zeta_n) : \mathbb{Q}] = 2^m.$$

Thus, G is a *2-group,* and hence there must exist a chain of subgroups

$$G = G_m > G_{m-1} > \ldots > G_0 = 1$$

with $[G_{i+1} : G_i] = 2$ for all i. Now by the *Fundamental theorem of Galois theory,* if K_i is the fixed field of G_i, then

$$\mathbb{Q} = K_m \subseteq K_{m-1} \subseteq \ldots \subseteq K_0 = \mathbb{Q}(\zeta_n),$$

and $[K_i : K_{i+1}] = [G_{i+1} : G_i] = 2$. Therefore, ζ_n is indeed constructible.

5.3.6. We have now shown that a regular n-gon is constructible if and only if $\phi(n)$ is a power of 2. It only remains to show that the integers which satisfy this are precisely the integers in the given form.

Let the prime factorization of n be

$$n = 2^\alpha \cdot p_1^{\alpha_1} \cdot p_2^{\alpha_2} \cdot \ldots \cdot p_k^{\alpha_k},$$

where $p_1, p_2, \ldots, p_k$ are distinct odd primes, $\alpha_i \geq 1$ for all i, and $\alpha \geq 0$.

Then by the formula for $\phi(n)$ we have

$$\phi(n) = 2^{\alpha-1} \cdot p_1^{\alpha_1-1} \cdot p_2^{\alpha_2-1} \cdot \ldots \cdot p_n^{\alpha_n-1} \cdot (p_1 - 1)(p_2 - 1)\ldots(p_n - 1),$$

or, if $\alpha = 0$,

$$\phi(n) = p_1^{\alpha_1-1} p_2^{\alpha_2-1} \ldots p_n^{\alpha_n-1}(p_1 - 1)(p_2 - 1)\ldots(p_n - 1).$$

Assume that $\phi(n) = 2^m$. As a power of 2 cannot be divisible by an odd prime, we cannot have $\alpha_i > 1$ for any i (otherwise we would have $p_i | 2^m$), so

$$\alpha_1 = \alpha_2 = \ldots = \alpha_n = 1.$$

Also, any divisor of a power of 2 must also be a power of 2, so $p_i - 1$ is a power of 2 for each i, from which is easily follows that each p_i is a Fermat prime. Hence, n has the desired form.

Conversely assume that

$$n = 2^a p_1 \cdot p_2 \cdot \ldots \cdot p_k,$$

where $p_1, p_2, \ldots, p_k$ are distinct Fermat primes, say, $p_i = 2^{2^{\beta_i}} + 1$. Then

$$\phi(n) = 2^{a-1} \cdot (p_1 - 1)(p_2 - 1)\ldots(p_n - 1) = 2^{a-1} \cdot 2^{2^{\beta_1}} \cdot 2^{2^{\beta_2}} \cdot \ldots \cdot 2^{2^{\beta_k}},$$

or, of $\alpha = 0$,

$$\phi(n) = (p_1 - 1)(p_2 - 1)\ldots(p_n - 1) = 2^{2^{\beta_1}} \cdot 2^{2^{\beta_2}} \cdot \ldots \cdot 2^{2^{\beta_k}}.$$

In either case this is a power of 2, as desired. This completes the sketch of the proof. □

The fool proof of the theorem and some additional information see, for example, in [Carm19], [Deza17], [Gaus01], [Gelf98], [Glea88], [KLS01], [Madd05], [Pier95], [Shko61], [Tsan10], [Want36].

Exercises

1. Construct a regular triangle with compass and straightedge.

2. Construct a regular hexagon with compass and straightedge.

3. Construct a square with compass and straightedge.

4. Construct a regular octagon with compass and straightedge.

5. Prove, that $\cos \frac{2\pi}{5} = \frac{1+\sqrt{5}}{4}$.

6. Find all solutions of the equation $\phi(x) = 8$; $\phi(x) = 9$; $\phi(x) = \frac{x}{2}$; $\phi(x) = x$.

7. Find first ten Pierpont primes.

8. Prove, that any Fermat prime is an Pierpont prime.

9. Prove, that all odd non-Fermat Pierpont primes have the form $6k + 1$.

10. Check, that 11 is a Pierpont prime of the second kind, i.e., a prime number of the form $2^u 3^v - 1$, $u, v \geq 0$.

11. Find first ten Pierpont primes of the second kind.

12. Prove, that any Mersenne prime is an Pierpont prime of the second kind.

5.4   Prime numbers records and Mersenne numbers

Prime numbers have long fascinated both amateurs and professional mathematicians.

At present, the research of new large primes is an important part of pure Mathematics and Computer Science. In the last few years, several distributed computing projects, including *PrimeGrid, Great Internet Mersenne Prime Search, Fermat Search, Twin Prime Search* etc., are hunting for very large prime numbers whilst also aiming to solve long-standing mathematical conjectures.

The PrimePages: Prime number research and record

5.4.1. The *PrimePages* is a website about prime numbers maintained by C. Caldwell at the University of Tennessee at Martin.

The site maintains the list of the "5000 largest known primes", selected smaller primes of special forms, and many "top twenty" lists for primes of various forms.

The PrimePages has articles on primes and primality testing. It includes "The Prime Glossary" with articles on hundreds of glosses related to primes, and "Prime Curios!" with thousands of curios about specific numbers.

The database started as a list of *titanic primes* (primes with at least 1000 decimal digits) by S. Yates. In subsequent years, the whole *top-5000* has consisted of *gigantic primes* (primes with at least 10000 decimal digits).

Primes of special forms are kept on the current lists if they are titanic and in the *top-20* or *top-5* for their form.

Any record in these lists of the *top-20* (or *top-10*, or *top-5*) is a testament to the incredible amount of work put in by the programmers, project directors, and the tens of thousands of enthusiasts. In the table below we list the 20 largest known primes.

20 Largest Known Primes

	Primee	Number of digits	Year	Comment
1	$2^{82589933} - 1$	24862048	2018	51-th(?) Mersenne prime
2	$2^{77232917} - 1$	23249425	2018	50-th(?) Mersenne prime
3	$2^{74207281} - 1$	22338618	2016	49-th(?) Mersenne prime
4	$2^{57885161} - 1$	17425170	2013	48-th(?) Mersenne prime
5	$2^{43112609} - 1$	12978189	2008	47-th Mersenne prime
6	$2^{42643801} - 1$	12837064	2009	46-th Mersenne prime
7	$2^{37156667} - 1$	11185272	2008	45-th Mersenne prime
8	$2^{32582657} - 1$	9808358	2006	44-th Mersenne prime
9	$10223 \cdot 2^{31172165} + 1$	9383761	2016	
10	$2^{30402457} - 1$	9152052	2005	43-th Mersenne prime
11	$2^{25964951} - 1$	7816230	2005	42-th Mersenne prime
12	$2^{24036583} - 1$	7235733	2004	41-th Mersenne prime
13	$2^{20996011} - 1$	6320430	2003	40-th Mersenne prime

(*continued*)

(Continued)

	Primee	Number of digits	Year	Comment
14	$1059094^{1048576} + 1$	6317602	2018	Generalized Fermat prime
15	$919444^{1048576} + 1$	6253210	2017	Generalized Fermat prime
16	$168451 \cdot 2^{19375200} + 1$	5832522	2017	
17	$7 \cdot 2^{18233956} + 1$	5488969	2020	Divides Fermat number $F_{18233954}$
18	$123447^{1048576} - 123447^{524288} + 1$	5338805	2017	Generalized unique prime
19	$7 \cdot 6^{6772401} + 1$	5269954	2019	
20	$8508301 \cdot 2^{17016603} - 1$	5122515	2018	

See for an additional information [Prim20], [Wiki20].

Mersenne primes and GIMPS

5.4.2. The largest known prime has almost always been a Mersenne prime.

Why? Because the way the largest numbers N are proven prime is based on the factorizations of either $N + 1$ or $N - 1$. For Mersenne mumbers the factorization of $N + 1$ is as trivial as possible; it is a power of two.

At present there are few practical uses for this new large primes, prompting some to ask "why search for these large primes"?

Those same doubts existed a few decades ago until important cryptography algorithms were developed based on prime numbers.

5.4.3. The *Great Internet Mersenne Prime Search* (GIMPS) is a collaborative project of volunteers who use freely available software to search for Mersenne prime numbers.

GIMPS is said to be one of the first large scale distributed computing projects over the Internet for research purposes.

It has had a virtual lock on the largest known prime, because its excellent free software is easy to install and maintain, requiring little of the user.

GIMPS was founded in early 1996 by G. Woltman to discover new world record size Mersenne primes.

GIMPS is registered as *Mersenne Research, Inc.* with S. Kurowski as Executive Vice President and board director.

The name for the project was coined by L. Welsh, one of its earlier searchers and the co-discoverer of the 29-th Mersenne prime.

Within a few months, several dozen people had joined, and over a thousand by the end of the first year.

J. Armengaud, a participant, discovered the primality of $M_{1398269}$ on November 13, 1996.

In 1997 S. Kurowski enabled GIMPS to automatically harness the power of thousands of ordinary computers to search for these "needles in a haystack".

Most GIMPS members join the search for the thrill of possibly discovering a record-setting, rare, and historic new Mersenne prime.

5.4.4. As of 2020, the project has found a total of seventeen Mersenne primes, fifteen of which were the largest known prime number at their respective times of discovery.

In 1996, J. Armengaud *et al.* discovered the 35-th Mersenne prime $M_{1398269}$ in France.

In 1999, Nayan N. Hajratwala has discovered the first known million-digit prime number using software written by G. Woltman and the distributed computing technology and services of Kurowski's company.

This prime number, $2^{6972593} - 1$, is the 38-th Mersenne prime. It contains 2098960 digits.

Nayan Hajratwala used a Pentium II IBM Aptiva computer running part-time for 111 days to prove the number prime.

Running uninterrupted it would take about three weeks to test the primality of this number.

There are several prizes offered by the Electronic Frontier Foundation for record primes.

This record, passed one million digits in 1999, earning a US$50000 prize.

In 2008, the record passed ten million digits (12978189 digits of $M_{43112609}$, the 47-th Mersenne prime), earning a US$100000 prize and a Cooperative Computing Award from the Electronic Frontier Foundation. *Time* (an American weekly news magazine), called it the 29-th top invention of 2008.

Both the US$50000 and the US$100000 prizes were won by participation in GIMPS.

As of 2020, the largest known prime is $2^{82589933} - 1$ ($M_{82589933}$). It has 24862048 digits. It was discovered in 2018 by P. Laroche, one of thousands of volunteers using free GIMPS software available at *www.mersenne.org/download/*.

The new prime number $M_{82589933}$ is calculated by multiplying together 82589933 two's, and then subtracting one.

It is more than one and a half million digits larger than the previous record prime number $2^{74207281} - 1$, $M_{14207281}$ (having 23249425 digits).

It is only the 51-st known Mersenne prime ever discovered, each increasingly more difficult to find.

For many years, P. Laroche had used GIMPS software as a free "stress test" for his computer builds. On the time, he started prime hunting on his media server to "give back" to the project. After less than four months and on just his fourth try, he discovered the new prime number.

By way of comparison, some GIMPS participants have searched for more than 20 years with tens of thousands of attempts but no success. Thus proving that even the "little guy" can compete against those with lots of computing resources.

The primality proof took twelve days of non-stop computing. To prove there were no errors in the prime discovery process, the new prime was independently verified using three different programs on three different hardware configurations.

The number $M_{82589933}$ has 24862048 decimal digits. To help visualize the size of this number, if it were to be saved to disk, the resulting text file would be nearly 25 megabytes long (most books in plain text format clock in under two megabytes). A standard word processor layout (50 lines per page, 75 digits per line) would require 6629 pages to display it.

GIMPS is also coordinating its long-range search efforts for primes of 100 million digits and larger and will split the Electronic Frontier Foundation's US$150000 prize with a winning participant. Additional prize is being offered for the first prime number found with at least one billion digits.

5.4.5. GIMPS has also been extremely lucky over the last 15 years. The largest known prime $M_{82589933}$ is GIMPS' 12-th prime discovery between $2^{20000000} - 1$ and $2^{85000000} - 1$, triple the expected number of new primes in this interval (see Chapter 3, section 3.6).

One reason to search for new primes is to match actual results with expected results. This anomaly is not necessarily evidence that existing theories on the distribution of Mersenne primes is incorrect. However, if the trend continues it may be worth further investigation.

On December 4, 2020, the project passed a major milestone after all exponents below 100 million were checked at least once.

5.4.6. There is a unique history to the arithmetic algorithms underlying the GIMPS project.

In the early 1990's, R. Crandall, Apple Distinguished Scientist, discovered ways to double the speed of what are called *convolutions*, essentially big multiplication operations.

The method is applicable not only to prime searching but other aspects of computation.

During that work he also patented the *Fast Elliptic Encryption system*, now owned by Apple Computer, which uses Mersenne primes to quickly encrypt and decrypt messages.

G. Woltman implemented Crandall's algorithm in assembly language, thereby producing a prime-search program of unprecedented efficiency.

In general, GIMPS project relies primarily on the *Lucas–Lehmer primality test* as it is an algorithm that is both specialized for testing Mersenne primes and particularly efficient on binary computer architectures.

There is also a *trial division* phase, used to rapidly eliminate many Mersenne numbers with small factors.

Pollard's $(p-1)$-algorithm is also used to search for smooth factors.

In 2017, GIMPS adopted the *Fermat primality test* as an alternative option for primality testing.

5.4.7. The search for more Mersenne primes is already under way. There may be smaller, as yet undiscovered Mersenne primes, and there almost certainly are larger Mersenne primes waiting to be found.

Anyone with a reasonably powerful personal computer can join GIMPS and become a big prime hunter, and possibly earn a cash research discovery award.

All the necessary software can be downloaded for free at *www.mersenne.org/download/*.

For an additional information see [DeKo18], [Ribe89], [Ribe96], [SlPl95]; for useful connections see [GIMPS20], [Prim20], [Sloa20], [Wiki20].

Exercises

1. Prove, that F_{73} has more than $24 \cdot 10^{20}$ digits.

2. Prove, that F_{1945} has more than 10^{582} digits.

3. Using the table of Mersenne primes, find the first Mersenne titanic prime number.

4. Using the table of Mersenne primes, find the first Mersenne gigantic prime number. (A gigantic prime is a prime number with at least 10000 decimal digits.)

5. Using the table of Mersenne primes, find the first Mersenne gigantic number.

6. Using the table of Mersenne primes, find the first Mersenne megaprime. (A megaprime is a prime number with at least one million decimal digits.)

7. Using a table of primes, find the first pair of titanic twin primes.

8. Using Prime Page, find the biggest known pair of twin primes.

5.5   Mersenne and Fermat numbers in Cryptography

Public-key Cryptography and large primes

5.5.1. *Cryptography*, or *Cryptology* (from Ancient Greek's *kryptós* ("hidden, secret") and *graphein* ("to write"), or *logia* ("study"), respectively) is the practice and study of techniques for secure communication in the presence of third parties, called *adversaries*.

More generally, Cryptography is about constructing and analyzing protocols that prevent third parties or the public from reading private messages.

Various aspects in Information Security such as Data Confidentiality, Data Integrity, Authentication, and Non-Repudiation are central to modern Cryptography.

Applications of Cryptography include Electronic Commerce, Chip-based Payment Cards, Digital Currencies, Computer Passwords, and Military Communications.

Cryptography plays a major role in Digital Rights Management and Copyright Infringement disputes in regard to digital media.

Modern Cryptography exists at the intersection of the disciplines of Mathematics, Computer Science, Electrical Engineering, Communication Science, and Physics.

It is heavily based on mathematical theory and Computer Science practice; cryptographic algorithms are designed around computational hardness assumptions, making such algorithms hard to break in actual practice by any adversary. While it is theoretically possible to break into a well-designed such system, it is infeasible in actual practice to do so.

Such schemes, if well designed, are therefore termed "computationally secure"; theoretical advances, e.g., improvements in *integer factorization algorithms* and, therefore, in study and search of new "good" big numbers, and faster computing technology require these designs to be continually re-evaluated, and if necessary, adapted.

In addition to being aware of cryptographic history, cryptographic algorithm and system designers must also sensibly consider probable future developments while working on their designs. For instance, continuous improvements in computer processing power have increased the scope of brute-force attacks, so when specifying key lengths, the required key lengths are similarly advancing. The potential effects of Quantum Computing are already being considered by some cryptographic system designers developing post-quantum Cryptography.

5.5.2. Cryptosystems all involve a *key*: a piece of information necessary to encrypt information.

This can be *symmetric*, for example, in the form of a *one-time pad*, such as the string of binary bits used by the *Vernam cipher*. This cryptosystem belongs to information-theoretically secure schemes that provably cannot be broken even with unlimited computing power.

There are very few cryptosystems that are proven to be unconditionally secure; the one-time pad is one, and was proven to be so by C. Shannon.

However, these schemes are much more difficult to use in practice than the best theoretically breakable but computationally secure schemes.

Symmetric-key Cryptography refers to encryption methods in which both the sender and receiver share the same key (or, less commonly, in which their keys are different, but related in an easily computable way).

So, symmetric-key cryptosystems use the same key for encryption and decryption of a message, although a message or group of messages can have a different key than others.

A significant disadvantage of symmetric ciphers is the key management necessary to use them securely. Each distinct pair of communicating parties must, ideally, share a different key, and perhaps for each ciphertext exchanged as well. The number of keys required increases as the square of the number of network members, which very quickly requires complex key management schemes to keep them all consistent and secret.

Contemporary symmetric-key ciphers are implemented as either *block ciphers* or *stream ciphers*. A block cipher enciphers input in blocks of plaintext as opposed to individual characters, the input form used by a stream cipher.

Symmetric-key Cryptography was the only kind of encryption publicly known until 1976.

5.5.3. Extensive open academic research into Cryptography began only in the mid-1970's.

In 1976, W. Diffie and M. Hellman proposed the notion of *Public-key* (also, more generally, called *Asymmetric-key*) *Cryptography* in which two different but mathematically related keys are used: a *public key* and a *private key*.

A public-key system is so constructed that calculation of one key (the "private key") is computationally infeasible from the other (the "public key"), even though they are necessarily related.

Instead, both keys are generated secretly, as an interrelated pair.

The historian D. Kahn described Public-key Cryptography as "...the most revolutionary new concept in the field since polyalphabetic substitution emerged in the Renaissance."

In public-key cryptosystems, the public key may be freely distributed, while its paired private key must remain secret.

In a public-key encryption system, the public key is used for encryption, while the private or secret key is used for decryption.

W. Diffie and M. Hellman showed that Public-key Cryptography was indeed possible by presenting the *Diffie–Hellman key exchange protocol* (*Diffie–Hellman algorithm*, DHA), a solution that is now widely used in secure communications to allow two parties to secretly agree on a shared encryption key.

Diffie and Hellman's publication sparked widespread academic efforts in finding a practical public-key encryption system.

This race was finally won in 1978 by R. Rivest, A. Shamir, and L. Adleman, whose solution has since become known as the *RSA algorithm*.

The Diffie–Hellman and RSA algorithms, in addition to being the first publicly known examples of high quality public-key algorithms, have been among the most widely used. Other asymmetric-key algorithms include the *Cramer–Shoup cryptosystem*, *ElGamal encryption*, and various *elliptic curve techniques*.

Public-key Cryptography is also used for implementing *digital signature algorithms* (DSA).

A digital signature is reminiscent of an ordinary signature; they both have the characteristic of being easy for a user to produce, but difficult for anyone else to forge. Digital signatures can also be permanently tied to the content of the message being signed; they cannot then be moved from one document to another, for any attempt will be detectable.

In digital signature schemes, there are two algorithms: one for signing, in which a secret key is used to process the message (or a *hash* of the message, or both), and one for verification, in which the matching public key is used with the message to check the validity of the signature.

RSA and DSA are two of the most popular digital signature schemes. Digital signatures are central to the operation of public-key infrastructures and many network security schemes.

5.5.4. Public-key algorithms are most often based on the computational complexity of "hard" mathematical problems (often from Number Theory), that are easy to state but have been found difficult to solve.

Most modern cryptographic techniques can only keep their keys secret if certain mathematical problems are intractable, such as the *integer factorization* or the *discrete logarithm problems*, so there are deep connections with abstract Mathematics.

For example, the *integer factorization problem*, i.e., the infeasibility of factoring extremely large integers, is the basis for believing that RSA is secure. The hardness of DHA and DSA are related to the *discrete logarithm problem*, i.e., on the assumption that the computation of discrete logarithm over carefully chosen groups has no efficient solution.

The security of Elliptic Curve Cryptography is based on number-theoretical problems involving elliptic curves.

But even so proofs of unbreakability are unavailable since the underlying mathematical problems remain open.

Because of the difficulty of the underlying problems, most public-key algorithms involve operations such as modular multiplication and exponentiation, which are much more computationally expensive than the techniques used in most block ciphers, especially with typical key sizes. As a result, public-key cryptosystems are commonly hybrid cryptosystems, in which a fast high-quality symmetric-key encryption algorithm is used for the message itself, while the relevant symmetric key is sent with the message, but encrypted using a public-key algorithm.

Similarly, hybrid signature schemes are often used, in which a cryptographic hash function is computed, and only the resulting hash is digitally signed.

For references and additional information see, for example, [Beuk88], [Bras88], [DeKo18], [DiHe76], [Kerc83].

$(N-1)$-based primaliy tests

5.5.5. The well-known deterministic primality test is the *Lucas test*; for a positive integer N, it requires that the all prime factors of the number $N-1$ be already known. In other words, the test uses the full factorization of $N-1$ to prove that a positive integer N is prime. Formally, the Fermat numbers are the best candidates for such kind of $(N-1)$-*based primality tests*, as in this case $N-1$ is just a power of two.

The Lucas test is the basis of the *Pratt certificate* that gives a concise verification that N is prime.

The algorithm of the test is thus.

- *Let $n > 1$ be an odd positive integer, and suppose there exist number $1 < a < n$, such that:*

- $a^{n-1} \equiv 1 (mod\ n)$;

- $a^{\frac{n-1}{q}} \not\equiv 1 (mod\ n)$ *for any prime* $q|(n-1)$.

Then n is prime. Otherwise n is composite.

□ In fact, by multiplicative order's properties, we obtain from the first condition, that $ord_n\ a|(n-1)$, and from the second condition it follows, that $ord_n\ a = n-1$. As $ord_n\ a|\phi(n)$, where $\phi(n)$ is the *Euler's totient function*, then $n-1|\phi(n)$; as $\phi(n) < n$ for $n \neq 1$, then $\phi(n) = n-1$, i.e., $n \in P$.

Conversely, if n is prime, then there exists a primitive root g modulo n. It holds $ord_n\ g = n-1$, and both equivalences will hold for any such primitive root g prime modulo n. □

Note that if there exists a positive integer $1 < a < n$ such that the first equivalence fails, a is a *Fermat witness* for the compositeness of n.

5.5.6. The *Pocklington–Lehmer primality test* is a deterministic primality test devised by H.C. Pocklington and D.H. Lehmer. The test uses a partial factorization of $N - 1$ to prove that an integer N is prime.

It produces a *primality certificate* to be found with less effort than the Lucas primality test, which requires the full factorization of $N-1$.

The basic version of the test is thus.

• *Let $n > 1$ be an integer, and suppose there exist numbers a and p such that:*

• $a^{n-1} \equiv 1 (mod\ n)$;

• p *is prime*; $p|(n-1)$; $p > \sqrt{n} - 1$;

• $gcd(a^{\frac{n-1}{p}} - 1, n) = 1.$

Then n is prime.

□ Suppose n is not prime. Then there must be a prime q, $q \le \sqrt{n}$, that divides n. Since

$$p > \sqrt{n} - 1 \ge q - 1, \text{ it holds } p > q - 1,$$

and since p is prime, $gcd(p, q - 1) = 1$. Thus, there must exist an integer u, a multiplicative inverse of p modulo $q-1$, with the property that

$$up \equiv 1 (mod\ q - 1).$$

Therefore, by the *Fermat's little theorem*, it holds

$$a^{up} \equiv a (mod\ q).$$

As $q|n$ and $a^{n-1} \equiv 1 (mod\ n)$, we get that $a^{n-1} \equiv 1 (mod\ q)$. Now this implies

$$1 \equiv (a^{n-1})^u \equiv a^{u(n-1)} \equiv a^{up\frac{n-1}{p}} \equiv (a^{up})^{\frac{n-1}{p}} \equiv a^{\frac{n-1}{p}} (mod\ q).$$

This shows that q divides $a^{\frac{n-1}{p}} - 1$. As $q|(a^{\frac{n-1}{p}} - 1)$ and $q|n$, it holds that $q|gcd(a^{\frac{n-1}{p}} - 1, n)$, and, therefore, $gcd(a^{\frac{n-1}{p}} - 1, n) \ne 1$; a contradiction.

On the other hand, if n is a prime number, then there exists a primitive root g modulo n. It is a generator of the group Z/nZ^*. Such a generator has order $|Z/nZ^*| = n - 1$. So,

$$g^{n-1} \equiv 1 (mod\ n), \quad \text{and}$$

$$g^{\frac{n-1}{p}} \not\equiv 1 (mod\ n) \quad \text{for any prime } p|(n-1).$$

It means, that $n \nmid (g^{\frac{n-1}{p}} - 1)$, and, for a prime n, we get that $gcd(g^{\frac{n-1}{p}} - 1, n) = 1$. $\square$

Given n, if p and a can be found which satisfy the conditions of the theorem, then n is prime. Moreover, the pair (p, a) constitute a primality certificate which can be quickly verified to satisfy the conditions of the theorem, confirming n as prime.

The main difficulty is finding a "good" value of p. First, it is usually difficult to find a large prime factor of a large number. Second, for many primes n, such a p does not exist. For example, $n = 17$ (as well as any Fermat number) has no suitable p because $n - 1 = 2^4$.

Note, that first equation is simply a Fermat primality test. If we find any value of a, not divisible by n, such that equation is false, we may immediately conclude that n is not prime.

For example, let $n = 35$. With $a = 2$, we find that $a^{n-1} \equiv 9 (mod\ n)$. This is enough to prove that n is not prime.

If $n = 229$, consider a decomposition

$$229 = 19 \cdot 12 + 1,$$

i.e., put $p = 19$. Let $a = 2$. It is easy to check, that

$$a^{n-1} \equiv 2^{228} \equiv 1 (mod\ 229), \quad \text{and} \quad gcd(a^{\frac{n-1}{p}}, n) = gcd(2^{12}, 229) = 1.$$

Therefore, 229 is a prime nuber.

The algorithm of the *generalized Pocklington test*, examples and some additional information see, for example, in [Bras88], [DeKo18], [DiHe76], [Kerc83], [Salo90].

$(N + 1)$-based primality tests

5.5.7. Define $(N+1)$-*based primality test* as a deterministic primality test working under conditions that full or partially decomposition of number $N + 1$ is known. Formally, the Mersenne numbers are the best candidates for such kind of primality tests, as in this case $N + 1$ is just a power of two.

For example, the following algorithm can be applied to the number n, if all prime divisors of $n + 1$ are known.

- *Given odd positive integer $n > 1$, let integer p and q are relatively prime; let the sequence $\{U_i\}_i$ is defined by the recurrence $U_0 = 0$, $U_1 = 1$, and*

$$U_{i+1} = p \cdot U_i - q \cdot U_{i-1} \ \text{for} \ i \geq 1.$$

Let

- $p^2 - 4q$ *is a quadratic nonresidue modulo n;*

- $U_{n+1} \equiv 0 (mod \ n)$;

- $U_{\frac{n+1}{p}} \not\equiv 0 (mod \ n)$ *for any prime divisor p of $n + 1$.*

Then n is a prime.

For example, let $n = 350657$. In this case, for $n + 1$ it holds:

$$350658 = 2 \cdot 3^3 \cdot 7 \cdot 11^2 \cdot 23.$$

Let $p = 3$, $q = 5$; in this case $p^2 - 4q = 3^2 - 4 \cdot 5 = -11$. Easy to see, that $\left(\frac{-11}{n}\right) = -1$.

Consider the sequence $\{U_i\}_i$ modulo n:

$$U_0 = 0, U_1 = 1, U_2 = 3, ...,$$

$$..., U_{n-1} \equiv 280525 (mod \ n), U_n \equiv 350656 (mod \ n), U_{n+1} \equiv 0 (mod \, n).$$

At last, it holds

$$U_{\frac{n}{2}} \equiv 7281 (mod \ n), U_{\frac{n}{3}} \equiv 155139 (mod \ n), U_{\frac{n}{7}} \equiv 299210 (mod \ n),$$

$$U_{\frac{n}{11}} \equiv 306723 (mod \ n), U_{\frac{n}{23}} \equiv 51824 (mod \ n).$$

So, all three conditions are true, and, hence, the number $n = 350657$ is prime.

5.5.8. This primality test is based on the properties of so-called *Lucas sequences*, named after a French mathematician É. Lucas.

Given integers P, Q, the *Lucas sequences* $U_n(P,Q)$ and $V_n(P,Q)$ are certain constant-recursive integer sequences that satisfy the recurrence relation

$$x_n = P \cdot x_{n-1} - Q \cdot x_{n-2}.$$

More exactly, given two integer parameters P and Q, the *Lucas sequence of the first kind* $U_n(P,Q)$ is defined by the recurrence relation

$$U_0(P,Q) = 0, U_1(P,Q) = 1,$$
$$U_n(P,Q) = P \cdot U_{n-1}(P,Q) - Q \cdot U_{n-2}(P,Q) \text{ for } n > 1,$$

while the *Lucas sequence of the second kind* $V_n(P,Q)$ are defined by the recurrence relation

$$V_0(P,Q) = 2, V_1(P,Q) = P,$$
$$V_n(P,Q) = P \cdot V_{n-1}(P,Q) - Q \cdot V_{n-2}(P,Q) \text{ for } n > 1.$$

Famous examples of Lucas sequences include *Fibonacci numbers, Mersenne numbers, Pell numbers, Lucas numbers, Jacobsthal numbers,* and a superset of *Fermat numbers.*

In fact, the sequence $U_n(1,-1)$ gives the *Fibonacci numbers*, while $V_n(1,-1)$ gives *Lucas numbers*.

The sequence $U_n(2,-1)$ form the set of *Pell numbers*, while $V_n(2,-1)$ gives *Pell-Lucas numbers* (or *companion Pell numbers*).

The sequence $U_n(3,2)$ gives *Mersenne numbers* $2^n - 1$, while $V_n(3,2)$ produces numbers of the form $2^n + 1$, which include the set of *Fermat numbers.*

It is not hard to show, that for $n > 0$ it holds

$$U_n(P,Q) = \frac{P \cdot U_{n-1}(P,Q) + V_{n-1}(P,Q)}{2},$$
$$V_n(P,Q) = \frac{(P^2 - 4Q) \cdot U_{n-1}(P,Q) + P \cdot V_{n-1}(P,Q)}{2}.$$

Moreover, the *characteristic equation* of the recurrence relation for Lucas sequences $U_n(P,Q)$ and $V_n(P,Q)$ is:

$$x^2 - Px + Q = 0.$$

It has the discriminant $D = P^2 - 4Q$ and the roots

$$a = \frac{P + \sqrt{D}}{2} \quad \text{and} \quad b = \frac{P - \sqrt{D}}{2}.$$

Thus,

$$a + b = P, \quad ab = \frac{1}{4}(P^2 - D) = Q, \quad a - b = \sqrt{D}.$$

5.5.9. There exists the following property of the sequence $U_n(P, Q)$.
- *Let n be a positive integer and let $\left(\frac{D}{n}\right)$ be the Jacobi symbol. Let*

$$\delta(n) = n - \left(\frac{D}{n}\right).$$

If n is a prime such that $\gcd(n, Q) = 1$, then the following congruence condition holds:

$$U_{\delta(n)} \equiv 0 (mod\ n).$$

If this congruence does not hold, then n is not prime. If n is composite, then this congruence usually does not hold.

These are the key facts that make Lucas sequences useful in primality testing.

If D is a quadratic residue modulo n, then $\left(\frac{D}{n}\right) = -1$, and $\delta(n) = n + 1$. So, we obtain two first conditions of the primality test above. In order to do this text deterministic, we add the third condition.

Note, that the *Lucas-Lehmer test* for Mersenne primes is a particular case of the test above.

□ In fact, in order to prove the Lucas-Lehmer test (see Chapter 3, Section 3.4), we considered the sequences $u_r = \frac{a^r - b^r}{a - b}$, and $v_r = a^r + b^r$, where $a = 1 + \sqrt{3}$, and $b = 1 - \sqrt{3}$, i.e., $a + b = 2$, and $ab = -2$. In other words, $u_r = U_r(2, -2)$, and $v_r = V_r(2, -2)$. □

5.5.10. However, the third condition seems too strong and requires the information about all prime divisors of n.

If we drop this condition, we obtain a probabilistic primality test.
- *Given integers P and Q, where $P > 0$ and $D = P^2 - 4Q$, let $U_k(P, Q)$ and $V_k(P, Q)$ be the corresponding Lucas sequences. Let n*

be a positive integer and let $\left(\frac{D}{n}\right)$ be the Jacobi symbol. Let

$$\delta(n) = n - \left(\frac{D}{n}\right).$$

If

$$U_{\delta(n)} \equiv 0 (mod\ n),$$

then n is, possible prime; if

$$U_{\delta(n)} \not\equiv 0 (mod\ n),$$

then n is composite.

A *Lucas probable prime* for a given (P, Q)-pair is any positive integer n for which the congruence $U_{\delta(n)} \equiv 0(mod\ n)$ above is true. A *Lucas pseudoprime* for a given (P, Q)-pair is a composite integer n for which the congruence $U_{\delta(n)} \equiv 0(mod\ n)$ is true.

In particular, a *Fibonacci pseudoprime* is defined as a composite number n not divisible by 5 for which congruence $U_{\delta(n)} \equiv 0(mod\ n)$ holds with $P = 1$ and $Q = -1$.

By this definition, the Fibonacci pseudoprimes form a sequence 323, 377, 1891, 3827, 4181, 5777, 6601, 6721, 8149, 10877, ... (sequence A081264 in the OEIS).

A *Pell pseudoprime* may be defined as a composite number n for which congruence above is true with $P = 2$ and $Q = -1$. The first Pelle pseudoprimes are then 35, 169, 385, 779, 899, 961, 1121, 1189, 2419, ...

5.5.11. The Lucas probabilistic primality test is most useful if D is chosen such that the Jacobi symbol $\left(\frac{D}{n}\right)$ is -1. In his case we obtain

$$U_{n+1} \equiv 0(mod\ n).$$

This is especially important when combining a Lucas test with a strong pseudoprime test, such as the *Baillie-PSW primality test*. Typically implementations will use a parameter selection method that ensures this condition (e.g., the *Selfridge method*).

As in the case with any other probabilistic primality test, if we perform additional Lucas tests with different D, P and Q, then unless one of the tests proves that n is composite, we gain more confidence that n is prime.

Before embarking on a probable prime test, one usually verifies that n, the number to be tested for primality, is odd, is not a perfect square, and is not divisible by any small prime less than some convenient limit.

The proofs of the properties above and some additional information see, for example, in [Brot64], [Buch09], [Cohn64], [Cohn65], [DeKo18].

Factorization of composite Mersenne numbers

5.5.12. *Integer factorization* is the process of determining which prime numbers divide a given positive integer. Doing this quickly has applications in Cryptography. The difficulty depends on both the size and form of the number and its prime factors; it is currently very difficult to factorize large *semiprimes*, i.e., products of two large primes (and, indeed, most numbers which have no small factors).

Since they are prime numbers, Mersenne primes are divisible only by 1 and by themselves. However, not all Mersenne numbers are Mersenne primes, and the composite Mersenne numbers may be factored non-trivially.

Mersenne numbers are very good test cases for the *special number field sieve* (SNFS) algorithm.

The *special number field sieve* (SNFS) is a *integer factorization* algorithm, which is efficient for integers of the form $r^e \pm s$, where r and s are small.

For Mersenne numbers, $r = 2$, and $s = 1$; it is a simplest possible case. So often the largest number factorized with this algorithm has been a Mersenne number.

It is also efficient for any integers which can be represented as a polynomial with small coefficients. This includes integers of the more general form $ar^e \pm bs^f$, and also for many integers whose binary representation has low *Hamming weight* (sum of binary digits).

The reason for this is as follows: the SNFS performs sieving in two different fields. The first field is usually the rationals. The second is a higher degree field. The efficiency of the algorithm strongly depends on the norms of certain elements in these fields. When an integer can be represented as a polynomial with small coefficients, the norms that arise are much smaller than those that arise when an integer is represented by a general polynomial.

The special number field sieve can factorize numbers with more than one large factor. If a number has only one very large factor then other algorithms can factorize larger numbers by first finding small factors and then making a primality test on the cofactor.

The SNFS has been used extensively by NFSNET (National Science Foundation Network, US), NFS@Home (a research project that uses Internet-connected computers to do the factorization of large integers), and others to factorise numbers of the *Cunningham project* (a project, started in 1925, to factor numbers of the form $b^n \pm 1$). For some time the records for integer factorization have been numbers factored by SNFS.

As of 2020, the number $2^{1193} - 1$ is the record-holder, having been factored with a variant of the special number field sieve that allows the factorization of several numbers at once.

Moreover, the largest factorization with probable prime factors allowed is

$$2^{7313983} - 1 = 305492080276193 \cdot q,$$

where q is an 2201714-digit probable prime. It was discovered in 2018 by O. Kruse.

However, the Mersenne number $M_{1277} = 2^{1277} - 1$ is the smallest composite Mersenne number with no known factors; it has no prime factors below 2^{67}.

Prime exponents p of composite Mersenne numbers.are 11, 23, 29, 37, 41, 43, 47, 53, 59, 67, ... (sequence A054723 in the OEIS).

For such p, prime factors of $M_p = 2^p - 1$, ordered by increasing p, then by increasing size of the factors, are 23, 89; 47, 178481; 233, 1103, 2089; 223, 616318177; 13367, 164511353; ... (sequence A244453 in the OEIS).

See [Bras88], [DeKo18], [DiHe76], [Kerc83], [Prim20], [Salo90], [Wiki20].

Pseudorandom Number Generation and Fermat numbers

5.5.13. *Pseudorandomness* measures the extent to which a sequence of numbers, "though produced by a completely deterministic and repeatable process, appear to be patternless."

A *pseudorandom number generator* (PRNG) is an algorithm for generating a sequence of numbers whose properties approximate the properties of sequences of random numbers.

The PRNG-generated sequence is not truly random, because it is completely determined by an initial value, called the *PRNG's seed* (which may include truly random values).

Although sequences that are closer to truly random can be generated using hardware random number generators, pseudorandom number generators are important in practice for their speed in number generation and their reproducibility.

PRNGs are central in applications such as Simulations, Electronic Games, and Cryptography.

A *linear congruential generator* (LCG) is an algorithm that yields a sequence of pseudo-randomized numbers calculated with a discontinuous piecewise linear equation. The method represents one of the oldest and best-known pseudorandom number generator algorithms. The theory behind them is relatively easy to understand, and they are easily implemented and fast, especially on computer hardware which can provide modular arithmetic by storage-bit truncation.

The generator is defined by the recurrence relation

$$X_{n+1} \equiv aX_n + c(mod\ m),$$

where X is the sequence of pseudorandom values, and m, $0 < m$ (the *modulus*), $a, 0 < a < m$ (the *multiplier*) $c, 0 \leq c < m$ (the *increment*) X_0, $0 \leq X_0 < m$ (the *seed* or *start value*) are integer constants that specify the generator. If $c = 0$, the generator is often called a *multiplicative congruential generator* (MCG), or *Lehmer RNG*.

Fermat primes are particularly useful in generating pseudorandom sequences of numbers in the range $1, ..., N$, where N is a power of 2.

The most common method used is to take any seed value between 1 and $P - 1$, where P is a Fermat prime. Now multiply this by a number A, which is greater than the square root of P and is a primitive root modulo P (i.e., it is not a quadratic residue). Then take the result modulo P. The result is the new value for the PRNG:

$$V_{j+1} \equiv A \cdot V_j \,(mod\ P).$$

This is useful in Computer Science since most data structures have members with 2^X possible values.

For example, a byte has $256 = 2^8$ possible values $0, 1, 2, ...,$ 255. Therefore, to fill a byte or bytes with random values, a random number generator which produces values 1–256 can be used.

Very large Fermat primes are of particular interest in Data Encryption for this reason.

This method produces only pseudorandom values as, after $P - 1$ repetitions, the sequence repeats. A poorly chosen multiplier can result in the sequence repeating sooner than $P - 1$.

Some additional information see, for example, in [Bras88] [DeKo18], [Prim20], [Salo90], [Wiki20].

Exercises

1. Check the primality of 71 using the Lucas primality test.

2. Check the primality of 27457 using the Pocklington–Lehmer primality test.

3. Check the primality of 27457 using the Generalized Pocklington test.

4. Construct first ten elements of the sequence $U_n(1, -1)$; check, that it gives the Fibonacci numbers.

5. Construct first ten elements of the sequence $V_n(1, -1)$; check, that it gives the Lucas numbers.

6. Construct first ten elements of the sequence $U_n(3, 2)$; check, that it gives the Mersenne numbers.

7. Construct first ten elements of the sequence $V_n(3,2)$; check, that it gives the numbers $2^n + 1$. Prove, that $F_n = V_{2^n}(3,2)$.

8. Check the property "if $n|m$, then U_m divides U_n" for the first ten elements of the sequences $U_n(1,-1)$ and $U_n(3,2)$. Prove this property.

9. Check the property "if $n|m$ is odd, then V_m divides V_n" for the first ten elements of the sequences $V_n(1,-1)$ and $V_n(3,2)$. Prove this property.

5.6   Open problems

5.6.1. *Computational Complexity Theory* focuses on classifying *computational problems* according to their resource usage, and relating these classes to each other.

In contemporary language, a *computational problem* is a task which is solvable by a computer, i.e., by mechanical application of mathematical steps, such as an *algorithm*.

A problem is regarded as inherently difficult if its solution requires significant resources. The theory formalizes this intuition, by introducing mathematical models of computation to study these problems and quantifying their *computational complexity*, i.e., the amount of resources needed to solve them, such as *time* and *storage*.

P versus *NP* problem

5.6.2. A major unsolved problem in Computer Science is *P versus NP problem* (or $P \neq NP$ *problem*).

It is one of the seven Millennium Prize Problems selected by the Clay Mathematics Institute, each of which carries a US$1000000 prize for the first correct solution.

It asks *whether every problem whose solution can be quickly verified can also be solved quickly.*

The informal term *quickly*, used above, means the existence of an algorithm solving the task that runs in *polynomial time*, such that the time to complete the task varies as a polynomial function on the size of the input to the algorithm (as opposed to, say *exponential time*).

The general class of questions for which some algorithm can provide an answer in polynomial time is called *class P* (or just *P*).

For some questions, there is no known way to find an answer quickly, but if one is provided with information showing what the answer is, it is possible to verify the answer quickly. The class of questions for which an answer can be verified in polynomial time is called *class NP* (or just *NP*), which stands for *nondeterministic polynomial time*.

An answer to the *P versus NP* question would determine whether problems that can be verified in polynomial time can also be solved in polynomial time. If it turned out that $P \neq NP$, which is widely believed, it would mean that there are problems in *NP* that are harder to compute than to verify: they could not be solved in polynomial time, but the answer could be verified in polynomial time.

Aside from being an important problem in Computational Theory, a proof either way would have profound implications for Mathematics, Cryptography, Algorithm Research, Artificial Intelligence, Game Theory, Multimedia Processing, Philosophy, Economics and many other fields.

5.6.3. The complexity class P is often seen as a mathematical abstraction modeling those computational tasks that admit an efficient algorithm. It contains all decision problems that can be solved by a *deterministic Turing machine* using a polynomial amount of computation time, or polynomial time.

The complexity class *NP*, on the other hand, contains many problems that people would like to solve efficiently, but for which no efficient algorithm is known. It is the set of decision problems solvable in polynomial time by a *non-deterministic Turing machine*.

An equivalent definition of *NP* is the set of decision problems for which the problem instances, where the answer is "yes", have proofs verifiable in polynomial time by a deterministic Turing machine.

These two definitions are equivalent because the algorithm based on the Turing machine consists of two phases, the first of which consists of a guess about the solution, which is generated in a non-deterministic way, while the second phase consists of a deterministic algorithm that verifies if the guess is a solution to the problem.

Since deterministic Turing machines are special non-deterministic Turing machines, it is easily observed that each problem in P is also member of the class *NP*.

NP-complete problems are a set of problems to each of which any other *NP-problem* can be reduced in polynomial time and whose solution may still be verified in polynomial time. That is, any *NP*-problem can be transformed into any of the *NP*-complete problems. Informally, an *NP*-complete problem is an *NP*-problem that is at least as "tough" as any other problem in *NP*.

If $P \neq NP$ then there exist problems in *NP* that are neither in P nor *NP*-complete. Such problems are called *NP-intermediate problems*. The *integer factorization problem* and the *discrete logarithm problem* are examples of problems believed to be *NP*-intermediate. They are some of the very few *NP*-problems not known to be in P or to be *NP*-complete.

For an additional information see, for example, [Bras88] [DeKo18], [Kobl87], [Salo90], [Wiki20].

Can integer factorization be done in polynomial time on a classical (non-quantum) computer?

5.6.4. The *integer factorization problem* is the computational problem of determining the prime factorization of a given integer.

No efficient integer factorization algorithm is known, and this fact forms the basis of several modern cryptographic systems, such as the RSA algorithm.

No algorithm has been published that can factor all integers in polynomial time, that is, that can factor an b-bit number n in time $O(b^k)$ for some constant k. Neither the existence nor non-existence of such algorithms has been proved, but it is generally suspected that they do not exist and hence that the problem is not in class P.

The problem is clearly in class *NP*, but it is generally suspected that it is not *NP*-complete, though this has not been proven.

The best known algorithm for integer factorization is the *general number field sieve* (GNFS), which takes time

$$O\left(e^{\left(\sqrt[3]{\frac{64}{9}}\right)\sqrt[3]{(\log n)}\sqrt[3]{(\log \log n)^2}}\right)$$

to factor an odd integer n.

However, the best known quantum algorithm for this problem, *Shor's algorithm*, does run in polynomial time. This is almost exponentially faster than the general number field sieve, which works in sub-exponential time.

Unfortunately, this fact doesn't say much about where the problem lies with respect to non-quantum complexity classes.

At the same time, the inverse problem, integer multiplication, is not difficult. It is easy to shows that such an algorithm of the multiplication of two d-digit numbers x and y, $y \leq x$, is implemented in $O(d^k)$ operations for some fixed $k \leq 2$, so the complexity of computing $x \cdot y$ is given by

$$O\big((\log x)^k\big).$$

So, today no efficient integer factorization algorithm is known, and this fact forms the basis of several modern cryptographic systems, such as the RSA algorithm.

In contrast, the decision of problem *Is n a composite number?* (or equivalently *Is n a prime number?*) appears to be much easier than the problem of specifying factors of n. This *composite/prime problem* can be solved in polynomial time in the number b of digits of n with the *Agrawal–Kayal–Saxena primality test* (2002).

In addition, there are several probabilistic algorithms that can test primality very quickly in practice if one is willing to accept a vanishingly small possibility of error.

The ease of primality testing is a crucial part of the RSA algorithm, as it is necessary to find large prime numbers to start with.

For an additional information see, for example, [Bras88], [DeKo18], [Kobl87], [Salo90], [Wiki20].

Is Public-key Cryptography possible?

5.6.5. We don't know for sure that Public-key Cryptography is safe. It could be, for example, that RSA can be broken in polynomial time if factoring can be done efficiently.

In Cryptography, the *RSA problem* summarizes the task of performing an RSA private key operation given only the public key.

The *RSA algorithm* raises a message to an exponent, modulo a composite number N whose factors are not known. Thus, the task can be neatly described as finding the e-th roots of an arbitrary number modulo N.

For large RSA key sizes (in excess of 1024 bits), no efficient method for solving this problem is known; if an efficient method is ever developed, it would threaten the current or eventual security of RSA-based cryptosystems, both for Public-key encryption and digital signatures.

More specifically, the RSA problem is to efficiently compute P given an RSA public key (N, e) and a ciphertext

$$C \equiv P^e (mod\ N).$$

The structure of the RSA public key requires that

- $N = p \cdot q$, $p, q \in P$ (N be a product of two large prime numbers, i.e., a large semiprime);

- $2 < e < \phi(N)$;

- $gcd(e, \phi(N)) = 1$ (e be coprime $\phi(N)$);

- $0 \leq C < N$ (C be a smallest nonnegative residue modulo N, chosen randomly).

To specify the problem with complete precision, one must also specify how N and e are generated, which will depend on the precise means of RSA random key pair generation in use.

The most efficient method known to solve the RSA problem is by first factoring the modulus N, a task believed to be impractical if N is sufficiently large, due to the *integer factorization problem*.

The RSA key setup routine

$$ed \equiv 1(mod \ \phi(N)), \ 1 < d < \phi(N),$$

already turns the public exponent e, with this prime factorization, into the private exponent d, and so exactly the same algorithm allows anyone who factors N to obtain the private key. Any C can then be decrypted with the private key.

Just as there are no proofs that integer factorization is computationally difficult, there are also no proofs that the RSA problem is similarly difficult.

By the above method, the RSA problem is at least as easy as factoring, but it might well be easier.

Indeed, there is strong evidence pointing to this conclusion: that a method to break the RSA algorithm cannot be converted necessarily into a method for factoring large semi-primes.

This is perhaps easiest to see by the sheer overkill of the factoring approach: the RSA problem asks us to decrypt one arbitrary ciphertext, whereas the factoring method reveals the private key: thus decrypting all arbitrary ciphertexts, and it also allows one to perform arbitrary RSA private key encryptions.

Along these same lines, finding the decryption exponent d indeed is computationally equivalent to factoring N, even though the RSA problem does not ask for d.

A different, unrelated problem with RSA is that it can be broken by quantum computers. This is an unrelated problem since the definition of a secure Public-key cryptosystem only requires that the cryptosystem not be breakable by classical (non-quantum) computers.

For additional information see, for example, [Bras88], [DeKo18], [Kobl87], [Salo90], [Wiki20].

Can the discrete logarithm be computed in polynomial time?

5.6.6. In the field $\mathbb{R}$ of real numbers, the $\log_b a$ is a number x such that $b^x = a$, for given numbers $a > 0$ and $b > 0, b \neq 1$.

Analogously, in any group G, powers b^k can be defined for all integers k, and the *discrete logarithm* $\log_b a$ is an integer k such that $b^k = a$. In Number Theory, the more commonly used term is *index*.

Discrete logarithms are quickly computable in a few special cases.

For example, in the group of the integers modulo p under addition, the power b^k becomes a product bk, and equality means congruence modulo p in the integers. The extended Euclidean algorithm finds k quickly.

However, no efficient method is known for computing them in general.

A general algorithm for computing $\log_b a$ in a finite group G is to raise b to larger and larger powers k until the desired a is found. This algorithm is sometimes called *trial multiplication*. It requires running time linear in the size of the group G and thus exponential in the number of digits in the size of the group. Therefore, it is an exponential-time algorithm, practical only for small groups G.

However, there is an efficient (polynomial) quantum algorithm due to Peter Shor.

Several important algorithms in Public-key Cryptography, including digital signature algorithm (DSA), base their security on the assumption that the discrete logarithm problem over carefully chosen groups has no efficient solution.

While computing discrete logarithms and factoring integers are distinct problems, they share some properties:

- both problems seem to be difficult (no efficient algorithms are known for non-quantum computers);

- for both problems efficient algorithms on quantum computers are known;

- algorithms from one problem are often adapted to the other;

- the difficulty of both problems has been used to construct various cryptographic systems.

At the same time, the inverse problem of discrete exponentiation is not difficult. It can be computed efficiently using exponentiation by squaring, for example.

A brief analysis shows that such an algorithm uses $\lfloor \log_2 n \rfloor$ squarings and at most $\lfloor \log_2 n \rfloor$ multiplications, where $\lfloor \ \rfloor$ denotes the floor function.

Each squaring results in approximately double the number of digits of the previous, and so, as multiplication of two d-digit numbers is implemented in $O(d^k)$ operations for some fixed k, then the complexity of computing x^n is given by

$$\sum_{i=0}^{O(\log n)} \left(2^i O(\log x) \right)^k = O\left((n \log x)^k \right).$$

This asymmetry is analogous to the one between integer factorization and integer multiplication: the complexity of computing $x \cdot y$, $y \le x$ is given by

$$O\left((\log x)^k \right).$$

Both asymmetries (and other possibly one-way functions) have been exploited in the construction of cryptographic systems.

For additional information see, for example, [Buch09], [Bras88], [DeKo13], [DeKo18], [Kobl87], [Salo90], [Wiki20].

Exercises

1. Prove, that the complicity of an algorithm of the addition $x + y$ of two positive integer numbers x, y, $y \le x$, is $O(\log x)$.

2. Prove, that the complicity of an algorithm of the subtraction $x - y$ of two positive integer numbers x, y, $y \le x$, is $O(\log x)$.

3. Prove, that the complicity of an algorithm of the multiplication $x \cdot y$ of two positive integer numbers x, y, $y \le x$, is $O(\log^2 x)$.

4. How many operations of multiplication can be used for calculation of 2^{100}? Check, that it can be reduced to 8 instead of 99.

5. Prove, that an algorithm of the exponentiation x^n, $x, n \in \mathbb{N}$, requires $O(\log n)$ multiplications.

6. Prove, that the complexity of an algorithm of the exponentiation x^n modulo N, $x, n, N \in \mathbb{N}$, is $O(\log^3 N)$.

7. Prove, that the complicity of an algorithm of the finding of $gcd(x,y)$ of two positive integer numbers x, y, $y \leq x$, is $O(\log^3 x)$.

8. Prove, that the complicity of an algorithm of the finding of the factorial $x!$ of a positive integer x is $O(n^2 \log^2 x)$.

Chapter 5: References

[IrRo90], [Arno38], [Apos86], [BePa01], [BiBa70], [Bras88], [Brot64], [Beuk88], [Buch09], [Chan70], [Cohn64], [Cohn65], [Dave99], [Dede63], [DeKo18], [Dick27], [Dick05], [Ehrm67], [Eule48], [Gard89], [Gelf98], [Gill64], [Glea88], [GKP94], [HaWr79], [Ingh32], [Ivic85], [Kahn67], [Kerc83], [Kara83], [KaVo92], [Knut68], [Knut76], [Kobl87], [Kost82], [Lagr70], [Land09], [Lege30], [Lege79], [LiNi96], [Luca75], [BrMo75], [Moto83], [Nico26], [Hogg69], [Poll74], [Prim20], [Shko61], [Sing00], [SlPl95], [Salo90], [Stra16], [Titc87], [Weis99], [Wiki20].

Chapter 6
Zoo of Numbers

In this chapter we collected some remarkable individual special numbers, related to Mersenne and Fermat numbers (see [DeDe12], [Deza17], [Deza18], [LeLi83], [SlPl95], [Weis99], [Weis20], [Wiki20], etc.).

- **2**: the only even prime.

- **3**: the first odd prime; the first Fermat prime, F_0. The first Mersenne prime, M_2; the only Fibonacci prime that is also a triangular number.

- **5**: the second Fermat prime, F_1; the only prime digit in which a perfect square can end; the only prime whose square is composed of only prime digits.

- **6**: the first perfect number; the only mean $\frac{5+7}{2}$ between a pair of twin primes which is triangular; the only known even number n such that both numbers $n^n - (n+1)$ and $n^n + (n+1)$ are primes (in fact, primes 46649 and 46663); the largest known number n such that there are n integers for which all pairwise sums are perfect squares.

- **7**: the third Mersenne prime, M_3; the only prime p such that $p+1$ is a perfect cube; the smallest centered hexagonal prime; the smallest *cuban number*, i.e., a number of the form $(n+1)^3 - n^3$; the only prime equal to the difference of the product and the sum of the two previous primes; the only prime that is member of two pairs of *cousin primes*; the biggest (besides 4 and 5) known solution of the *Brocard's problem*: to find integers n such that $n! + 1$ is a perfect square; the largest known prime that is not the sum of a triangular number, a square, and a cube, all of them positive.

- **8**: the largest Fibonacci number of the form $p + 1$ or $p - 1$ for a prime p; the largest composite number such that all its proper divisors plus 1 are primes.

- **9**: the only known composite number n such that both $2^n + n^2$ and $2^n - n^2$ are primes (in fact, the primes 431 and 593).

- **10**: the only number with the property that the sum as well as the difference of its prime divisors are primes ($2 + 5 = 7$, and $5 - 2 = 3$).

- **11**: the smallest prime repunit; the smallest prime strobogrammatic number; the only prime ($2 \cdot 3 + 5$) of the form $p_1 \cdot p_2 + p_3$, where p_1, p_2, p_3 are consecutive primes; the only prime that can be expressed ($2^3 + 3 = 3^2 + 2$) by two consecutive primes in the forms $p^q + q$ and $q^p + p$; the largest integer that cannot be expressed as a sum of at least two distinct primes; the largest number which is not expressible as sum of two composite numbers.

- **12**: the only number n such that $n \pm 1$, $\frac{n}{2} \pm 1$ and $\frac{n}{3} \pm 1$ are all primes.

- **13**: the only prime that can divide two successive integers of the form $n^2 + 3$; the only prime sum ($2^2 + 3^2$) of squares of two consecutive primes; the only Fibonacci prime sum of squares of two consecutive Fibonacci primes; the prime whose square is equal to the sum of squares of all digits in which primes end: $13^2 = 1^2 + 2^2 + 3^2 + 5^2 + 7^2 + 9^2$; the largest known Fibonacci prime with index of a Fibonacci prime; the smallest prime having exactly one representation as a sum of squares greater than one (31, its reversal, is the largest one).

- **15**: the smallest Mersenne composite, M_4; the product of two consecutive Fermat primes, 3 and 5 (thus, a regular polygon with 15 sides is constructible with compass and unmarked straightedge);

the smallest number that can be factorized using Shor's quantum algorithm.

- **16**: the largest known integer n for which $2^n + 1$ (in fact, F_4) is prime; the only integer of the form $n^m = m^n$ for distinct integers n and m: $16 = 2^4 = 4^2$.

- **17**: the third Fermat prime, F_2; the only prime that is the average $(\frac{13+21}{2})$ of two consecutive Fibonacci numbers; the only prime that is the sum $(17 = 2 + 3 + 5 + 7)$ of 4 consequtive primes; the only prime of the form $p^q + q^p$, where p and q are primes: $17 = 2^3 + 3^2$; the only number n having n partitions into prime parts; the largest, if Goldbach's conjecture is true, integer which is not the sum of three distinct primes; the only prime number of the form $p^2 + 8$; the smallest prime whose sum of the digits is a cubic number; the only known prime that is equal to the sum of digits of its cube: $17^3 = 4913$, and $4 + 9 + 1 + 3 = 17$; the smallest prime that is the *quartan*, i.e., the sum of two biquadratic numbers: $17 = 1^4 + 2^4$; 17^2 can be expressed as the sum of $1, 2, 3, 4, 5, 6, 7, 8$ distinct squares.

- **23**: the smallest odd prime that is not a twin prime; the largest integer n such that no factor of a binomial coefficient $\binom{n}{k}$ is a perfect square; the biggest, besides $2, 3, 5, 7, 11$, prime which is uniquely expressible as a sum of at most four squares.

- **24**: the sum of the twin primes 11 and 13; the only integer $n > 1$ such that $\sum_{i=1}^{n} i^2$ is a perfect square: $\sum_{i=1}^{24} i^2 = 70^2$.

- **25**: the third Cullen number $3 \cdot 2^3 + 1$; the smallest Fermat pseudoprime modulo 7, i.e., satisfying the congruence $7^n \equiv 7 (mod\ n)$; the only perfect square of the form $k^3 - 2$.

- **26**: the largest integer n such that the segment $[m, m + 100]$ contains n primes (it happens only for $m = 2$).

- **28**: the second perfect number; the only perfect number of the form $n^k + m^k$ with $k > 1$: $28 = 3^3 + 1^3$.

- **29**: the only two-digital prime whose square is the sum of squares of two consecutive two-digital numbers: $29^2 = 20^2 + 21^2$; the smallest prime equal to the sum of three consecutive squares: $29 = 2^2 + 3^2 + 4^2$; the smallest multi-digital prime which on adding its reverse gives perfect square: $29 + 92 = 11^2$; the only *non-titanic prime* (i.e., with less than 1000 decimal digits) of form $p^p + 2$.

- **30**: the only integer n, besides $0, 3, 8, 10, 18$ and 24, such that the number of its divisors is equal to the number of integers $k, 0 < k < n$, which are coprime with n; the largest integer n such that all integers $k, 0 < k < n$, which are coprime with n, are also prime.

- **31**: the third Mersenne prime, M_5; the smallest prime that can be represented as a sum of two triangular numbers in two different ways: $31 = 21 + 10 = 28 + 3$; the smallest prime that can be represented as a sum of two triangular numbers with prime indices; there are only 31 numbers that cannot be expressed as a sum of distinct squares; $3 + 5 + 7 + 11 + \cdots + 89 = 31^2$, i.e., a sum of the first 31 odd primes is a square of prime.

- **33**: the smallest odd repdigit that is not a prime number; the largest integer which is not a sum of distinct triangular numbers.

- **36**: the smallest perfect square expressible as a sum of four consecutive primes which are also two couples of prime twins: $36 = 5 + 7 + 11 + 13$; the smallest triangular number whose sum of divisors, as well as sum of its proper divisors, is also a triangular number.

- **41**: the only known prime number that is not a sum of two triangular numbers and a nonnegative cube; the smallest prime whose cube can be written as a sum of three cubes in two ways: $41^3 = 40^3 + 17^3 + 2^3 = 33^3 + 32^3 + 6^3$; a sum of two consecutive squares: $4^2 + 5^2$.

- **42**: the smallest number n such that n^2 is the mean of cubed twin primes, $42^2 = \frac{11^3 + 13^3}{2}$.

- **65**: the only number which gives a square of prime on adding as well as subtracting its reverse from it: $65 + 56 = 11^2$, $65 - 56 = 3^2$; the only number which is the difference $(3^4 - 2^4)$ of two biquadratic numbers with prime indices; $(65!)^2 + 1$ is prime.

- **67**: the smallest multi-digital prime whose square 4489 and cube 300763 consist of different digits.

- **81**: the only known square n such that $n \cdot 2^n - 1$ is a prime.

- **83**: the largest known prime that can be expressed as a sum of three positive triangular numbers in exactly one way; the smallest prime whose square 6889 is a strobogrammatic number; the only prime equal to a sum of squares of odd primes: $83 = 3^2 + 5^2 + 7^2$; the only prime of the form $p^4 + 2$, where p is a prime.

- **89**: the smallest positive integer whose square 7921 and cube 704969 are likewise primes upon reversal.

- **100**: the smallest perfect square whose summation of the differences between itself and each of its digits, where each difference is raised to the power of the corresponding digit, is equal to a prime: $101 = (100 - 1)^1 + (100 - 0)^0 + (100 - 0)^0$ is a prime.

- **101**: the smallest 3-digital strobogrammatic prime; the smallest 3-digital palindrome prime.

- **109**: the smallest number (coincidentally prime) that has more distinct digits than its square 11881; a strobogrammatic prime, which is equal to the square root of the strobogrammatic number 11881 (or $118 - 8 - 1$).

- **113**: the smallest prime which is a sum of three biquadratic numbers with prime indices: $113 = 2^4 + 2^4 + 3^4$.

- **121**: the only perfect square of the form $1 + p + p^2 + p^3 + p^4$, $p \in P$: $121 = 1 + 3 + 3^3 + 3^3 + 3^4$; the only, besides 4, perfect square of the form $n^3 - 4$ ($121 = 5^3 - 4$).

- **127**: the 4-th Mersenne prime, M_7; the exponent for 12-th Mersenne prime $M_{127} = 2^{127} - 1$ (it is the largest prime ever discovered by hand calculations, as well as the largest known double Mersenne prime); the first nice Friedman number (the representation $127 = 2^7 - 1$ uses all digits 1, 2, 7); the smallest prime that can be written as the sum of the first two or more odd primes ($127 = 3 + 5 + 7 + 11 + 13 + 17 + 19 + 23 + 29$).

- **144**: the largest, besides 0 and 1, perfect square, which is a Fibonacci number; the sum of a twin prime pair $(71; 73)$.

- **149**: the only known prime in the concatenate square sequence.

- **169**: the largest Pell number (i.e., a member of the recurrent sequence $P_{n+2} = 2P_{n+1} + P_n$, $P_0 = 0, P_1 = 1$), which is a perfect square; the smallest composite Pell number with a prime index.

- **173**: the largest known prime whose square 29929 and cube 5177717 consist of different digits.

- **191**: a palindromic prime whose square 36481 is a distinct-digital number whose first two digits, central digit, and last two digits, are perfect squares.

- **196**: the smallest candidate *Lychrel number*, i.e., a natural number which cannot form a palindrome through the iterative process of repeatedly reversing its base 10 digits and adding the resulting numbers; in fact, 196 does not yield a palindrome after 700 000 000 iterations.

- **211**: the largest known prime that cannot be written as a sum of a prime and a positive triangular number.

- **216**: the smallest cubic number, which is a sum of three cubic numbers: $216 = 6^3 = 5^3 + 4^3 + 3^3$.

- **229**: replacing each digit of prime 229 with its square, respectively its cube, results in two new primes (4481 and 88729) with a palindromic difference of 84248; coincidentally, $229 + 4481 + 88729$ is palindromic as well.

- **239**: a Chen prime (241 is a prime); a Sophie Germain prime (479 is a prime); the largest, besides 23, integer, which is not a sum of less than 9 cubic numbers: $239 = 2 \cdot 4^3 + 4 \cdot 3^3 + 3 \cdot 1^3$.

- **257**: the 4-th Fermat prime, F_3; the largest prime in a sequence of fifteen primes of the form $2n + 17$, where n runs through the first 15 triangular numbers.

- **277**: the square 76729 of the prime 277 is the smallest square with 5 or more digits that is the concatenation of three primes (7, 67, and 29).

- **289**: the square of the sum of the first four primes: $289 = (2 + 3 + 5 + 7)^2$.

- **343**: the only, cubic number (7^3) besides 1, such that the sum of its divisors is a perfect square: $1 + 7 + 7^2 + 7^3 = 20^2$; a *Friedman number*, as is the result of a non-trivial expression using all its own digits, $343 = (3 + 4)^3$.

- **367**: the largest number (in fact, prime) whose square 134689 has strictly increasing digits.

- **400**: the only known square of the form $1 + k + k^2 + k^3$, where $k \in \mathbb{N}\backslash\{1\}$ (in fact, $k = 7$).

- **407**: the largest, besides 1, 153, 370 and 371, integer, which is the sum of cubes of their decimal digits.

- **463**: the smallest multi-digital prime such that both a sum of digits and product of digits of its square remain squares.

- **496**: the third perfect number; the smallest triangular number such that the sum of the cubes of its digits is prime: $4^3 + 9^3 + 6^3 = 1009$.

- **514**: 19-th centered triangle number; a *constructible number*, because $514 = 2 \cdot 257$ is a power of two times the product of Fermat primes.

- **541**: the 100-th prime number, the 10-th hexagonal star (i.e., Star of David) number.

- **576**: the only known perfect square represented as a difference between a squared sum of consecutive primes and the sum of their squares: $576 = 24^2 = (2+3+5+7+11)^2 - (2^2 + 3^2 + 5^2 + 7^2 + 11^2)$; it is the only such case for all primes up to $2 \cdot 10^9$.

- **613**: a prime which gives a mathematical enigma in the story *Number of the End* by Jason Earls: *bring the first digit back to* *get* 136, *it is triangular, now bring the first digit of that back to get* 361, *it is a square*; the square of $613 = 375769$ is the largest known perfect square that divides a number of the form $n! + 1$, which happens when $n = 229$, another prime.

- **631**: a prime which is the reverse concatenation of the first three triangular numbers.
  A *constructable number* because $640 = 2^7 \cdot 5$ is a power of two times product of Fermat primes.

- **641**: the prime divisor of the smallest Fermat composite, F_5: $F_5 = 4294967297 = 641 \cdot 6700417$.

- **691**: the only known prime which is a square 169 when turned upside down and another square 196 when reversed; moreover, 169 and 196 are the smallest consecutive squares using the same digits.

- **701**: the smallest prime whose square (491401) contains all of square digits only; it is equal to $5^4 + 4^3 + 3^2 + 2^1 + 1^0$.

- **727**: the first prime whose square (528529) can be represented as the concatenation of two consecutive numbers.

- **773**: replacing each digit of prime 773 with its square respectively its cube results in two new primes 49499 and 34334327; this latter number is also an *emirp* (72343343 is a prime).

- **786**: the largest known number n such that the binomial coefficient $\binom{2n}{n}$ is not divisible by the square of an odd prime.

- **900**: the smallest perfect square which is a sum of different primes by using all the ten digits: $900 = 503 + 241 + 89 + 67 = 509 + 283 + 61 + 47$.

- **919**: a prime, which is also the 18-th centered hexagonal number. One can use the digits of 919 to write $18 = (9 \cdot 1) + 9 = 9 + (1 \cdot 9)$.

- **2047**: the first Mersenne composite with prime index, M_{11}.

- **65537**: the biggest known Fermat prime, F_4.

- **4294967297**: the smallest Fermat composite (in fact, semi-prime), F_5.

- 10^{100}: one *googol*; in decimal notation, it is written as the digit 1 followed by one hundred zeroes; a googol is approximately 70!; 1 googol $\approx 2^{332.19280949}$.

- $2^{2^9} + 1 = 2^{512} + 1$: the first Fermat number, F_9, greater than googol, i.e., having more 100 digits (in fact, 154 digits).

- **2317006...230657** $= 2^{2^{11}} + 1 = 2^{2048} + 1$: the biggest completely factored Fermat number, F_{11}; it has 617 digits; it is a product of five prime numbers with 6, 6, 21, 22 and 564 decimal digits.

- **1000...0007** $= 10^{999} + 7$: the smallest *titanic prime*, i.e., a prime of at least 1000 decimal digits.

- $2^{4253} - 1$: the smallest Mersenne titanic prime, M_{4253}; 19-th Mersenne prime; it has 1281 digits.

- $2^{2^{33}} + 1$: the smallest Fermat number, F_{33}, with unknown character.

- **1000...00033603** $= 10^{9999} + 33603$: the smallest *gigantic prime*, i.e., a prime number with at least 10000 decimal digits; it has exactly 10000 digits.

- $2^{44497} - 1$: the smallest Mersenne giganic prime, $M_{44,497}$; 27-th Mersenne prime; it has 13395 digits.

- $191273 \cdot 2^{3321908} - 1$: the smallest known *megaprime*, i.e., a prime with at least one million decimal digits; it has exactly 1000000 digits.

- $2^{6972593} - 1$: the smallest Mersenne megaprime, $M_{6972593}$; 38-th Mersenne prime; it has 2098960 digits.

- $2^{82589933} - 1$: the biggest known Mersenne prime, $M_{82589933}$; 51-th known Mersenne prime; it has 24862048 digits.

- $10^{10^{100}} = 10^{googol}$: one *googolplex*; written out in ordinary decimal notation, it is 1 followed by 10^{100} zeroes, that is, a unity followed by a googol zeroes.

- $2^{2^{334}} + 1$: the first Fermat number, F_{334}, greater than googoplex.

- $2^{2^{18233954}} + 1$: the largest known composite Fermat number, $F_{18233954}$; it has the prime factor $p = 7 \cdot 2^{18233956} + 1$.

Chapter 7

Mini Dictionary

In this Chapter, a Mini Dictionary, i.e., the list of all special numbers, relative to Mersenne and Fermat numbers, is represented (see [CoGu96], [DeDe12], [Deza17], [Deza18], [Weis20], [Wiki20], etc.).

- $n \in \mathbb{N}$, $\sigma(n) > 2n$ — *abundant number*: positive integers, for which their sum of proper divisors is greater than the number itself.

- (m, n), $m, n \in \mathbb{N}$, $\sigma(n) = \sigma(m) = n + m$ — *amicable numbers*: a pair of positive integers which are the sum of each other's proper divisors.

- $B(n) = \sum_{k=0}^{n-1} \binom{n-1}{k} B(k)$ with $B(0) = 1$ — *Bell number*.

- $B_n = -\frac{1}{n+1} \sum_{k=1}^{n} \binom{n+1}{k+1} B_{n-k}$ with $B_0 = 0$ — *Bernoulli number*.

- $BC(n) = n^4$, $n = 1, 2, 3, \ldots$ — *biquadratic number*.

- $\binom{n}{m}$, $n = 0, 1, 2, \ldots$, $m = 0, 1, \ldots, n$ — *binomial coefficient*; they form the *Pascal's triangle* — number triangle, the sides of which are formed by 1, and any inner entry is obtained by adding the two entries diagonally above.

- $n \in S$, $a^{n-1} \equiv 1 (mod\ n)$ for any $a \in \mathbb{Z}$, a coprime to n — *Carmichael number*: numbers n that are Fermat pseudoprimes to any base a coprime to n.

- $Ca(n) = 4^n - 2^{n+1} - 1$, $n = 1, 2, 3, \ldots$ — *Carol number*.

- $C_n = \frac{1}{n}\binom{2n-2}{n-1}$, $n = 1, 2, 3, \ldots$ — *Catalan number*.

- $p \in P$, such that $p + 2 \in P$, or $p + 2 = q_1 \cdot q_2$, $q_1, q_2 \in P$ — *Chen prime*: prime numbers p, such that $p + 2$ are either primes or semiprimes (a product of two primes).

- $n = 2^r \cdot p_1 \cdot \ldots \cdot p_k$, where $r \in \mathbb{Z}$, $r \geq 0$, and $p_1 < \ldots < p_k$ are distinct Fermat primes — *constructible number*: line segment of length n can be constructed with compass and straightedge in a finite number of steps.

- $(p, p + 4)$, $p, p + 4 \in P$ — *cousin primes*.

- $(n + 1)^3 - n^3$, $n = 1, 2, 3, \ldots$ — *cuban number*.

- $C(n) = n^3$, $n = 1, 2, 3, \ldots$ — *cubic number* (or *cube number*).

- $Cu(n) = n \cdot 2^n + 1$, $n = 1, 2, 3, \ldots$ — *Cullen number*.

- $n \in \mathbb{N}$, $\sigma(n) < 2n$ — *deficient number*: positive integers, for which their sum of proper divisors is less than the number itself.

- 2, 3, 7, 23, 37, 53, ... — *digit prime*: prime numbers that has only prime digits 2, 3, 5, or 7.

- $D(n) = \frac{n(9n^2 - 9n + 2)}{2}$, $n = 1, 2, 3, \ldots$ — *dodecahedral number*.

- $M_{M_n} = 2^{2^n - 1} - 1$, $n = 1, 2, 3, \ldots$ — *double Mersenne number*.

- $(1 + \omega)^n - 1$, where $\omega = \frac{-1 + i\sqrt{3}}{2}$, $i^2 = -1$, $n = 1, 2, 3, \ldots$ — *Eisenstein Mersenne number*.

- 13, 17, 31, 37, 71, ... — *emirp*: prime numbers that result in different primes when its decimal digits are reversed.

- $F_n = 2^{2^n} + 1$, $n = 0, 1, 2, \ldots$ — *Fermat number*.

- $n \in S$, n is odd, $a^{n-1} \equiv 1 \pmod{n}$ — *Fermat pseudoprime* to base a.

- $u_{n+2} = u_{n+1} + u_n$, $u_1 = u_2 = 1$ — *Fibonacci number*.

- 25, 121, 125, 126, 127, ... — *Friedman number*: positive integers, represented as the result of a non-trivial expression using all its own digits (e.g., $25 = 5^2$; $121 = 11^2$; $343 = (3 + 4)^3$, etc.).

- $(1+i)^n - 1$, where $i^2 = -1$, $n = 1, 2, 3, \ldots$ — *Gaussian Mersenne numbers.*

- $F_n(a, b) = a^{2^n} + b^{2^n}$, where a and b are coprime integers, such that $a > b > 0$, while $n = 0, 1, 2, \ldots$ — *generalized Fermat number.*

- $M_n(a, b) = \frac{a^n - b^n}{a - b}$, where a and b are coprime integers, such that $a > 1$, and $-a < b < a$, while $n = 1, 2, 3, \ldots$ — *generalized Mersenne numbers*; an important special case is $M_n(b, 1) = M_n(b) = \frac{b^n - 1}{b - 1}$.

- $GCu(n) = n \cdot b^n + 1$, where $n + 2 > b$ — *generalized Cullen number.*

- $GW_n = n \cdot b^n - 1$, where $n + 2 > b$ — *generalized Woodall number.*

- $p \in P$, $p > 10^{9999}$ — *gigantic prime*: prime numbers with at least 10000 decimal digits.

- $Gn(n) = 2n - 1$, $n = 0, 1, 2, \ldots$ — *gnomonic number.*

- $n \in \mathbb{N}$, $n = p + q$, $p, q \in P \backslash \{2\}$ — *Goldbach number*: even positive integers that can be expressed as the sum of two odd primes.

- $\frac{a^{2^n} + 1}{2}$, where $a > 1$ is an odd positive integer, while $n = 0, 1, 2, \ldots$ — *half generalized Fermat number to an odd base a.*

- $O_H(n) = \frac{1}{3}(2n - 1)(2n^2 - 2n + 3)$, $n = 1, 2, 3, \ldots$ — *Haüy octahedral number.*

- $RD_H(n) = (2n - 1)(8n^2 - 14n + 7)$, $n = 1, 2, 3, \ldots$ — *Haüy rhombic dodecahedral number.*

- $n \in \mathbb{N}$, $n = 1 + k(\sigma(n) - n - 1)$ — *k-hyperperfect number*, $k \in \mathbb{N}$; *hyperperfect number* — positive integers which are k-hyperperfect numbers for some positive integer k.

- $I(n) = \frac{n(5n^2 - 5n + 2)}{2}$, $n = 1, 2, 3, \ldots$ — *icosahedral number.*

- $C^k(n) = n^k$, $n = 1, 2, 3, \ldots$ — *k-dimensional hypercube number*, $k \in \mathbb{N}$, $k > 1$.

- $O^k(n) = \sum_{j=0}^{k-1} (-1)^j \binom{k-1}{j} 2^{k-j-1} S_3^{k-j}(n)$ — *k-dimensional hyper-octahedron number*, $k \in \mathbb{N}$, $k > 1$.

- $S_3^k(n) = S_3^k(n) = \frac{n(n+1)\ldots(n+k-1)}{k!}$, $n = 1, 2, 3, \ldots$ — *k-dimensional hypertetrahedron number*, $k \in \mathbb{N}$, $k > 1$.

- $S_m^k(n) = \frac{n(n+1)\ldots(n+k-2)((m-2)n-m+k+2)}{k!}$, $n = 1, 2, 3, \ldots$ — *k-dimensional m-gonal pyramidal number*, $k \in \mathbb{N}$, $k > 1$.

- $N^k(n) = (n+1)^{k+1} - n^{k+1}$, $n = 1, 2, 3, \ldots$ — *k-dimensional nexus number*, $k \in \mathbb{N}$, $k > 1$.

- $Ky(n) = 4^n + 2^{n+1} - 1$, $n = 1, 2, 3, \ldots$ — *Kynéa number*.

- $L_{n+2} = L_{n+1} + L_n$, $L_0 = 2$, $L_1 = 1$ — *Lucas number*.

- $p \in P$, $p > 10^{999999}$ — *megaprime*: prime numbers with at least one million decimal digits.

- $M_n = 2^n - 1$, $n = 1, 2, 3, \ldots$ — *Mersenne number*.

- $MF(p, r) = \frac{2^{p^r} - 1}{2^{p^{r-1}} - 1}$ — *Mersenne-Fermat number*; when $r = 1$, they are Mersenne numbers; when $p = 2$, they are Fermat numbers.

- $S_m(n) = \frac{n((m-2)n-m+4)}{2}$, $n = 1, 2, 3, \ldots$ — *m-gonal number*.

- $S_m^3(n) = \frac{n(n+1)((m-2)n-m+5)}{6}$ $n = 1, 2, 3, \ldots$ — *m-gonal pyramidal number*.

- $P_m(n) = m \cdot n^2 - m \cdot n + 1$, $n = 1, 2, 3, \ldots$ — *m-gram number*; in particular, $P_6(n) = S(n) = 6n^2 - 6n + 1$ — *star number*.

- $MN_n = MN_{n-1} + \sum_{i=0}^{n-2} MN_i MN_{n-2-i}$, $MN_0 = MN_1 = 1$ — *Motzkin number*.

- $n \in \mathbb{N}$, $\sigma(n) = kn$ — *k-perfect number* (or *k-fold perfect number*): the sum of all positive divisors of n is equal to kn, $k \in \mathbb{N}$, $k > 1$; *multiply perfect number* — numbers that are k-perfect for some positive integer $k > 1$.

- 1, 6, 28, 140, 270, 496, ... — *Ore number* (or *harmonic divisor number*): positive integers whose divisors have a harmonic mean that is an integer (e.g., the harmonic mean $\frac{4}{\frac{1}{1}+\frac{1}{2}+\frac{1}{3}+\frac{1}{6}}$ of divisors 1, 2, 3, 6 of 6 is an integer 2).

- $(abc...cba)_{10}$ — *palindromic number* (or *numeral palindrome*, or *numeric palindrome*): positive integers that remain the same when their decimal digits (in general, digits in any positional number system) are reversed (e.g., 16461).

- 2, 3, 5, 7, 11, 13, ... — *permutable prime* (or *absolute prime*): prime numbers which remain primes on every rearrangement of their decimal digits; in general, digits in any positional number system (e.g., 337 is a permutable prime because each of 337, 373 and 733 is prime).

- $P_{n+2} = 2P_{n+1} + P_n$, $P_0 = 0$, $P_1 = 1$ — *Pell number.*

- $PL_{n+2} = 2PL_{n+1} + PL_n$, $PL_0 = 2$, $PL_1 = 2$ — *Pell-Lucas number* (or *companion Pell number*).

- 3, 5, 6, 7, 9, ... — *pernicious number*: positive integers such that the *Hamming weights* of their binary representations (i.e., the sums of binary digits) are primes (e.g., $3 = 11_2$, and $1 + 1 = 2$, $2 \in P$).

- $P_5(n) = 5n^2 - 5n + 1$, $n = 1, 2, 3, ...$ — *pentagram number.*

- $2^u 3^v + 1 \in P$, where $u, v \in \mathbb{Z}$, $u, v \geq 0$ — *Pierpont prime.*

- $2^u 3^v - 1 \in P$, where $u, v \in \mathbb{Z}$, $u, v \geq 0$ — *Pierpont prime of the second kind.*

- $n \in S$, n is odd, $2^{n-1} \equiv 1 (mod\ n)$ — *Poulet number* (or *Sarrus number*): Fermat pseudoprime to base 2.

- 1, 2, 4, 6, 8, 12, ... — *practical number* (or *panarithmic number*): positive integers n such that all smaller positive integers can be represented as sums of distinct divisors of n.

- $P(n) = n(n+1)$, $n = 1, 2, 3, ...$ — *pronic number.*

- $k \cdot 2^n + 1$, $k \in \mathbb{N}$, k is odd, $k < 2^n$ — *Proth number.*

- $O(n) = \frac{n(2n^2+1)}{3}$, $n = 1, 2, 3, ...$ — *octahedral number.*

- $n^4 + m^4$, $n, m = 1, 2, 3, ...$ — *quartan*: sums of two biquadratic numbers.

- $n \in \mathbb{N}$, $n\frac{u(u+1)}{2} = 2^v - 1$, $u \geq 0$, $v \geq 0$ — *Ramanujan-Nagell number*: the numbers 0, 1, 3, 15, 4095, which are the only non-negative integers which have simultaneously the form $\frac{u(u+1)}{2}$, $u = 0, 1, 2, ...$, and the form $2^v - 1$, $v = 0, 1, 2, ...$

- $(aaa...aaa)_{10}$ — *repdigit* (or *monodigit*): positive integers composed of repeated instances of the same digit in decimal system (in general, in any positional number system).

- $(111...111)_{10}$ — *repunit*: positive integers containing only the digit 1 in decimal system (in general, in any positional number system); $R_n = \frac{10^n - 1}{9}$.

- $RD(n) = 4n^3 - 6n^2 + 4n - 1$, $n = 1, 2, 3, ...$ — *rhombic dodecahedral number*.

- 509203, 762701, 777149, 790841, 992077, ... — *Riesel number*: odd positive integers k such that $k \cdot 2^n - 1$ is composite for all natural numbers n.

- $q = 2p + 1 \in P$, where $p \in P$ — *safe prime*; q is the prime number $2p + 1$ associated with a *Sophie Germain prime* p.

- 6, 12, 18, 20, 24, 28, ... — *semiperfect number*: positive integers that are equal to the sum of all or some of their proper divisors; abundant numbers which are not semiperfect are called *weird number*.

- $p \cdot q$, $p, q \in P$ — *semiprime*: positive integers, which are product of two primes.

- $(p, p + 6)$, $p, p + 6 \in P$ — *sexy primes*.

- 78557, 271129, 271577, 322523, 327739, ... — *Sierpiński number*: odd positive integers k such that $k \cdot 2^n + 1$ are composite for all natural numbers n.

- $p \in P$, such that $2p + 1 \in P$ — *Sophie Germain prime*; the prime number $q = 2p + 1$, associated with p, is called *safe prime*.

- $S(n) = 6n^2 - 6n + 1$, $n = 1, 2, 3, ...$ — *star number*.

- $SO(n) = n(2n^2 - 1)$, $n = 1, 2, 3, \ldots$ — *stella octangula number*.

- $S(n, m) = \frac{1}{m!} \sum_{i=0}^{m} (-1)^i \binom{m}{i} (m - i)^n$, $n = 1, 2, 3, \ldots,$ $1 \leq m \leq n$ — *Stirling numbers of the second kind.*

- 0, 1, 8, 11, 69, 88, 96, 101,... — *strobogrammatic number*: positive integers whose numeral is rotationally symmetric, so that it appears the same when rotated 180 degrees.

- $p \in P$, $p > 10^{999}$ — *titanic prime*: prime numbers of at least 1000 decimal digits.

- $Tr(n, k) = n + (n+1) + \ldots + (n + (k-1))$, $n, k \in \mathbb{N} \backslash \{1\}$ — *trapecoidal number*.

- $Q(b, n) = \frac{b^n + 1}{b + 1}$, $b \geq 2$, $n = 1, 2, 3, \ldots$ — *Wagstaff numbers base b.*

- $\frac{2^p + 1}{3} \in P$, where $p \in P$ — *Wagstaff primes.*

- 70, 836, 4030, 5830, 7192, ... — *weird numbers*: abundant numbers which are not semi-perfect.

- $W_n = n \cdot 2^n - 1$, $n = 1, 2, 3, \ldots$ — *Woodall number* (or *Cullen number of the second kind*).

Chapter 8
Exercises

In this Chapter we represent some interesting problems concerning Mersenne and Fermat numbers, and give (sketches of) their solutions (see references to Chapters 2, 3, 4).

Problems, connected with Mersenne numbers

1. Prove the following linear recurrent relation with constant coefficients of the second order for the Mersenne numbers:

$$M_{n+2} = 3M_{m+1} - 2M_n, \ M_1 = 1, M_2 = 3.$$

2. Using recurrent relation $M_{n+2} = 3M_{m+1} - 2M_n$, $M_1 = 1$, $M_2 = 3$, prove that the generating function for the sequence $M_1, M_2, M_3, \ldots, M_n, \ldots$ of Mersenne numbers has the form $f(x) = \frac{1}{2x^2-3x+1}$, i.e.,

$$\frac{1}{2x^2 - 3x + 1} = M_1 + M_2 x + M_3 x^2 + \cdots$$

$$+ M_{m+1} x^n + \ldots, \ |x| < 0.5.$$

3. Using only elementary arguments, prove, that the generating function of the sequence $M_1, M_2, M_3, \ldots, M_n, \ldots$ of Mersenne numbers has the form $\frac{1}{(1-x)(1-2x)}$, and the corresponding series is convergent for $|x| \leq 0.5$.

4. Find the value of the sum $M_1 + M_2 \cdot x + M_3 \cdot x^2$ for $x = 1, 2, 3, 100, 1000$.

5. Prove, that $M(x) = \sum_{M_n \leq x} 1 = \frac{e^\gamma}{\ln 2} \ln \ln x$, where γ is the Euler constant.

6. Prove, that if p is a prime, then all Mersenne numbers M_p are primes or pseudoprimes to the base 2.

7. Prove, that any odd positive integer n divides finitely many Mersenne numbers.

8. Prove the infiniteness of primes, using Mersenne numbers.

9. Check, that $M_n \equiv -1 (mod\ 8)$ for $n \geq 3$.

10. Prove that $M_n \neq x^2 + y^2 + z^2$, $x, y, z \in \mathbb{Z}$, for $n \geq 3$.

11. Prove that $M_n = x^2 + y^2 + z^2 + t^2$, $x, y, z, t \in \mathbb{Z}$, for $n \geq 1$.

12. Prove, that for any positive integer $m \leq 20$ there exists a Mersenne number M_n, such that the first decimal digits of M_n coincide with the decimal representation of the number m.

13. Prove, that for any positive integer s, the last s digits of the numbers M_n, $n = 1, 2, ...$, form an infinite periodical sequence with the period, having $4 \cdot 5^{s-1}$ elements.

14. The *Tower of Hanoi* (or *Tower of Brahma, Lucas' Tower*) is a mathematical puzzle. It consists of three rods and a number of disks of different sizes, which can slide onto any rod. The puzzle starts with the disks in a neat stack in ascending order of size on one rod, the smallest at the top, thus making a conical shape. The objective of the puzzle is to move the entire stack to another rod, obeying the following simple rules: only one disk can be moved at a time; each move consists of taking the upper disk from one of the stacks and placing it on top of another stack or on an empty rod; no larger disk may be placed on top of a smaller disk. Check, that with 3 disks, the puzzle can be solved in $M_3 = 7$ moves. Prove, that the minimal number of moves required to solve a Tower of Hanoi puzzle with n disks (and three rods) is $M_n = 2^n - 1$.

15. Prove, that the last digit of any Mersenne prime greater 3 is 1 or 7.

16. Prove, that the last digit of any even perfect number is 6 or 8.

17. Write the first four perfect numbers 6, 28, 496, 8128 in binary; prove, that any even perfect number has in binary the form $111...111000...00_2$, where after p unities we have $p - 1$ zeros.

18. Find all even numbers less than 1000 of the form $x^3 + 1$. Check, that the only perfect number of such form is 28.

19. Prove, that the number of divisors of a perfect number is even.

20. A *Pierpont prime of the second kind* is a prime number of the form $2^u 3^v - 1$, $u, v \geq 0$. Find first ten Pierpont primes of the second kind. Prove, that any Mersenne prime is a Pierpont prime of the second kind. Check, that the largest known Pierpont prime is $2^{82589933} - 1$.

21. A *Pierpont prime of the second kind* $2^u 3^v - 1$, $u, v \geq 0$, with $v = 0$ is of the form $2^u - 1$ and is therefore a Mersenne prime for $u \geq 2$. Prove, that any non-Mersenne Pierpont prime of the second kind greater than 3 have the form $6k - 1$.

22. Consider the numbers of the form $k \cdot 2^n - 1$, with $k < 2^n$. Find first ten of such numbers. Find the first ten of primes of such form. Prove, that all Mersenne primes have such form.

23. Given number N of the form $k \cdot 2^n - 1$, with $k < 2^n$, prove the *Lucas–Lehmer–Riesel primality test*: N *is prime if and only if it divides* U_{n-2}, *where a sequence* $U_0, U_1, ..., U_i, ...U_{n-2}$ *is defined for all* $i > 0$ *by* $U_i = U_{i-1}^2 - 2$, *and starting point* U_0 *depending on the value of* k. Check the following starting conditions:

- if $k = 1$ and $n \equiv \pm1 (mod\ 4)$ is odd, one can take $U_0 = 4$;
- if $k = 1$ and $n \equiv 3 (mod\ 4)$, one can take $U_0 = 3$;
- if $k = 3$, $n \equiv 0, 3 (mod\ 4)$, one can take $U_0 = 5778$;
- if $k \equiv 1, 5 (mod\ 6)$ and $N \not\equiv 0 (mod\ 3)$, one can take $U_0 = (2 + \sqrt{3})^k + (2 - \sqrt{3})^k$.

24. A sequence $p, 2p+1, 2(2p+1)+1, \ldots$ in which all of the numbers are primes, is called a *Cunningham chain of the first kind*. Find several Cunningham chains of the first kind. Prove, that every term of such a sequence except the last is a *Sophie Germain prime*, and every term except the first is a *safe prime*.

25. A sequence $p, 2p-1, 2(2p-1)-1, \ldots$ in which all of the numbers are primes is called a *Cunningham chain of the second kind*. Find several Cunningham chains of the second kind.

26. Prove, that for $n \geq 2$ it holds $u_{n+5} \geq 10 u_n$, where u_n is the n-th Fibonacci number.

27. Prove, that for $l \in \mathbb{N}$ and for any $n \geq 2$ we have $u_{n+5l} \geq 10^l u_n$, where u_n is the n-th Fibonacci number.

28. Prove, that starting with $W_4 = 63$ and $W_5 = 159$, every sixth Woodall number $W_n = n \cdot 2^n - 1$ is divisible by 3.

29. Prove, that for a positive integer m, the Woodall number W_{2^m} may be prime only if $2^m + m$ is prime.

30. Using the formula $Q(b,n) = \frac{b^n+1}{b+1}$, $b \geq 2$, find first ten *Wagstaff numbers base b* for $b = 2, 3, 4, 5, 6, 7, 8$. For which specific values of b all $Q(b,n)$ (with a possible exception for very small n) are composite?

31. Check, that *Wagstaff numbers base b* $b \geq 2$, defined as $Q(b,n) = \frac{b^n+1}{b+1}$, can be considered, for odd indexes n, as the *repunits with negative base* $-b$.

32. Check, that $M_7 = 127$ is a *palindromic prime* in nonary and binary.

Problems, connected with Fermat numbers

1. Check, that the numbers v_n of the form $2^n + 1$ can be obtained by the recurrence relation $v_{n+1} = 2v_n - 1$, $v_1 = 3$.

2. Prove that the numbers v_n of the form $2^n + 1$ can be obtained by the linear recurrence relation of second order with constant coefficients:

$$v_{n+2} = 3v_{n+1} - 2v_n, \quad v_1 = 3, v_2 = 5.$$

3. Using recurrent relation $v_{n+2} = 3v_{m+1} - 2v_n, \ v_1 = 3, v_2 = 5$, prove that the generating function for the sequence $v_1, v_2, v_3, \ldots, v_n, \ldots$ of numbers $2^n + 1$ has the form $f(x) = \frac{3x-4}{2x^2-3x+1}$, i.e.,

$$\frac{3 - 4x}{2x^2 - 3x + 1} = v_1 + v_2 x + v_3 x^2 + \cdots + v_{n+1} x^n + \ldots, \quad |x| < 0.5.$$

4. Using only elementary arguments, prove, that the generating function of the sequence $v_1, v_2, v_3, \ldots, v_n, \ldots$ of numbers $2^n + 1$ has the form $\frac{3-4x}{(1-x)(1-2x)}$, and the corresponding series is convergent for $|x| \le 0.5$.

5. Let $q = p^m$, $m > 0$, be a power of an odd prime p. Prove, that the Fermat number F_n is divisible by q if and only if $ord_q\, 2 = 2^{n+1}$.

6. Prove, that any divisor $d > 1$ of a Fermat number F_n, $n > 1$, is of the form $k \cdot 2^{n+2} + 1$, where k is a positive integer.

7. Prove, that the set of all quadratic nonresidues of a Fermat prime is equal to the set of all its primitive roots.

8. Let the largest prime factor of the Fermat number F_n be $P(F_n)$. Check, that $P(F_n) \ge 2^{n+2}(4n + 9) + 1$.

9. Check, that in view of the known facts about the factors of F_n, Fermat's question, "whether $(2k)^{2^m} + 1$ is always a prime except when divisible by an F_n", is without further particular interest.

10. Prove the infiniteness of primes, using Fermat numbers.

11. Prove the infiniteness of primes in the arithmetic progression $4k + 1$, $k \in \mathbb{N}$.

12. Given $t \in \mathbb{N}$, prove the infiniteness of primes in the arithmetic progression $2^t k + 1$, $k \in \mathbb{N}$.

13. Given $p \in P, t \in \mathbb{N}$, prove the infiniteness of primes in the arithmetic progression $2p^t k + 1$.

14. Prove, that for n-th prime number p_n, $n \geq 7$, it holds $p_n \leq F_{n-5}$.

15. Prove, that $\pi(x) \geq \log_2 \log_2 x + 4$ for any $x \geq 17$.

16. Prove, that every Fermat number F_n, $n > 1$, has infinitely many representations in the form $x^2 - 2y^2$, where x and y are both positive integers.

17. Prove, that $F_n = p^2 + q^2 + z^2$ if and only if $n = 2$.

18. Prove, that $F_n = x^2 + y^2 + z^2 + t^2$, $x, y, z, t \in \mathbb{Z}$, for any $n \geq 0$.

19. Prove, that every Fermat number F_n, $n > 0$, is of the form $6m - 1$.

20. Prove, that any prime of the form $n^n + 1$ is a Fermat prime.

21. Find the first digits in the decimal expansion of the sum $\sum_{n \geq 0} \frac{1}{F_n} = \sum_{n \geq 0} \frac{1}{2^{2^n} + 1}$ of the reciprocals of all the Fermat numbers.

22. Prove, that F_{73} has more than $24 \cdot 10^{20}$ digits.

23. Check, that $F_8 < 10^{100}$, but $F_9 > 10^{100}$.

24. Prove, that F_{334} has more than $10^{10^{100}}$ digits.

25. Prove, that F_{1945} has more than $24 \cdot 10^{582}$ digits.

26. Prove, that the Fermat numbers sit in the Pascal's triangle only in the trivial way: if $F_m = \binom{n}{k}$ for some $n \geq 2k \geq 2$, then $k = 1$.

27. Prove, that there exist non-trivial solutions of $F_m(a) = \binom{n}{k}$, where $F_m(a) = a^{2^m} + 1$, $a > 1$, is a generalized Fermat number.

28. Find all regular n-gons, $n \leq 1000$, which can be constructed by Euclidean methods.

29. Gauss proved that we can subdivide a circle into n parts using a ruler (an unmarked straightedge) and a compass (which draws circles) if and only if n is a power of two times a product of distinct Fermat primes. Later Pierpont proved (1895, [Pier95]) that we can divide a circle into n parts using origami (paper

folding) if and only if n is a product of a power of two times a power of three times a product of distinct primes of the form $2^u 3^v + 1$, $u, v \geq 0$. These primes are called *Pierpont primes*.

Find the first ten Pierpont primes. Find also, the first ten regular polygons, which can be constructible using origami. Prove, that any classical constructible polygon can be constructed using origami.

30. A *Pierpont prime* $2^u 3^v + 1$, $u, v \geq 0$, with $v = 0$ is of the form $2^u + 1$ and is therefore a Fermat prime for $u > 0$. Prove, that any odd non-Fermat Pierpont prime have the form $6k + 1$.

31. A *Proth number* is a number $k \cdot 2^n + 1$, where k and n are positive integers, k is odd, and $2^n > k$. Find the first ten Proth numbers. Prove, that any Fermat number is a Proth number.

32. A *Proth prime* is a Proth number $k \cdot 2^n + 1$ (where k and n are positive integers, k is odd, and $2^n > k$), that is prime. Find the first ten Proth primes.

33. Prove the *Proth theorem*: if p is a Proth number (i.e., a number of the form $k \cdot 2^n + 1$, k is odd, $k < 2^n$), and if there exists an integer a for which

$$a^{\frac{p-1}{2}} \equiv -1 \pmod{p},$$

then p is prime. Check, that 5, 13, 17 are primes, using the Proth theorem.

34. Prove, that for all $n \geq 2$ it holds $F_n + 8, F_n^2 + 8, F_n^3 + 8 \in S$; find other possible shifts of Fermat numbers, such that for all $n \geq n_0$ it holds $F_n + K \in S$; $F_n^2 + T \in S$, $F_n^3 + R \in S$.

35. Check, that the third Fermat prime, $F_2 = 17$: is the average of two consecutive Fibonacci numbers; is the sum of four consecutive primes; is a prime of the form $p^q + q^p$, where p and q are primes; has 17 partitions into prime parts; is an integer which is not the sum of three distinct primes; is a prime number of the form $p^2 + 8$; is the smallest prime whose sum of the digits is a cubic number; is a prime that is equal to the sum of digits of its cube; is the smallest prime that is a *quartan*, i.e., the sum of two

biquadratic numbers; moreover, its square can be expressed as the sum of $1, 2, 3, 4, 5, 6, 7, 8$ distinct squares.

Other problems

1. Given two integer parameters P and Q, the *Lucas sequences of the first kind* $U_n(P, Q)$ and *of the second kind* $V_n(P, Q)$ are defined by the following recurrence relations:

$$U_0(P, Q) = 0, \ U_1(P, Q) = 1,$$
$$U_n(P, Q) = P \cdot U_{n-1}(P, Q) - Q \cdot U_{n-2}(P, Q) \ \text{for } n > 1;$$
$$V_0(P, Q) = 2, \ V_1(P, Q) = P,$$
$$V_n(P, Q) = P \cdot V_{n-1}(P, Q) - Q \cdot V_{n-2}(P, Q) \ \text{for } n > 1.$$

   Check, that the sequence $U_n(1, -1)$ gives the Fibonacci numbers; the sequence $V_n(1, -1)$ gives the Lucas numbers.

2. Check, that the sequence $U_n(2, -1)$ gives the Pell numbers; the sequence $V_n(2, -1)$ gives the Pell-Lucas numbers.

3. Check that the sequence $U_n(3, 2)$ gives the Mersenne numbers $2^n - 1$; the sequence $V_n(3, 2)$ gives the numbers of the form $2^n + 1$, which includes the Fermat numbers.

4. Prove that for $n > 0$, it holds

$$U_n(P, Q) = \frac{P \cdot U_{n-1}(P, Q) + V_{n-1}(P, Q)}{2},$$
$$V_n(P, Q) = \frac{(P^2 - 4Q) \cdot U_{n-1}(P, Q) + P \cdot V_{n-1}(P, Q)}{2}.$$

5. Prove, that the terms of Lucas sequences can be expressed as follows: $U_n = \frac{a^n - b^n}{a - b}$, $V_n = a^n + b^n$, where $a = \frac{P + \sqrt{P^2 - 4Q}}{2}$, and $b = \frac{P - \sqrt{P^2 - 4Q}}{2}$.

6. A Sierpiński number is an odd natural number k such that $k \cdot 2^n + 1$ is composite for all natural numbers n. Check, that all odd numbers less 100 are not Sierpiński numbers.

7. A *Riesel number* is an odd natural number k such that $k \cdot 2^n - 1$ is composite for all natural numbers n. Check, that all odd numbers less than 100 are not Riesel numbers.

8. Let p be a prime such that $p \equiv 3(mod\ 4)$. Prove, that the Fermat number F_p is prime if and only if $M_p^{\frac{F_p-1}{2}} \equiv -1(mod\ F_p)$, where M_p is the associated Mersenne number.

9. Let p be a prime such that $p \equiv 3, 5(mod\ 8)$. Prove, that the Fermat number F_p is prime if and only if $M_p^{\frac{F_{p+1}-1}{2}} \equiv -1(mod\ F_{p+1})$, where M_p is the associated Mersenne number.

10. Prove, that every *permutable prime* (i.e., prime number which remains prime on every rearrangement of their decimal digits) is a *near-repdigit*, that is, it is a permutation of the integer $B_n(a, b) = (aaa...aab)_{10}$, where a and b are distinct digits from the set $\{1, 3, 7, 9\}$. Find the first ten permutable primes. Find also, the first non-trivial permutable Mersenne prime, and find first non-trivial permutable Fermat prime.

11. An *emirp* (*prime* spelled backwards) is a prime number that results in a different prime when its decimal digits are reversed. Find the first ten emirps. Find a Mersenne prime, which is an emirp. Find a Fermt prime which is an emirp.

12. A prime number p is called a *Chen prime* if $p + 2$ is either a prime or a product of two primes (a semi-prime). Find the first ten Chen primes. Find the first ten Chen primes which are not the lower member of a pair of twin primes. Find the smallest prime, which is not a Chen prime.

Solutions: Problems, connected with Mersenne numbers

1. In fact, $M_{n+1} = 2M_n + 1$. So, $M_{n+2} = 2M_{n+1} + 1$. Then

$$M_{n+2} - M_{n+1} = 2M_{m+1} - 2M_n, \quad \text{and}$$
$$M_{n+2} = 3M_{m+1} - 2M_n.$$

2. In order to obtain the above formula, let us consider two polynomials

$$f(x) = a_0 + a_1 x + \cdots + a_m x^m \text{ and } g(x) = b_0 + b_1 x + \cdots + b_n x^n$$

with real coefficients and $m < n$. It follows (see, for example, [DeMo10], [Plou92]), that *the rational function $\frac{f(x)}{g(x)}$ is the generating function of the sequence $c_0, c_1, c_2, ..., c_n, ...,$ which is a solution of the linear recurrent equation $b_0 c_{n+k} + b_1 c_{n+k-1} + \cdots + b_n c_k = 0$ of n-th order with coefficients $b_0, b_1, ..., b_n$.*

In fact, one has the decomposition $\frac{f(x)}{g(x)} = c_0 + c_1 x + c_2 x^2 + \cdots + c_n x^n + \cdots$, if $|x| < r$, and $r = \min_{1 \le i \le n} |x_i|$, where $x_1, ..., x_n$ are the roots of the polynomial $g(x)$. It yields, that the rational function $\frac{f(x)}{g(x)}$ is the generating function of the obtained sequence $c_0, c_1, c_2, ..., c_n, ...$ Moreover, one gets $f(x) = g(x)(c_0 + c_1 x + c_2 x^2 + \cdots + c_n x^n + \cdots)$. In other words, it holds

$$a_0 + a_1 x + a_2 x^2 + \cdots + a_m x^m = (b_0 + b_1 x + b_2 x^2 + \cdots + b_n x^n)$$
$$\times (c_0 + c_1 x + c_2 x^2 + \cdots + c_n x^n + \cdots).$$

It is easy to check now the following equalities:

$$a_0 = b_0 c_0, \quad a_1 = b_0 c_1 + b_1 c_0, \quad a_2 = b_0 c_2 + b_1 c_1 + b_2 c_0, ...,$$

$$a_m = b_0 c_m + \cdots + b_m c_0,$$

$$0 = b_0 c_{m+1} + \cdots + b_{m+1} c_0, ..., \quad 0 = b_0 c_n + \cdots + b_n c_0,$$

$$0 = b_0 c_{n+1} + \cdots + b_n c_1, ... \ .$$

So, the sequence $c_0, c_1, ..., c_n, ...$ is a solution of the *linear recurrent equation*

$$b_0 c_{n+k} + \cdots + b_n c_k = 0$$

of n-th order with coefficients $b_0, ..., b_n$. Moreover, one can find the first n elements of this sequence using the first n above equalities: $a_0 = b_0 c_0$, $a_1 = b_0 c_1 + b_1 c_0$, ..., $a_{n-1} = \sum_{k=0}^{n-1} b_k c_{n-1-k}$.

On the other hand, let the sequence $c_0, c_1, c_2, ..., c_n, ...$ be a solution of a linear recurrent equation $b_0 c_{n+k} + \cdots + b_n c_k = 0$ of n-th order with coefficients $b_0, ..., b_n$. Let us define numbers

$a_0, ..., a_{n-1}$ by the formulas $a_i = \sum_{k=0}^{i} b_k c_{i-k}$, $i = 0, 1, 2, ...,$ $n - 1$, using the *initial values* $c_0, c_1, ..., c_n$ of the given sequence. It yields the equality

$$a_0 + a_1 x + a_2 x^2 + \cdots + a_{n-1} x^{n-1}$$
$$= (b_0 + b_1 x + b_2 x^2 + \cdots + b_n x^n)$$
$$\times (c_0 + c_1 x + c_2 x^2 + \cdots + c_n x^n + ...).$$

In other words, one gets

$$\frac{a_0 + a_1 x + a_2 x^2 + \cdots + a_{n-1} x^{n-1}}{b_0 + b_1 x + b_2 x^2 + \cdots + b_n x^n}$$
$$= c_0 + c_1 x + c_2 x^2 + ... + c_n x^n + ...$$

So, the generating function of the sequence $c_0, c_1, ..., c_n, ...$ has the form $\frac{f(x)}{g(x)}$, where

$$g(x) = b_0 + b_1 x + \cdots + b_n x^n, \quad \text{and} \quad f(x) = a_0 + \cdots + a_{n-1} x^{n-1},$$

with $a_0 = b_0 c_0$, $a_1 = b_0 c_1 + b_1 c_0$, ..., $a_{n-1} = b_0 c_{n-1} + b_1 c_{n-2} + ... + b_{n-1} c_0$.

So, the characteristic equation for the recurrence $M_{n+2} = 3M_{m+1} - 2M_n$ is $x^2 = 3x - 2$, i.e., $x^2 - 3x + 2 = 0$. Then the generating function of the sequence $M_1, M_2, M_3, ..., M_n,$ ... of Mersenne numbers has the form $f(x) = \frac{h(x)}{g(x)}$, where $g(x) = 2x^2 - 3x + 1$, and $f(x) = a_0 + a_1 x$, where $a_0 = 1 \cdot 1 = 1$, and $a_1 = 1 \cdot 3 + (-3) \cdot 1 = 0$. Thus, $f(x) \equiv 1$. We obtain $f(x) = \frac{1}{2x^2 - 3x + 1} = \frac{1}{2(x-1)(x-0.5)}$, i.e.,

$$\frac{1}{2x^2 - 3x + 1} = M_1 + M_2 x + M_3 x^2 + ... M_{m+1} x^n + ...$$

As $\min\{1, 0.5\} = 0.5$, we get $|x| < 0.5$.

3. Consider an elementary proof, using the sums of the infinite geometric progressions: $1 + 2x + (2x)^2 + \cdots = \frac{1}{1-2x}$ for $|2x| < 1$, and $1 + x + x^2 + \cdots = \frac{1}{1-x}$ for $|x| < 1$. So, for $|x| < 0.5$ it

holds

$$M_1 + M_2 x + \cdots + M_n x^{n-1} + \cdots = (2^1 - 1) + (2^2 - 1)x$$
$$+ (2^3 - 1)x^2 + \ldots + (2^n - 1)x^{n-1} + \cdots$$
$$= -(1 + x + x^2 + \cdots + x^{n-1} + \cdots) + 2(1 + 2x + (2x)^2 + \cdots$$
$$+ (2x)^{n-1} + \cdots)$$
$$= -\frac{1}{1-x} + \frac{2}{1-2x} = \frac{1}{(1-x)(1-2x)}.$$

4. For example, $M_1 + M_2 \cdot 100 + M_3 \cdot 100^2 = 70301$; it is easy to find the first three Mersenne numbers in this representation.

5. One should use the estimation $2^n - 1 \le x < 2^{n+1} - 1$; see Chapter 5, Section 5.1.

6. Let $M_p = 2^p - 1$ be a Mersenne number, where p is a prime. If M_p is a composite, then p is odd. By the Fermat's little theorem, $\frac{M_p - 1}{2} = 2^{p-1} - 1 \equiv 0 \pmod p$. So $\frac{M_p - 1}{2} = kp$ for some positive integer k. Hence, $M_p = 2^p - 1 | (2^{kp} - 1) = 2^{\frac{M_p - 1}{2}} - 1$. It is equivalent to say that $2^{\frac{M_p - 1}{2}} \equiv 1 \pmod{M_p}$, which implies that $2^{M_p - 1} \equiv 1 \pmod{M_p}$.

7. Consider the numbers $M_{k\phi(n)}$.

8. Let $q_i | M_{p_i}$; as $\gcd(M_{p_i}, M_{p_j}) = 1$, then $q_i \ne q_j$; if the set P of primes is finite, then $P = \{q_1, ..., q_k\}$, but $p_1 = 2$, while all q_i are odd; a contradiction.

9. In fact, for $n \ge 3$ it holds

$$M_n \equiv 2^n - 1 \equiv 0 - 1 \equiv -1 \pmod 8.$$

10. It is well-known, that numbers of the form $8t + 7$ cannot be represented as a sum of three perfect squares. In fact, any integer a has the form $8t + r$, where $r \in \{0, \pm 1, \pm 2, \pm 3, 4\}$. Then $a^2 = 8k + r$, where $r \in \{0, 1, 4\}$. Checking all 27 possible combinations $0 + 0 + 0 \equiv 0 \pmod 8$, $0 + 0 + 1 \equiv 1 \pmod 8$, $0 + 0 + 4 \equiv 4 \pmod 8$, ..., $4 + 4 + 4 \equiv 4 \pmod 8$, we get, that $x^2 + y^2 + z^2 = 8l + r$, where $r \in \{0, 1, 2, 3, 4, 5, 6\}$. So, $x^2 + y^2 + z^2 \ne 8l + 7$.

As $M_n = 8l + 7$ for $n \geq 3$, we get that $M_n \neq x^2 + y^2 + z^2$ for $n \geq 3$.

11. It is well-known, that any positive integer can be represented as a sum of four integer squares (it is the *Lagrange's four-square theorem*).

12. It is true for any positive integer m. It is easy to check for small m (see, for example, the table of known Mersenne primes). For general proof see, for example, [CFHLZ17].

13. It is easy to check for small s. Thus, for $s = 1$ it was proven in Chapter 3, Section 3.2.

    In general, the period is equal to the multiplicative order of 2 modulo 5^s, which is $\phi(5^s) = 4 \cdot 5^{s-1}$.

    For an additional information see, for example, [CFHLZ17].

14. The objective of the puzzle is to move the entire stack to another rod, obeying the following simple rules: only one disk can be moved at a time; each move consists of taking the upper disk from one of the stacks and placing it on top of another stack or on an empty rod; no larger disk may be placed on top of a smaller disk.

    Check, that with 3 disks, the puzzle can be solved in $M_3 = 7$ moves.

    Prove, that the minimal number of moves required to solve a Tower of Hanoi puzzle with n disks (and three rods) is $M_n = 2^n - 1$.

    A simple solution for the toy puzzle is to alternate moves between the smallest piece and a non-smallest piece. When moving the smallest piece, always move it to the next position in the same direction (to the right if the starting number of pieces is even, to the left if the starting number of pieces is odd). If there is no tower position in the chosen direction, move the piece to the opposite end, but then continue to move in the correct direction. For example, if you started with three pieces, you would move the smallest piece to the opposite end, then continue in the left direction after that. When the turn is to move the non-smallest piece, there is only one legal move. Doing this will complete the puzzle in the fewest moves. As a result of the above procedure, the number h_n of moves required to solve the puzzle of n disks on three rods is given by the recurrence relation $h_n = 2h_{n-1}+1$ with $h_1 = 1$. Solving gives $h_n = 2^n - 1$, i.e., the Mersenne number M_n.

For three rods, the proof that the above solution is minimal can be achieved using the Lucas correspondence which relates Pascal's triangle to the *Hanoi graph*.

While algorithms are known for transferring disks on four rods, none has been proved minimal.

15. In fact, if $M_p \in P, M_p > 3$, then $p \in P, p > 2$. For any odd prime p it holds $p \equiv \pm 1 (mod\ 4)$. If $p \equiv 1 (mod\ 4)$, then $p = 4k+1$, and

$$M_p \equiv 2^p - 1 \equiv 2^{4k+1} - 1$$
$$\equiv 2 \cdot (2^4)^k - 1 \equiv 2 \cdot 6^k - 1 \equiv 2 \cdot 6 - 1 \equiv 1 (mod\ 10).$$

If $p \equiv -1 (mod\ 4)$, then $p = 4k+3$, and

$$M_p \equiv 2^p - 1 \equiv 2^{4k+3} - 1$$
$$\equiv 2^3 \cdot (2^4)^k - 1 \equiv 8 \cdot 6^k - 1 \equiv 8 \cdot 6 - 1 \equiv 7 (mod\ 10).$$

16. It is true for the first perfect number 6. Otherwise, any even perfect number has the form $2^{p-1}(2^p - 1)$ with odd prime p. For any odd prime p it holds $p \equiv \pm 1 (mod\ 4)$.
    If $p \equiv 1 (mod\ 4)$, then $p = 4k+1$, and

$$2^{p-1}(2^p - 1) \equiv 2^{4k}(2^{4k+1} - 1)$$
$$\equiv (2^4)^k \cdot ((2^4)^k - 1) \equiv 6^k \cdot (2 \cdot 6^k - 1)$$
$$\equiv 6(2 \cdot 6 - 1) \equiv 6 (mod\ 10).$$

If $p \equiv -1 (mod\ 4)$, then $p = 4k+3$, and

$$2^{p-1}(2^p - 1 \equiv 2^{4k+2}(2^{4k+3} - 1))$$
$$\equiv 2^2 \cdot (2^4)^k \cdot (2 \cdot (2^4)^k - 1) \equiv 4 \cdot 6^k(8 \cdot 6^k - 1)$$
$$\equiv 4 \cdot 6(8 \cdot 6 - 1) \equiv 4 \cdot 7 \equiv 8 (mod\ 10).$$

See the first perfect numbers 6, 28, 496, 8128, 33550336, 8589869056, 137438691328, ... (sequence A000396 in the OEIS); all of them end with either the digit 6 or the digit 8; but the digits do not alternate: 6, 8, 6, 8, 6, 6, 8, ...

17. It is obviously:

$$2^{k-1}(2^k - 1) = 2^{k-1}(1 + 2 + 2^2 + \cdots + 2^{k-1})$$
$$= 2^{2k-2} + 2^{2k-3} + \cdots + 2^{k-1}.$$

In particular, $6 = 110_2$, $28 = 11100_2$, $496 = 111110000_2$, $8128 = 1111111000000_2$.

18. It is proven, that the only even perfect number of the form $x^3 + 1$ is 28 (Makowski, 1962). Moreover, 28 is also the only even perfect number that is a sum of two positive cubes of integers (Gallardo, 2010).

19. The number of divisors of a perfect number N (whether even or odd) must be even, because N cannot be a perfect square.

20. These numbers are 2, 3, 5, 7, 11, 17, 23, 31, 47, 53, ... (sequence A005105 in the OEIS). The largest known primes of this type are Mersenne primes; currently the largest known is $2^{82589933} - 1$. The largest known Pierpont prime of the second kind that is not a Mersenne prime, is $3 \cdot 2^{11895718} - 1$. It was found by *PrimeGrid*.

21. A Pierpont prime of the second kind with $v = 0$ is of the form $2^u - 1$ and is therefore a Mersenne prime. If v is positive, then u must also be positive, because a number of the form $3^v - 1$, greater 2, would be even and therefore non-prime. Therefore, the odd non-Mersenne Pierpont primes all have the form $6k - 1$.

22. For $k = 1$ we obtain Mersenne numbers. So, any Mersenne number has the form $k \cdot 2^n - 1$, with $k < 2^n$.

23. The *Lucas–Lehmer–Riesel test* is a particular case of group-order primality testing; we demonstrate that some number is prime by showing that some group has the order that it would have were that number prime, and we do this by finding an element of that group of precisely the right order.

For Lucas-style tests on a number N, we work in the multiplicative group of a quadratic extension of the integers modulo N; if N is prime, the order of this multiplicative group is $N^2 - 1$, it has a subgroup of order $N + 1$, and we try to find a generator for that subgroup.

We start off by trying to find a non-iterative expression for the U_i. Following the model of the Lucas–Lehmer test, put $U_i = a^{2^i} + a^{-2^i}$, and by induction we have $U_i = U_{i-1}^2 - 2$.

So we can looking at the 2^i-th term of the sequence $V_i = a^i + a^{-i}$. If a satisfies a quadratic equation, this is a Lucas sequence, and has an expression of the form $V_i = \alpha V_{i-1} + \beta V_{i-2}$. Really, we're looking at the $(k \cdot 2^i)$-th term of a different sequence, but since decimations (take every k-th term starting with 0) of a Lucas sequence are themselves Lucas sequences, we can deal with the factor k by picking a different starting point.

24. It follows that $p_i = 2^{i-1}p_1 + (2^{i-1} - 1)$, where $p_1 = p$. By setting $a = \frac{p+1}{2}$, $p_i = 2^i a - 1$.

Examples of (complete) Cunningham chains of the first kind include these:

- 2, 5, 11, 23, 47 (the next number would be 95, but that is not prime);

- 3, 7 (the next number would be 15, but that is not prime);

- 29, 59 (the next number would be $119 = 7 \cdot 17$, but that is not prime);

- 41, 83, 167 (the next number would be 335, but that is not prime);

- 89, 179, 359, 719, 1439, 2879 (the next number would be $5759 = 13 \cdot 443$, but that is not prime).

As of 2020, the longest known Cunningham chain of the first kind is of length 17, with starting prime 2759832934171386593519, discovered by J. Wroblewski in 2008.

Extending the conjecture that there exist infinitely many Sophie Germain primes, it has also been conjectured that arbitrarily long Cunningham chains of the first kind exist, although infinite chains are known to be impossible.

25. It follows that $p_i = 2^{i-1}p_1 - (2^{i-1} - 1)$, where $p_1 = p$. By setting $a = \frac{p-1}{2}$, $p_i = 2^i a + 1$.

   Examples of (complete) Cunningham chains of the second kind include these:

   - 2, 3, 5 (the next number would be 9, but that is not prime);

   - 7, 13 (the next number would be 25, but that is not prime);

   - 19, 37, 73 (the next number would be 145, but that is not prime);

   - 31, 61 (the next number would be 121 = 112, but that is not prime).

   As of 2020, the longest known Cunningham chain of the second kind is of length 19, with starting prime 79910197721 667870187016101, discovered by Chermoni and Wroblewski in 2014.

   It has been conjectured that arbitrarily long Cunningham chains of the second kind exist.

26. In fact, if $n = 2$, then $u_7 = 13 \geq 1 - u_2 = 10$. If $n \geq 3$, then $u_{n+5} = u_{n+4} + u_{n+3} = 2u_{n+3} + u_{n+2} = 3u_{n+2} + 2u_{n+1} = 5u_{n+1} + 3u_n = 8u_n + 5u_{n-1} > 8u_n + 4u_{n-1} = 8u_n + 2(u_{n-1} + u_{n-1}) > 8u_n + 2(u_{n-1} + u_{n-2}) = 10u_n$.

27. In fact, for $l = 1$ we obtained this result easily. Going form l to $l + 1$, we get: $4u_{n+5(l+1)} = u_{(n+5l)+5} \geq 10u_{n+5l} \geq 10 \cdot 10^l u_n = 10^{l+1}u_{n+1}$.

28. It is easy to check, that for $n \equiv 4, 5 \pmod 6$ it holds the $W_n \equiv 0 \pmod 3$. For example, if $n \equiv 5 \pmod 6$, then

$$W_n \equiv 5 \cdot (2^6)^k \cdot 2^5 - 1 \equiv (-1) \cdot 1 \cdot (-1) - 1 \equiv 0 \pmod 3.$$

   Thus, in order for $W_n = n \cdot 2^n - 1$ to be prime, the index n cannot be congruent to 4 or 5 modulo 6.

29. In fact, $W_{2^m} = 2^m \cdot 2^{2^m} - 1 = 2^{m+2^m} - 1 = M_{m+2^m}$.

As of 2020, the only known primes that are both Woodall primes and Mersenne primes are $W_2 = M_3 = 7$, and $W_{512} = M_{521}$.

30. For some specific values of b, all $Q(b,n)$ (with a possible exception for very small n) are composite because of an algebraic factorization. Specifically, if b has the form of a perfect power with odd exponent (like 8, 27, 32, 64, 125, 128, 216, 243, 343, 512, 729, 1000, ... (sequence A070265 in the OEIS)), then the fact that $x^m + 1$, m odd, is divisible by $x + 1$ shows that $Q(a^m, n)$ is divisible by $a^n + 1$ in these special cases.

    Another case is $b = 4k^4$, with k positive integer (like 4, 64, 324, 1024, 2500, 5184, ... (sequence A141046 in the OEIS)), where we have the aurifeuillean factorization.

    However, when b does not admit an algebraic factorization, it is conjectured that an infinite number of n values make $Q(b,n)$ prime (all such n should be odd primes).

    For $b = 10$, the primes themselves have the following appearance:

$$9091, 909091, 909090909090909091,$$

$$909090909090909090909090909091, \dots$$

(sequence A097209 in the OEIS); these n are 5, 7, 19, 31, 53, 67, 293, 641, 2137, 3011, 268207, ... (sequence A001562 in the OEIS).

31. It is easy to see, that, for odd n, it holds $\frac{b^n+1}{b+1} = \frac{(-b)^n-1}{(-b)-1} = R_n(-b)$. So, the *Wagstaff primes base b*, $b \geq 2$, are can be considered, for odd n, as generalized repunits with the negative base $-b$. In particular, the *generalized Wagstaff primes base b* are generalized repunit primes base $-b$. Smallest base b such that $R_p(-b)$ is prime (where p is the n-th prime), are 3, 2, 2, 2, 2, 2, 2, 2, 2, 7, ... (sequence A103795 in the OEIS).

    Smallest prime $p > 2$ such that $R_p(-b)$ is prime are (start with $b = 2$, 0 if no such p exists), are 3, 3, 3, 5, 3, 3, 0, 3, 5, 5, ... (sequence A084742 in the OEIS).

32. In binary, $127 = 1111111_2$; In nonary, $127 = 151_9$.

    Moreover, if is a *Friedman number* in base 10, since $127 = 2^7 - 1$, as well as in binary, since $1111111 = (1+1)^{111} - 1$.

Solutions: Problems, connected with Fermat numbers

1. In fact, $2 \cdot (2^n + 1) - 1 = (2^{n+1} + 2) - 1 = 2^{n+1} + 1$.

2. In fact, $v_{n+1} = 2v_n - 1$, $v_{n+2} = 2v_{n+1} - 1$. Then $v_{n+2} - v_{n+1} = 2v_{m+1} - 2v_n$, and $v_{n+2} = 3v_{m+1} - 2v_n$.

3. In order to obtain the above formula, let us consider two polynomials

$$f(x) = a_0 + a_1 x + \cdots + a_m x^m \quad \text{and} \quad g(x) = b_0 + b_1 x + \cdots + b_n x^n$$

with real coefficients and $m < n$. It follows (see, for example, [DeMo10], [Plou92]), that *the rational function* $\frac{f(x)}{g(x)}$ *is the generating function of the sequence* $c_0, c_1, c_2, ..., c_n, ...,$ *which is a solution of the linear recurrent equation* $b_0 c_{n+k} + b_1 c_{n+k-1} + \cdots + b_n c_k = 0$ *of n-th order with coefficients* $b_0, b_1, ..., b_n$.

In fact, one has the decomposition $\frac{f(x)}{g(x)} = c_0 + c_1 x + c_2 x^2 + \cdots + c_n x^n + ...$ if $|x| < r$, and $r = \min_{1 \le i \le n} |x_i|$, where $x_1, ..., x_n$ are the roots of the polynomial $g(x)$. It yields, that the rational function $\frac{f(x)}{g(x)}$ is the generating function of the obtained sequence $c_0, c_1, c_2, ..., c_n, ...$ Moreover, one gets $f(x) = g(x)(c_0 + c_1 x + c_2 x^2 + ... + c_n x^n + \cdots)$. In other words, it holds

$$a_0 + a_1 x + a_2 x^2 + \cdots + a_m x^m = (b_0 + b_1 x + b_2 x^2 + \cdots + b_n x^n)$$
$$\times (c_0 + c_1 x + c_2 x^2 + ... + c_n x^n + \cdots).$$

It is easy to check now the following equalities:

$$a_0 = b_0 c_0, \quad a_1 = b_0 c_1 + b_1 c_0, \quad a_2 = b_0 c_2 + b_1 c_1 + b_2 c_0, ...,$$
$$a_m = b_0 c_m + \cdots + b_m c_0,$$
$$0 = b_0 c_{m+1} + \cdots + b_{m+1} c_0, ..., \quad 0 = b_0 c_n + \cdots + b_n c_0,$$
$$0 = b_0 c_{n+1} + \cdots + b_n c_1, ...$$

So, the sequence $c_0, c_1, ..., c_n, ...$ is a solution of the *linear recurrent equation* $b_0 c_{n+k} + \cdots + b_n c_k = 0$ *of n-th order with coefficients* $b_0, ..., b_n$. Moreover, one can find the first n elements of this sequence using the first n above equalities: $a_0 = b_0 c_0$, $a_1 = b_0 c_1 + b_1 c_0$, ..., $a_{n-1} = \sum_{k=0}^{n-1} b_k c_{n-1-k}$.

On the other hand, let the sequence $c_0, c_1, c_2, ..., c_n, ...$ be a solution of a linear recurrent equation $b_0 c_{n+k} + \cdots + b_n c_k = 0$ of n-th order with coefficients $b_0, ..., b_n$. Let us define numbers $a_0, ..., a_{n-1}$ by the formulas $a_i = \sum_{k=0}^{i} b_k c_{i-k}$, $i = 0, 1, 2, ..., n - 1$, using the *initial values* $c_0, c_1, ..., c_n$ of the given sequence. It yields the equality

$$a_0 + a_1 x + a_2 x^2 + \cdots + a_{n-1} x^{n-1}$$
$$= (b_0 + b_1 x + b_2 x^2 + \cdots + b_n x^n)$$
$$\times (c_0 + c_1 x + c_2 x^2 + ... + c_n x^n + \cdots).$$

In other words, one gets

$$\frac{a_0 + a_1 x + a_2 x^2 + \cdots + a_{n-1} x^{n-1}}{b_0 + b_1 x + b_2 x^2 + \cdots + b_n x^n}$$
$$= c_0 + c_1 x + c_2 x^2 + ... + c_n x^n + \cdots .$$

So, the generating function of the sequence $c_0, c_1, ..., c_n, ...$ has the form $\frac{f(x)}{g(x)}$, where

$$g(x) = b_0 + b_1 x + \cdots + b_n x^n, \quad \text{and} \quad f(x) = a_0 + \cdots + a_{n-1} x^{n-1},$$

with $a_0 = b_0 c_0$, $a_1 = b_0 c_1 + b_1 c_0$, ..., $a_{n-1} = b_0 c_{n-1} + b_1 c_{n-2} + \cdots + b_{n-1} c_0$.

So, the characteristic equation for the recurrence $v_{n+2} = 3v_{m+1} - 2v_n$ is $x^2 = 3x - 2$, i.e., $x^2 - 3x + 2 = 0$. Then the generating function $f(x) = \frac{h(x)}{g(x)}$, where $g(x) = 2x^2 - 3x + 1$, and $f(x) = a_0 + a_1 x$, where $a_0 = 1 \cdot 3 = 3$, and $a_1 = 1 \cdot 5 + (-3) \cdot 3 = -4$. So, $f(x) = 3 - 4x$. We obtain $f(x) = \frac{3-4x}{2x^2-3x+1} = \frac{3-4x}{2(x-1)(x-0.5)}$, i.e.,

$$\frac{3 - 4x}{2x^2 - 3x + 1} = v_1 + v_2 x + v_3 x^2 + \cdots v_{n+1} x^n + \cdots .$$

As $\min\{1, 0.5\} = 0.5$, we get $|x| < 0.5$.

4. One can obtain an elementary proof of this fact, using the sums of infinite geometric progressions.

5. Let $q|F_n$. Then $q|(2^{2^n}+1)\cdot(2^{2^n}-1) = 2^{2^{n+1}}-1$, and, hence, $2^{2^{n+1}} \equiv 1 (mod\ q)$. It follows that $ord_q\, 2|2^{n+1}$, i.e., $2^{n+1} = k\cdot ord_q\, 2$ for some positive integer k. Thus, k is a power of 2 and so is $ord_q\, 2$. Let $e = ord_q\, 2 = 2^j$. If $j < n+1$, then we have $q|(2^{e2^{n-j}}-1) = 2^{2^n}-1$. But this is impossible because $q|(2^{2^n}+1)$, and $q \neq 2$. Hence, $j = n+1$ and so $ord_q\, 2 = 2^{n+1}$.

    Conversely, if we assume that $ord_q\, 2 = 2^{n+1}$, then $q|(2^{2^{n+1}} - 1) = (2^{2^n}+1)\cdot(2^{2^n}-1)$. Since q is an odd prime, q divides either $2^{2^n}+1$ or $2^{2^n}-1$. But q cannot divide $2^{2^n}-1$ because $2^n < ord_q\, 2$. Hence, $q|(2^{2^n}+1)$, i.e., $q|F_n$.

6. Consider the product $(k\cdot 2^{n+2}+1)(t\cdot 2^{m+2}+1)$. Without loss of generality, assume $m \geq n$. Then $(k\cdot 2^{n+2}+1)(t\cdot 2^{m+2}+1) = k\cdot t\cdot 2^{m+n+4}+k\cdot 2^{n+2}+t\cdot 2^{m+2}+1 = (k\cdot t\cdot 2^{m+2}+k+t\cdot 2^{m-n})\cdot 2^{n+2}+1$, which is also in the form of $k\cdot 2^{n+2}+1$. So, all divisors $d > 1$ of a Fermat number F_n, $n > 1$, have the form $k\cdot 2^{n+2}+1$.

7. First let a be a quadratic non-residue of the Fermat prime F_n and let $e = ord_{F_n}\, a$. According to the Fermat's little theorem, $a^{F_n-1} \equiv 1 (mod\ F_n)$, so $e|(F_n - 1) = 2^{2^n}$. It follows that $e = 2^k$ for some non-negative integer $k \leq 2^n$. On the other hand, by the Euler's criterion, $a^{\frac{F_n-1}{2}} \equiv -1 (mod\ F_n)$. Hence, if $k < 2^n$, then $2^k|2^{2^n-1}$, and so $2^{2^{n-1}} \equiv 1 (mod\ F_n)$, which is a contradiction. So, $k = 2^n$, i.e., $ord_{F_n}\, a = 2^n$. Therefore, a is a primitive root modulo F_n.

    Conversely, suppose that g is a primitive root modulo F_n. It follows that $g^{\frac{F_n-1}{2}} \not\equiv 1 (mod\ F_n)$, and by, the Euler's criterion, g cannot be a quadratic residue.

8. It is easy to check for small values of n (see, for example, Chapter 4, Section 4.1.)

    In general, it was proven by Grytczuk, Luca and Wójtowicz in 2001.

9. See the paper *Fermat numbers $F_n = 2^{2^n}+1$* of R.D. Carmichael, 1919, [Carm19].

10. Let q_n be the least prime divisor of F_n. As $gcd(F_m, F_n) = 1$ for $n \neq m$, then $q_n \neq q_m$, and we obtain infinitely many primes $q_0, q_1, ..., q_k, ...$.

11. Note, that any prime divisor of F_n, $n \geq 1$, has the form $4k + 1$, and use the well-known property: $gcd(F_m, F_n) = 1$ for $n \neq m$.

12. For $t = 2$ it is obviously. For $t = 2$ it was proven above. Let $t > 2$. In this case consider the numbers $F_t, F_{t+1}, ..., F_{t+l}, ...$ Each such number F_{t+l}, $l \geq 0$, has a prime divisor of the form $k \cdot 2^{t+l+2} + 1 = s \cdot 2^t + 1$.

13. If $p = 2$, the situation is trivial; see that problem above. For $p \in P$, $p > 2$, see, for example, [Step01].

14. Let q_n be the smallest prime divisor of F_n. Then $q_0 = 3 = p_2$, $q_1 = 5 = p_3$, $q_2 = 17 = p_7$, and we obtain, that $p_n \leq q_{n-5} \leq F_{n-5}$ for all $n \geq 7$.

15. Let $x \geq 17$, and $x \in [p_n, p_{n+1})$. Then $\pi(x) = \pi(p_n) = n$, $n \geq 7$. Moreover, $x \leq p_{n+1} \leq 2^{2^{n-4}} + 1$, i.e., $x \leq 2^{2^{n-4}}$, or $\log_2 \log_2 x \leq n - 4$. Now we obtain, that $\pi(x) = n \geq \log_2 \log_2 x + 4$.

16. First, $(x_0, y_0) = (F_{n-1}, F_{n-2}-1)$ gives one such representation. Now notice that

$$(3x + 4y)^2 - 2(2x + 3y)^2$$
$$= 9x^2 + 24xy + 16y^2 - 8x^2 - 24xy - 18y^2 = x^2 - 2y^2.$$

If x and y are both positive, then $3x + 4y > x$ and $2x + 3y > y$ are also positive. This means that we can find (x_i, y_i) recursively by setting $(x_i, y_i) = (3x_{i-1} + 4y_{i-1}, 2x_{i-1} + 3y_{i-1})$. The set of all points (x_m, y_m) we find will be infinite, and each point will give a representation for F_n in the desired form.

17. See Chapter 4, Section 4.2.

18. It is true due to the *Lagrange four-square theorem*.

19. It is equivalent to show that $F_n + 1$ is divisible by 6. In fact, we have $F_n + 1 = (3F_1 \cdot \cdot F_n + 2) + 1 = 3(F_1 \cdot ... \cdot F_n + 1)$, where $F_1 \cdot ... \cdot F_n + 1$ is an even number.

20. If $n^n + 1$ is prime, there exists an integer m such that the exponent n has the form $n = 2^{2^m}$. The equation $n^n + 1 = F(2^m + m)$ holds in that case.

21. In fact, $\sum_{n \geq 0} \frac{1}{F_n} = \sum_{n \geq 0} \frac{1}{2^{2^n}+1} = 0,59606...$. More exactly, the decimal expansion of $\sum_{n \geq 0} \frac{1}{2^{2^n}+1}$ is given by the sequence $5, 9, 6, 0, 6, 3, 1, 7, 2, 1, ...$ (sequence A051158 in the OEIS). It is known, that $\sum_{n \geq 0} \frac{1}{2^{2^n}+1}$ is irrational. (Solomon and Golomb, 1963).

22. In fact, $2^{10} = 1024 > 10^3$, $2^{2^{73}} + 1 > 2^{2^{73}} > (2^{8(2^{10})})^7 > 2^{8 \cdot 10^{21}} = 2^{80 \cdot 10^{20}} > (2^{10})^{8 \cdot 10^{20}} > 10^{24 \cdot 10^{20}}$.

23. So, F_9 is the first Fermat number with more that 100 digits, i.e., the first Fermat prime, greater than *googol*.

24. So, F_{334} is a Fermat number with more that 10^{100} digits (in fact, the first such Fermat number), i.e., the first Fermat number, greater than *googoplex*.

25. It can be proven using similar reasons.

26. See [Luca01].

27. This fact can be illustrated by the example $F_1(3) = \binom{5}{3}$.

28. For $11 < n \leq 1000$, there exist 54 such numbers.

29. The first few Pierpont primes are 2, 3, 5, 7, 13, 17, 19, 37, 73, 97, ... (sequence A005109 in the OEIS). As any Fermat prime is a Pierpont prime (with $v = 0$, $u = 2^n$), then any constructible polygon can be constructed using origami.

30. A Pierpont prime with $v = 0$ is of the form $2^u + 1$, If $u = 0$, we obtain 2, if $u > 0$, we obtain a Fermat prime. If v is positive then u must also be positive, because a number of the form $3v + 1$ would be even and therefore non-prime. Therefore, the odd non-Fermat Pierpont primes all have the form $6k + 1$.

31. The first ten Proth numbers are 3, 5, 9, 13, 17, 25, 33, 41, 49, 57, ... (sequence A080075 in the OEIS). Without the condition that $2^n > k$, all odd integers larger than 1 would be Proth numbers.

32. The first ten Proth primes are 3, 5, 13, 17, 41, 97, 113, 193, 241, 257, ... (sequence A080076 in the OEIS).

33. The proof of this theorem uses the *Pocklington-Lehmer primality test*, and closely resembles the proof of Pépin's test (see, for example, [Ribe89], [Ribe96]).

    For example, for $p = 3 = 1 \cdot 2^1 + 1$, we have that $2^{\frac{3-1}{2}} + 1 = 3$ is divisible by 3, so 3 is prime.

    For $p = 5 = 1 \cdot 2^2 + 1$, we have that $3^{\frac{5-1}{2}} + 1 = 10$ is divisible by 5, so 5 is prime.

    For $p = 13 = 3 \cdot 2^2 + 1$, we have that $5^{\frac{13-1}{2}} + 1 = 15626$ is divisible by 13, so 13 is prime.

    For $p = 9$, which is not prime, there is no a such that $a^{9-12} + 1$ is divisible by 9.

34. Note, for example, that for all $n \geq 1$, it holds $2^{2^n} + 1 \equiv 2 (mod\ 3)$; for all $n \geq 2$, it holds $2^{2^n} + 1 \equiv 2 (mod\ 5)$, etc.

35. See Chapter 6, Zoo of Numbers.

Solutions: Other problems

1. It follows by definition of the Fibonacci sequence: the recurrence has the form $u_n = 1 \cdot u_{n-1} + (-1) \cdot u_{n-2}$, and $u_0 = u_1 = 1$. As for the Lucas sequence, the recurrence is the same, while $L_0 = 2, L_1 = 1$.

2. It follows by definition of the Pell sequence: the recurrence has the form $P_n = 2 \cdot P_{n-1} + (-1) \cdot P_{n-2}$, and $P_0 = 0, P_1 = 1$. As for the Pell-Lucas sequence, the recurrence is the same, while $PL_0 = 2, PL_1 = 2$.

3. We checked this fact before. In fact, we have shown, that a common recurrence for the numbers of the form $2^n \pm 1$ is $v_{n+2} = 3v_{n+1} + 2v_n$, while the starting values are $2^0 - 1 = 0$, $2^1 - 1 = 1$; $2^0 + 1 = 2$, $2^1 + 1 = 3$.

4. It follows from the definitions.

5. The characteristic equation of the recurrence relation for Lucas sequences $U_n(P, Q)$ and $V_n(P, Q)$ is: $x^2 - Px + Q = 0$. It has the discriminant $D = P^2 - 4Q$ and the roots $a = \frac{P+\sqrt{D}}{2}$ and $b = \frac{P-\sqrt{D}}{2}$. Thus: $a + b = P$, $ab = \frac{1}{4}(P^2 - D) = Q$, $a - b = \sqrt{D}$.

Note that the sequence a^n and the sequence b^n also satisfy the given recurrence relation. However these might not be integer sequences.

When $D \neq 0$, a and b are distinct and one quickly verifies that $a^n = \frac{V_n + U_n\sqrt{D}}{2}$; $b^n = \frac{V_n - U_n\sqrt{D}}{2}$. It follows that the terms of Lucas sequences can be expressed in terms of a and b as follows

$$U_n = \frac{a^n - b^n}{a - b} = \frac{a^n - b^n}{\sqrt{D}}, \ V_n = a^n + b^n.$$

The case $D = 0$ occurs exactly when $P = 2S$ and $Q = S^2$ for some integer S, so that $a = b = S$. In this case one easily finds that $U_n(P, Q) = U_n(2S, S^2) = nS^{n-1}$, $V_n(P, Q) = V_n(2S, S^2) = 2S^n$.

6. In fact, for $k = 1$ we get the numbers $2^n + 1$; we know exactly five primes of such form; they are Fermat primes. For $k = 3$ we get a prime on the first step: $3 \cdot 2^1 + 1 = 7$, etc. In 1960, W. Sierpiński proved that there are infinitely many odd integers k such that $k \cdot 2^n + 1$ is composite for all natural numbers n (see Chapter 4, Section 4.3).

   The sequence of currently known Sierpiński numbers begins with 78557, 271129, 271577, ... (sequence A076336 in the OEIS). It is unknown, if 78557 is the smallest Sierpiński number.

7. In fact, for $k = 1$ we get the numbers $2^n - 1$; we know exactly 51 primes of such form; they are Mersenne primes. For $k = 3$ we get a prime on the first step: $3 \cdot 2^1 - 1 = 5$, etc.

   In 1956, H. Riesel showed that there are an infinite number of integers k such that $k \cdot 2^n - 1$ is not prime for any integer n. He showed that the number 509203 has this property, as does 509203 plus any positive integer multiple of 11184810.

   The sequence of currently known Riesel numbers begins with: 509203, 762701, 777149, 790841, 992077, 1106681, 1247173, 1254341, 1330207, ... (sequence A101036 in the OEIS). The *Riesel problem* consists in determining the smallest Riesel number.

8. It suffices to show that Jacobi symbol $\left(\frac{M_p}{F_p}\right) = -1$. We have $2^{2^p - 2} \equiv 1 (mod \ M_p)$, and multiplying 2 to both sides we get

$2^{2^p-1} \equiv 2(mod\ M_p)$. This implies that $F_p = 2 \cdot 2^{2^p-1} + 1 \equiv 5(mod\ M_p)$.

Moreover, since $p \equiv 3(mod\ 4)$, $M_p = 2^p-1 = 2^{4k+3} - 1 = 8 \cdot 2^{4k} - 1 \equiv 3 \cdot 1-1 = 2(mod\ 5)$. Thus, $(\frac{M_p}{F_p}) = (\frac{F_p}{M_p}) = (\frac{5}{M_p}) = (\frac{M_p}{5}) = (\frac{2}{5}) = -1$.

9. It sufficies to show that Jacobi symbol $(\frac{M_p}{F_{p+1}}) = -1$.

We can show that $F_p \equiv 5(mod\ M_p)$. Then $F_{p+1} = (F_p - 1)^2 + 1 = 4^2 + 1 = 17(mod\ M_p)$.

First we assume $p \equiv 3(mod\ 8)$, then $M_p = 2^{8k+3} - 1 = 8 \cdot 16^{2k} - 1 \equiv 8 - 1 = 7(mod\ 17)$. Hence, $(\frac{M_p}{F_{p+1}}) = (\frac{F_{p+1}}{M_p}) = (\frac{17}{M_p}) = (\frac{M_p}{17}) = (\frac{7}{17}) = (\frac{17}{7}) = (\frac{3}{7}) = -(\frac{7}{3}) = -(\frac{1}{3}) = -1$.

Now if we assume $p \equiv 5(mod\ 8)$, then $M_p = 2^{8k+5} - 1 \equiv 2^5 - 1 = -3(mod\ 17)$. Hence, $(\frac{M_p}{F_{p+1}}) = (\frac{M_p}{17}) = (\frac{-3}{17}) = (\frac{-1}{17})(\frac{3}{17}) = (\frac{3}{17}) = (\frac{17}{3}) = (\frac{2}{3}) = -1$.

10. See, for example 337. It is a permutable prime because each of 337, 373 and 733 is prime. The first few permutable primes are 2, 3, 5, 7, 11, 13, 17, 31, 37, 71, 73, 79, 97, 113, 131, 199, 311, 337, 373, 733, 919, 991, $R_{19} = 1111111111111111111$, ... (sequence A003459 in the OEIS).

11. The sequence of emirps begins from 13, 17, 31, 37, 71, 73, 79, 97, 107, 113, ... (sequence A006567 in the OEIS). So, the first mersenne emirp is $M_5 = 31$, while the first Fermat emirp is $F_2 = 17$.

As of 2020, the largest known emirp is $10^{10006} + 941992101 \cdot 10^{4999} + 1$, found by J. K. Andersen in 2007.

12. The Chen primes are named after Chen Jingrun, who proved in 1966 that there are infinitely many such primes. This result would also follow from the truth of the twin prime conjecture as the lower member of a pair of twin primes is by definition a Chen prime.

The *Chen's theorem* states that every sufficiently large even number can be written as the sum of either two primes, or a prime and a semi-prime. The Chen's theorem is a giant step towards the *Goldbach's conjecture*, and a remarkable result of the sieve methods.

The first few Chen primes are 2, 3, 5, 7, 11, 13, 17, 19, 23, 29, ... (sequence A109611 in the OEIS).

The first few Chen primes that are not the lower member of a pair of twin primes are 2, 7, 13, 19, 23, 31, 37, 47, 53, 67, ... (sequence A063637 in the OEIS).

The first few non-Chen primes are 43, 61, 73, 79, 97, 103, 151, 163, 173, 193, ... (sequence A102540 in the OEIS).

Bibliography

[Abra74]   Abramovitz V. *Bernoulli numbers*, Kvant, 6, 1974.

[AbSt72]   Abramowitz M. and Stegun I.A. *Handbook of Mathematical Functions with Formulas, Graphs, and Mathematical Tables*, New York: Dover Publications, 1972.

[AGP94]   Alford W.R., Granville A. and Pomerance C. *There are Infinitely Many Carmichael Numbers*, Annals of Mathematics, 139, 1994.

[Andr98]   Andrews G.E. *The Theory of Partitions*, Cambridge, England: Cambridge University Press, 1998.

[Anke57]   Ankeny N.C. *Sums of Three Squeres*, Proc. Amer. Math. Soc., 8, 1957.

[Anto85]   Antoniadis J.A. *Generalized Fibonacci Numbers and Some Diophantine Equations*, The Fibonacci Quarterly, 23.3, 1985.

[Apos86]   Apostol T.M. *Introduction to Analytic Number Theory*, Springer-Verlag New-York, 1986.

[Arno38]   Arnold I.V. *Theoretical Arithmrtics*, Moscow, 1938.

[Avan67]   Avanesov E.T. *Solution of a Problem on Figurate Numbers*, Acta Arithmetica, 12, 1967.

[BaCo87]   Ball W.W. R. and Coxeter H.S.M. *Mathematical Recreations and Essays*, 13-th ed., New York: Dover, 1987.

[BaWe76]   Ballew D. and Weger R. *Repdigit Triangular Numbers*, J. Rec. Math., 8(2), 1976.

[BSW89]   Bateman P.T., Selfridge J.L. and Wagstaff Jr.S.S. *The New Mersenne Conjecture*, American Mathematical Monthly, Mathematical Association of America, 96 (2), 1989.

[Bern99]   Bernhart F.R. *Catalan, Motzkin, and Riordan numbers*, Discrete Mathematics, 204 (1–3), 1999.

295

[Beuk88]   Beukers F. *On Oranges and Integral Points on Certain Plane Cubic Curves*, Nieuw Arch. Wisk., 6, 1988.

[BiBa70]   Birkhoff G. and Bartee T.C. *Modern Applied Algebra*, McGraw Hill Text, 1970.

[Bond93]   Bondarenko B.A. *Generalized Pascal's Triangles and Pyramids, Their Fractals, Graphs, and Applications*, Fibonacci Association, 1993.

[Bras88]   Brassard G. *Modern Cryptology*, Springer-Verlag New York, 1988.

[BrMo75]   Brillhart J., Morrison M.A. *A Method of Factoring and the Factorization of F_7*, Math. Comp., 29, 1975.

[Broc76]   Brocard H. *Question 166*, Nouv. Corres. Math., 2, 1876.

[Broc85]   Brocard H. *Question 1532*, Nouv. Ann. Math., 4, 1885.

[Brot64]   Brother U.A. *On Square Lucas Numbers*, The Fibonacci Quarterly, 2.1, 1964.

[Buch09]   Buchstab A.A. *Number Theory*, Moscow, 2009.

[BePa01]   Burnett S. and Paine S. *RSA Security's Official Guide to Cryptography*, New York: Osborne/McGraw-Hill, 2001.

[CFHLZ17]   Cai Z., Faust M., Hildebrand A.J., Li J. and Zhang Y. *Leading Digits of Mersenne Numbers*, University of Illinois, 2017.

[Carm19]   Carmichael R.D. *Fermat Numbers $F_n = 2^{2^n}+1$*, The American Mathematical Monthly, 26 (4), 1919.

[Cata44]   Catalan E. *Note Extraite d'une Lettre Adressée à l'Éditeur*, J. Reine Angew. Math., 27, 1844.

[Chan70]   Chandrasekharan K. *Arithmetical Functions*, Springer-Verlag Berlin Heidelberg, 1970.

[ChFa07]   Chen Y.-G. and Fang J.-H. *Triangular Numbers in Geometrical Progression*, Electronic Journal of Combinatorial Number Theory, 7, 2007.

[Cher39]   Chernick J. *On Fermat's Simple Theorem*, Bull. Amer. Math. Soc., 45 (4), 1939.

[Clau40]   Clausen T. *Lehrsatz aus einer Abhandlung über die Bernoullischen Zahlen*, Astron. Nachr., 17 (22), 1840.

[Cohn64]   Cohn J.H.E. *On Square Fibonacci Numbers*, J. London Math. Soc., 39, 1964.

[Cohn65]   Cohn J.H.E. *Lucas and Fibonacci Numbers and Some Diophantine Equations*, Proc. Glasgow Math. Assn., 7, 1965.

[CoGu96]   Conway J.H. and Guy R.K. *The Book of Numbers*, New York: Springer-Verlag, 1996.

[CoRo96]   Courant R. and Robbins H. *What is Mathematics?: An Elementary Approach to Ideas and Methods*, 2-nd ed., Oxford University Press, 1996.

[Cull05]   Cullen J. *Question 15897*, Educ. Times, 534, 1905.

[CuWu17]   Cunningham A.J.C. and Woodall H.J. *Factorization of $Q =$ $(2^q \mp q)$ and $(q \cdot 2^q \mp 1)$*, Messenger of Mathematics, 47, 1917.

[Dave47]   Davenport H. *The Geometry of Numbers*, Mathematical Gazette, 31, 1947.

[Dave99]   Davenport H. *The Higher Arithmetic: An Introduction to the Theory of Numbers*, Cambridge University Press, 8-th ed., 2008.

[Dede63]   Dedekind R. *Essays on the Theory of Numbers*, Cambridge University Press, 1963.

[DeMo10]   Deza E. and Model D. *Elements of Discrete Mathematics*, Moscow, URSS, 2010.

[DeDe12]   Deza E. and Deza M.M. *Figurate Numbers*, World Scientific, 2012.

[DeKo13]   Deza E.I. and Kotova L.V. *Collection of Problems on Number Theory*, Moscow, URSS, 2013.

[Deza17]   Deza E. *Special Positive Integer Numbers*, Moscow, URSS, 2017.

[Deza18]   Deza E. *Special Combinatorial Numbers*, Moscow, URSS, 2018.

[DeKo18]   Deza E.I. and Kotova L.V. *Introduction into Cryptography*, Moscow, URSS, 2018.

[Dick27]   Dickson L.E. *All Positive Integers are Sums of Values of a Quadratic Function of x*, Bull. Amer. Math. Soc., 33, 1927.

[Dick05]   Dickson L.E. *History of the Theory of Numbers*, New York: Dover, 2005.

[DiHe76]   Diffie W. and Hellman M.E. *New Directions in Cryptography*, Transactions on Information Theory, 22 (6), 1976.

[Diop74]   Diophantus *Arithmetics and the Book of Polygonal Numbers*, Moscow: Nauka, 1974.

[DiDe63]   Dirichlet J.P.G.L. and Dedekind R. *Vorlesungen Über Zahlentheorie*, F. Vieweg und sohn, 1863.

[DoSh77]   Donaghey R. and Shapiro L.W. *Motzkin Numbers*, Journal of Combinatorial Theory, Series A, 23 (3), 1977.

[Edva77]   Edwards H.M. *Fermat's Last Theorem: A Genetic Introduction to Algebraic Number Theory*, New York: Springer-Verlag, 1977.

[Ehrm67]   Ehrman J.R. *The Number of Prime Divisors of Certain Mersenne Numbers*, Mathematics of Computation, 21 (100), 1967.

[ErOb37]   Erdős P. and Obláth R. *Über diophantische Gleichungen der Form $n! = x^p \pm y^p$ und $n! \pm m! = x^p$*, Acta Szeged., 8, 1937.

[Eule48]   Euler L. *Introductio in Analysin Infinitorum*, 1, 2, Marc Michel Bousquet, 1748.

[Gard61]   Gardner M. *Second Book of Mathematical Puzzles and Diversions*, New York, 1961.

[Gard88]   Gardner M. *Time Travel and Other Mathematical Bewilderments*, New York: Freeman, 1988.

[Gard89]   Gardner M. *Penrose Tiles to Trapdoor Ciphers*, W.N. Freeman, New York, 1989.

[Gaus01]   Gauss C.F. *Disquisitiones Arithmeticae*, Leipzig, 1801.

[Gees20]   De Geest P. *Palindromic Numbers and Other Recreational Topics*, *http://www.worldofnumbers.com/*, 29.12.2020.

[Gelf98]   Gelfand I.M. *Lectures on Linear Algebra*, Courier Dover Publications, 1998.

[GIMPS20]   Great Internet Mersenne Prime Search (GIMPS) *https://ru.wikipedia.org/wiki/GIMPS*, 29.12.2020.

[Gill64]   Gillies D.B. *Three New Mersenne Primes and a Statistical Theory*, Mathematics of Computation, 18 (85), 1964.

[Glea88]   Gleason A.M. *Angle Trisection, the Heptagon, and the Triskaidecagon*, American Mathematical Monthly, 95 (3), 1988.

[GKP94]   Graham R.L., Knuth D.E. and Patashnik O. *Concrete Mathematics: A Foundation for Computer Science*, 2-nd ed., Reading, MA: Addison-Wesley, 1994.

[Goul85]   Gould H.W. *Catalan Numbers: Research Bibliography of Two Special Numbers Sequences*, Morgantown, 1985.

[Guy94]   Guy R.K. *Unsolved Problems in Number Theory*, 2-nd ed., New York: Springer-Verlag, 1994.

[HaWr79]   Hardy G.H. and Wright E.M. *An Introduction to the Theory of Numbers*, 5-th ed., Oxford, England: Clarendon Press, 1979.

[Hogg69]   Hoggatt V.E. Jr. *The Fibonacci and Lucas Numbers*, Boston, MA: Houghton Mifflin, 1969.

[Hons91]   Honsberger R. *More Mathematical Morsels*, Washington, DC: Math. Assoc. Amer., 1991.

[Hool76]   Hooley Ch. *Applications of Sieve Methods*, Cambridge Tracts in Mathematics, 70, Cambridge University Press, 1976.

[JoLo99]   Jones Ch. and Lord N. *Characterising Non-trapezoidal Numbers*, Math. Gazette, 83 (497), 1999.

[John77]   Johnson A.W. *Absolute Primes*, Mathematics Magazine, 50, 1977.

[Ingh32]   Ingham A. *The Distribution of Prime Numbers*, Cambridge University Press, 1932.

[IrRo90]   Ireland K. and Rosen M. *A Classical Introduction to Modern Number Theory*, Springer-Verlag New York, 1990.

[Ivic85]   Ivić A. *The Riemann Zeta Function*, New York: John Wiley & Sons, 1985.

[Kahn67]   Kahn D. *The Codebreakers — The Story of Secret Writing*, New York: Charles Scribner's Sons, 1967.

[Kara83]   Karatsuba A. A. *Basic Analytic Number Theory*, Moscow: Nauka, 1983.

[KaVo92]   Karatsuba A.A. and Voronin S.M. *The Riemann zeta-function*, Berlin: Walter de Gruyter & Co., 1992.

[Kell20]   Keller W. *Prime Factors of Fermat Numbers*, *http://www.prothsearch.com/fermat.html*, 29.12.2020.

[Kerc83]   Kerckhoffs A. *La cryptographie militaire*, Journal des sciences militaires, 9, 1983.

[Kell95]   Keller W. *New Cullen Primes*, Mathematics of Computation, I, 1995.

[Khin97]   Khinchin A. *Continued Fractions*, Mineola, New-York: Dover Publications, 1997.

[Knut68]   Knuth D.E. *Fundamental Algorithms*, Addison-Weslay, 1968.

[Knut76]   Knuth D.E. *Mathematics and Computer Science. Coping with Finiteness*, Science, 1976.

[KGP94]   Knuth D., Graham R. and Patashnik O. *Concrete Mathematics: A Foundation for Computer Science*, Addison-Wesley, 1994.

[Kobl87]   Koblitz N. *A Course in Number Theory and Cryptography*, Graduate Texts in Math., 114, Springer-Verlag, New York, 1987.

[Kost82]   Kostrikin A. *Introduction to Algebra*, Springer Verlag, 1982.

[KLS01]   Krízek M., Luca F. and Somer L. *17 Lectures on Fermat Numbers: From Number Theory to Geometry*, CMS Books in Mathematics, 9, Springer-Verlag, New York, 2001.

[Lagr70]   Lagrange J.L. *Démonstration d'un Téorème d'Arithmétique*, Nouveaux Mémoires de l'Acad. Royale des Sci. et Belles-L. de Berlin, 1770.

[Land09]   Landau E. *Handbuch der Lehre von der Verteilung der Primzahlen*, Leipzig, Berlin: Teubner, 1909.

[Lege30]   Legendre A.-M. *Théorie des Nombres*, 3-th ed., Paris: Chez Firmin Didot Frères, Libraires, 1830.

[Lege79]   Legendre A.-M. *Théorie des Nombres*, 4-th ed., Paris: A. Blanchard, 1979.

[LiNi96]   Lidl R. and Niederreiter H. *Finite Fields*, Cambridge University Press, 1996.

[Line86]   Lines M.E. *A Number for your Though*, Bristol: Adam Hilger, 1986.

[LeLi83]   Le Lionnais F. *Les Nombres Remarquables*, Paris: Hermann, 1983.

[Luca75]   Lucas É. *Question 1180*, Nouv. Ann. Math., Ser. 2, 14, 1875.

[Luca00]   Luca F. *The Anti-Social Fermat Number*, American Mathematical Monthly, 107 (2), 2000.

[Luca01]   Luca F. *Fermat Numbers in the Pascal's Triangle*, Divulgaciones Matemáticas, 9 (2), 2001.

[Mada79]   Madachy J.S. *Madachy's Mathematical Recreations*, New York: Dover, 1979.

[Madd05]   Maddux F.C. *Fermat Numbers: Historical View with Applications Related to Fermat Primes*, Master's Theses. 2857. San Jose State University, 2005.

[Malc86]   Malcolm E. *A Number for Your Thoughts: Facts and Speculations about Number from Euclid to the latest Computers*, CRC Press, 1986.

[Mati93]   Matiyasevich Y. *Hilbert's 10-th Problem*, Foreword by Martin Davis and Hilary Putnam, The MIT Press, 1993.

[MaSt15]   Mazur B. and Stein W. *Prime Numbers and the Riemann Hypothesis*, Cambridge University Press, 2015.

[McDa98]   McDaniel W.L. *Pronic Fibonacci Numbers*, Fib. Quart., 36, 1998.

[McDa98a]   McDaniel W.L. *Pronic Lucas Numbers*, Fib. Quart., 36, 1998.

[MSC96]   Mitrinovic D.S., Sandor J. and Crstici B. *Handbook on Number Theory*, Kluwer Acad. Publ., 1996.

[Moto83]   Motohashi Y. *Lectures on Sieve Methods and Prime Number Theory*, Berlin: Springer-Verlag, 1983.

[Motz48]   Motzkin T.S. *Relations between Hypersurface Cross Ratios, and a Combinatorial Formula for Partitions of a Polygon, for Permanent Preponderance, and for Non-associative Products*, Bulletin of the American Mathematical Society, 54, 1948.

[Moze09]   Mozer L. *An Introduction to the Theory of Numbers*, The Trillia Group West Lafayette, IN, 2009.

[Nage51]   Nagell T. *Introduction to Number Theory*, New York: Wiley, 1951.

[Nage61]   Nagell T. *The Diophantine Equation $x^2 + 7 = 2^n$*, Ark. Mat., 30, 1961.

[Nico26]   Nicomachus of Geraza *Introduction in Arithmetic*, translated by Martin Luther D'Ooge, Macmillan, 1926.

[Ore48]   Ore O. *Number Theory and Its History*, Dover Publications, Inc., 1948.

[Pasc54]   Pascal B. *Traité du triangle arithmétique, avec quelques autres petits traitez sur la mesme matière*, Paris, 1654.

[Pier95]   Pierpont J. *On an Undemonstrated Theorem of the Disquisitiones Arithmetica*, Bulletin of the American Mathematical Society, 2 (3), 1895.

[Plou92]   Plouffe S. *1031 Generating Functions and Conjectures*, Université du Québec à Montréal, 1992.

[Plut78]   Plutarch *Platonic Questions*, in *The Morals*, Vol. 5, Boston: Little, Brown, and Co., 1878.

[Poll74]   Pollard J.M. *Theorems on Factorization and Primality Testing*, Proc. Cambridge Phil. Soc., 76, 1974.

[Prim20]   *The Prime Pages (prime number research, records and resources), https://primes.utm.edu/* 29.12.2020.

[RaTo57]   Rademacher H. and Toeplitz O. *The Enjoyment of Mathematics: Selections from Mathematics for the Amateur*, Princeton, NJ: Princeton University Press, 1957.

[Rama00]   Ramanujan S. *Collected Papers of Srinivasa Ramanujan*, Ed. Hardy G.H., Aiyar P. V. S. and Wilson B.M., Providence, RI: Amer. Math. Soc., 2000.

[Ribe89]   Ribenboim P. *The Book of Prime Number Records*, New York: Springer-Verlag, 1989.

[Ribe96]   Ribenboim P. *The New Book of Prime Numbers Records*, New-York: Springer-Verlag, 1996.

[Ries94]   Riesel H. *Prime Numbers and Computer Methods for Factorization*, 2-nd ed., Basel: Birkhouser, 1994.

[Salo90]   Salomaa A. *Public-Key Cryptography*, Springer-Verlag Berlin Heidelberg, 1990.

[Shan93]   Shanks D. *Solved and Unsolved Problems in Number Theory*, 4-th ed., New York: Chelsea, 1993.

[Shko61]   Shkolnik A.G. *The Problem of the Circle's Dividing*, Moscow, 1961.

[Sier64]   Sierpiński W. *Elementary Theory of Numbers*, Warszawa, 1964.

[Sier03]   Sierpiński W. *Pythagorean Triangles*, New York: Dover, 2003.

[Sing00]   Singh S. *The Code Book: The Secret History of Codes and Code-breaking*, London: Forth Estate, 2000.

[SlPl95]   Sloane N.J.A. and Plouffe S. *The Encyclopedia of Integer Sequences*, San Diego: Academic Press, 1995.

[Sloa20]   Sloane N.J.A. at all *On-line Encyclopedia of Integer Sequences*, *http://oeis.org/* 29.12.2020.

[Smit84]   Smith D.E. *A Source Book in Mathematics*, New York: Dover, 1984.

[Smit97]	Smith J. *Trapezoidal Numbers*, Math. in School, 11, 1997.
[Stau40]	von Staudt K.G.Ch. *Beweis eines Lehrsatzes, die Bernoullischen Zahlen betreffend*, Journal fur die reine und angewandte Mathematik, 21, 1840.
[Stan15]	Stanley R. *Catalan Numbers*, Cambridge Universiy Press, 2015.
[Step01]	Stepanova L.L. *Selected Chapters in Number Theory*, Moscow, MSPU, 2001.
[Stra16]	Strang G. *Linear Algebra and Its Applications*, Wellesley-Cambridge Press; 5-th ed., 2016.
[Stru87]	Struik D.J. *A Concise History of Mathematics*, Dover Publications, 1987.
[Titc87]	Titchmarsh E.Ch. *The Theory of the Riemann Zeta-Function*, Oxford University Press, 2-nd ed., 1987.
[Tsan10]	Tsang C. *Fermat Numbers*, University of Washington, 2010.
[Uspe76]	Uspensky V. A. *Pascal's Triangle: Certain Applications of Mechanics to Mathematics*, Moscow: Mir, 1976.
[Vile14]	Vilenkin N.Ia. *Combinatorics*, Academic Press, 2014.
[Vino03]	Vinogradov I.M. *Elements of Number Theory*, Mineola, NY: Dover Publications, 2003.
[Voro61]	Vorob'ev N.N. *Fibonacci Numbers*, New York: Blaisdell, 1961.
[Want36]	Wantzel M.L. *Recherches sur les Moyens de Reconnaître si un Problème de Géométrie Peut se Pésoudre avec la Règle et le Compas*, J. Math. pures appliq., 1, 1836.
[Weis99]	Weisstein E.W. *CRC Concise Encyclopedia of Mathematics*, CRC Press, 1999.
[Weis20]	Weisstein E.W. *MathWorld — A Wolfram Web Resource*, http://mathworld.wolfram.com/ 29.12.2020.
[Well86]	Wells D. *The Penguin Dictionary of Curious and Interesting Numbers*, Middlesex: Penguin Books, 1986.
[Wief09]	Wieferich A. *Beweis des Satzes, dass sich eine jede ganze Zahl als Summe von höchstens neun positiven Kuben darstellen lässt*, Math. Ann., 66, 1909.
[Wiki20]	*Wikipedia, the free Encyclopedia*, http://en.wikipedia.org, 29.12.2020.

Index

Printed in the United States
by Baker & Taylor Publisher Services